Science Communication in the World

Bernard Schiele • Michel Claessens • Shunke Shi
Editors

Science Communication in the World

Practices, Theories and Trends

Springer

Editors
Bernard Schiele
Department of Communication
University of Quebec at Montreal (UQAM)
P.O. Box 8888
Station Centre-Ville
Montreal, QC H3C 3P8
Canada

Michel Claessens
ITER, Route de Vinon sur Verdon
F-13115 St Paul-lez-Durance
France

Shunke Shi
China Research Institute
for Science Popularization (CRISP)
86 Xueyuan Nanlu, Haidian District
Beijing 100081
China

ISBN 978-94-007-4278-9 ISBN 978-94-007-4279-6 (eBook)
DOI 10.1007/978-94-007-4279-6
Springer Dordrecht Heidelberg New York London

Library of Congress Control Number: 2012936493

Printed on acid-free paper

Springer is part of Springer Science+Business Media (www.springer.com)

Contents

Part I National Overviews

1 **The 'Communicative Turn' in Contemporary Techno-science: Latin American Approaches and Global Tendencies** ... 3
Carmelo Polino and Yurij Castelfranchi

2 **The Evolution of Science Communication Research in Australia** ... 19
Jenni Metcalfe and Toss Gascoigne

3 **The Development of Science Communication Studies in Canada** ... 33
Bernard Schiele and Anik Landry

4 **Science Popularization Studies in China** ... 65
Fujun Ren, Lin Yin, and Honglin Li

5 **Policy Perspective on Science Popularization in China** ... 81
Shunke Shi and Huiliang Zhang

6 **Deliberation, Dialogue or Dissemination: Changing Objectives in the Communication of Science and Technology in Denmark** ... 95
Maja Horst

7 **Social Sciences and the Communication of Science and Technology in France: Implications, Experimentation and Critique** ... 109
Michèle Gellereau, Yves Jeanneret, and Joëlle Le Marec

8 **The Recent Public Understanding of Science Movement in Germany** ... 125
Markus Lehmkuhl

9 Public Understanding of Science: Glimpses of the Past and Roads Ahead ... 139
Gauhar Raza, Surjit Singh, and P.V.S. Kumar

10 Whose Science? What Knowledge? Science, Rationality and Literacy in Africa ... 151
Hester du Plessis

11 An Experience of Science Communication in Korea: The Space-Sharing Project with Mass Media ... 169
Sook-Kyoung Cho

12 From Science Popularization to Public Engagement: The History of Science Communication in Korea ... 181
Sung Kyum Cho and Ock Tae Kim

13 Spanish PCST and the European Science in Society Strategy ... 193
Vladimir de Semir

14 Science Museums and Cultural Images of Modernity: Scientific Communication, New Identities and Sociopolitical Constraints on Science Museums in Spain ... 211
Xavier Roigé

Part II Horizontal Issues

15 Slowly But Surely: How the European Union Promotes Science Communication ... 227
Michel Claessens

16 Vital and Vulnerable: Science Communication as a University Subject ... 241
Brian Trench

17 Visible Scientists, Media Coverage and National Identity: Nobel Laureates in the Italian Daily Press ... 259
Massimiano Bucchi

18 Engagement: The Key to the Communicative Effectiveness of Science and Ideas ... 269
Hak-Soo Kim

19 From Public to Policy ... 281
Jan Riise

20 Science Culture and Its Indicators ... 295
Martin W. Bauer

Index ... 313

Author Biography

Martin W. Bauer read Psychology and Economic History (Bern, Switzerland). He is currently Professor of Social Psychology & Research Methodology at the London School of Economics, where he directs the M.Sc. Social and Public Communication and edits the international journal *Public Understanding of Science.* A former Research Fellow at the Science Museum and a regular visiting professor in Brazil (UFRGS, Porto Alegre and CAMPINAS, Sao Paulo), he conducts PUS research, including theory building, comparative attitude surveys, media analysis and qualitative enquiries. Recent publications include *Journalism, science & society: Science communication between news and public relations* (Routledge, 2007, with M. Bucchi); *The culture of science: How the public relates to science across the globe* (New York, Routledge, 2012, with R. Shukla and N. Allum) and *Atoms, bytes and genes: Public resistance and socio-technical responses* (Routledge, in preparation). Martin's research papers have appeared in *Nature*, *Science*, *Nature–Biotechnology*, *Public Understanding of Science*, *Genetics & Society*, *Social Science Information*, *Journal for the Theory of Social Behaviour*, *Social Studies of Science*, *International Journal of Public Opinion Research*, *Science Communication* and *DIOGENE* (UNESCO).

Massimiano Bucchi is Professor of Science and Technology in Society at the University of Trento, Italy.

He has published several books, including *Science and the media* (London and New York, Routledge, 1998), *Science in society: An introduction to social studies of science* (London and New York, Routledge, 2004), *Handbook of public communication of science and technology* (with B. Trench, London and New York, Routledge, 2008) and *Beyond technocracy: Citizens, politics, technoscience* (New York, Springer, 2009), and essays in international journals such as *History and Philosophy of the Life Sciences*, *Nature, New Genetics and Society, Science* and *Public Understanding of Science.*

He has served as adviser and evaluator for several research and policy bodies, including the United States National Science Foundation, the Royal Society, the European Commission and the European Food Safety Auhority, and is a member of the editorial board of *Public Understanding of Science.*

Yurij Castelfranchi is a quantum physicist with a master's in science communication and a Ph.D. in sociology. He is Professor of Sociology at the Federal University of Minas Gerais, Belo Horizonte, Brazil. He is editor of the journal *Teoria & Sociedade* and co-editor of *Jcom* (the Journal of Science Communication). His main areas of research are the sociology of S&T, public communication of S&T, science journalism and public perception of S&T.

Sook-Kyoung Cho has been devoted to science culture and public understanding of science activities in Korea since studying at King's College of London (M.Sc. in history and philosophy of science) and Seoul National University (Ph.D. in the history of science). Since she joined KOFAC as a programme director in 2002, she has initiated and executed various new programmes, such as the Space-sharing Project, the ninth International Conference on Public Communication of Science and Technology, the Science Culture Symposia among Korea, China and Japan at AAAS annual conferences, and the Gwa-whal: Science Volunteers for College Students.

She also works on the academic side of science culture as the chair of the Science Culture Committee at the Korean Association for Science Education, as a committee member of the Korean History of Science Society and as a Scientific Committee member for the PCST Network. She has lectured in many universities and, after writing or translating more than ten books, she is now preparing a book on the history and practices of science culture activities in Korea during the past 40 years.

Sung Kyum Cho is a professor in the Department of Communication at Chungnam National University, South Korea. He has served as the Science/Health/Environment/Risk Division Chair of the Korean Society for Journalism and Communication Studies since 2009. He is the president of the Korean Association for Survey Research and has served as the chair of the editorial board of its official journal, *Survey Research.* He was also a member of the editorial board of the *Journal of ELSI Studies* from 2003 to 2008. He earned his Ph.D. in 1990 from Seoul National University. His recent publications include the papers 'How to obtain better survey results on life science issues', 'A variation on deliberative polling of the social perception of life science issues' and 'The challenges of governing biotechnology in Korea'.

Michel Claessens is currently Head of Communication at ITER Organization in Cadarache, France.

Michel was born in 1958 in Brussels. He has a Ph.D. in science and was a scientific researcher for 10 years (physical chemistry, medical imaging, biotechnology and pharmaceuticals). He has also been a freelance scientific journalist since 1980.

Michel joined the European Commission in 1994. He has been acting Head of the Communication Unit in the Research Directorate-General. He was also the editor-in-chief of the *research*eu* magazine of the European Commission. In April 2011, Michel was recruited at ITER Organization.

Michel also teaches science communication at the Free University of Brussels and is a member of the scientific committee of the international Publication Communication of Science and Technology Network. As a scientific journalist and writer, Michel Claessens has published 250 articles and 9 books covering science communication and several aspects of science and technology.

Fujun Ren is the General Director of the China Research Institute for Science Popularization (CRISP), Editor-in-Chief of the Chinese academic *Journal of Science Popularization*, Deputy Director of the Science and Technology Communication Research Center of the China Association for Science & Technology and Tsinghua University. He is also a professor of Beijing Normal University, the Chinese Science & Technology University and Harbin University of Science and Technology.

Mr. Ren gained his engineering doctoral degree in the graduate school of Harbin Institute of Technology before taking up his postdoctoral positions in Tsinghua University and Harbin Engineering University.

He has published more than 10 academic monographs and tertiary textbooks and more than 100 academic papers. He has also conducted 30 scientific research programmes at national and provincial levels.

Toss Gascoigne is the inaugural President of the PCST Network, an international group promoting the public communication of science and technology, and a founding member of Australian Science Communicators. He is a former executive director of Australian organizations representing researchers in the sciences and in the humanities and social sciences. He is interested in the linkages between researchers and government, and approaches that persuade governments to adopt evidence-based processes in determining policy.

He has edited a number of books on science communication and published on subjects ranging from the history of science communication in Australia to an examination on the status of science communication as a discipline.

Michèle Gellereau is Professor of Information and Communication Sciences at the University of Lille (France) and a member of the GERiiCO research centre (and its director from 2005 to 2010). Michèle has worked in several national academic research programmes and in interregional European programmes on culture, heritage, museology, science communication, informal education and cultural mediation. Many of her publications explore the question of the dissemination and appropriation of information and knowledge, cultural and social mediation in the public sphere, and cultural practices. Michèle is co-editor of the international academic journal *Etudes de communication (language, information, mediation)* edited by the University of Lille. She is member of several scientific committees of social and human scientific journals, conferences and the French National Agency for Research.

Honglin Li is a Research Associate at CRISP. She got her Ph.D. from Tsinghua University. Her research interests include theoretical and empirical studies of public understanding of science, as well as S&T policy. She has published 'The theoretical evolution of public understanding of science: Take Miller's system as a clue' (2010), 'Living science and public scientific literacy' (2009), 'The social context, theoretical origin and target direction of Miller's system' (2009) and 'On the basic research investment and its coordination mechanism in typical countries' (2008).

Maja Horst is Head of the Department of Media, Cognition and Communication at the University of Copenhagen, Denmark. Before taking up that position in 2011, she was associate professor in Science and Technology Studies at Copenhagen Business

School. Maja Horst has published in journals such as Public Understanding of Science; Social Studies of Science; Science, Technology and Human Values; Science and Public Policy; and Science and Engineering Ethics.

She has been engaged in a number of research projects on public understanding of science funded by the Danish research councils and in international collaborative research networks on dialogical science communication funded by the EU and Nordforsk (the Scandinavian research council). She has also been experimenting with dialogical communication of her own research in two spatial installations created in collaboration with designers and artists. On the basis of at work, she was awarded the Danish Science Minister's Communication Prize in 2009.

Huiliang Zhang is an Assistant Researcher at the China Research Institute. She has a master's degree in Mass Communications from the University of Leicester and a Ph.D. in Education from Zhejiang University. She works primarily on science education and science communication policy.

Yves Jeanneret is professor at Celsa, the school of information and communication sciences of Paris Sorbonne University, and a member of the GRIPIC research team (an interdisciplinary research group on information and communication processes) and the 'Concepts and languages' doctoral school of Paris Sorbonne University. He has published several books on the popularization of science, the social sharing of culture and knowledge, writing practices, and the epistemology of communication and semiotics, such as *Ecrire la science* (Presses universitaires de France, 1994), *Y a-t-il (vraiment) des technonologies de l'information* (Septentrion, 2000) and *Penser la trivialité* (Hermès, 2008). He is the editor of the journal *Communication & langages.*

Hak-Soo Kim is Professor, College of Communication, Sogang University and Fellow & Policy Studies Division Chair, Korean Academy of Science and Technology, Seoul, South Korea. He also directs the Academy for Scientific Culture as well as the Science Communication Laboratory of the same university. His research interests focus on theory construction, model-building, basic communication theory, and science communication.

Ock Tae Kim is a research fellow in the Department of Communication, Dongguk University, Seoul, and has an affiliation with the Seoul National University Institute of Communication Research. He has a Ph.D. in Telecommunications from Indiana University at Bloomington. His scholarship focuses on the intersections among media effects on individuals and society, communication technology, and science communication. His work has appeared in a number of conferences and journals in Korea and abroad.

P. V. S. Kumar is a social scientist with a Ph.D. in demography from IIT (Bombay). After specializing in the management of family planning and health programs, he worked for 3 years for the Indo-Global Social Services Society, an NGO dealing with rural and urban developmental issues, mainly monitoring and evaluating ongoing projects in sustainable livelihoods and the adult and continuing education sectors.

Later, he joined the National Institute of Science, Technology and Development Studies in New Delhi. He has 25 years of research experience in health, science, and technology and society. He has expertise in database design and implementation, and a particular interest and experience in developing ICT capabilities and infrastructure.

Anik Landry is coordinator of the Office of Internships at the Faculty of Education and a professional researcher at UQAM. She sometimes teaches at master's level in museum studies. She is an active member of RGEM/GREM (Research Group on Education and Museums), which conducts research on museum education. She has been doing research in that field for more than 10 years, and is focusing on the pedagogical evaluation of museum educational activities carried on the web. She has an extensive knowledge of the development of web-related communication technologies in the museum field.

Over the years, Anik has been doing research for IRCST/CIRST (Inter-university Research Center on Science and Technology), McGill University, Biosphere, the Society for the Promotion of Science and Technology, and various museums.

Markus Lehmkuhl is a science journalist and senior researcher at Research Center Jülich and at Science Journalism Division, Freie Universität Berlin. He is the chief editor of the internet journal *wpk-quarterly*, which aims to bridge the gap between theory and practice in science journalism. He has been a team leader in various international and national qualitative and quantitative studies on science in radio and TV in Europe, the elaboration of risk information, the perception of risk by mass media, science journalism in the local press, and the role of interpersonal influence in risk perception and judgement.

Lin Yin is Associate Professor and Assistant Director, Division of Theoretical Studies on Science Popularization, at CRISP. She is also a member of the Association of Chinese Science Writers. Her main research interest is in the history of science popularization, science writing and the development in S&T communication around the world. She has completed several research programmes, including 'Research on science popularization studies in China', 'Survey on popular science publications in China', and 'Looking into S&T educational activities for youth and adolescents in America'.

Joëlle Le Marec is professor of information and communication sciences at the University Paris Diderot (France), in scientific journalism and museology courses. She managed an interdisciplinary scientific community of sciences studies (the 'Enjeux et representations des Sciences, des technologies et de leurs usages' cluster) in Lyon for 4 years from 2007 to 2011. She also works with museums, libraries and all cultural institutions related to knowledge in society. She is very much interested in scientific sociability and in relations between universities and non-academic research.

Jenni Metcalfe is Director of Econnect Communication, which is Australia's earliest science communication consultancy, operating since 1995. Econnect conducts all phases of science communication, from research to practice, and then evaluation of that practice. Jenni lectures in science and medical journalism at the University of Queensland and has qualifications in science, journalism, education and communication. She has worked as a science communicator for 22 years and is a foundation member of Australian Science Communicators, of which she was president from 2005 to 2007. Jenni has been a member of the Public Communication of Science and Technology Network scientific committee since 1996. She is particularly interested in the interface between the theory and practice of science communication and is currently studying that topic as part of her Ph.D. with the University of Queensland.

Hester du Plessis (DLitt et Phil Philosophy) is a Senior Research Specialist in the Research Use and Impact Assessment unit at the Human Sciences Research Council of South Africa. Her area of research is in the public understanding of science. She works in collaboration with the Science Communication through Multi-Media research team at the National Institute of Science Communication and Information Resources of the Council for Scientific and Industrial Research in New Delhi, India. She is an Associate Researcher at the Sustainable Energy Technology and Research (SeTAR) Centre, Faculty of Science, University of Johannesburg. At the SeTAR Centre, she collaborates on a research programme on the public understanding of science as well as a research programme on affordable alternative energy as a key element for sustainable development. She is a Fellow at the Mapungubwe Institute for Strategic Reflection and a team member of 'The concept and application of transdisciplinarity in intellectual discourse and research' project.

Carmelo Polino is a researcher at Centro REDES (Argentina), an institution associated with the Conicyt (National Council of Science and Technology). He is also an associate researcher of the Network of Science and Technology Indicators and the Observatory of Science, Technology and Society. He holds a doctorate from the Universidad de Oviedo (Spain); a Master's of Communication of Science, Technology and Society, University of Salamanca (Spain); and a Master's of Science, Technology and Society, Universidad de Quilmes (Argentina). His research interests are related to the sociology of science communication, public perception of S&T and scientific culture.

Gauhar Raza is a scientist and the head of the Science Communication through Multimedia Division at the National Institute of Science Communication and Information Resources in New Delhi. His current interests include methodological questions that researchers face in the area of public understanding of science, especially in the third world. Raza has worked on issues related to indigenous knowledge systems, cultural aspects of public understanding of science and scientific temper. He has also produced a number of documentary films on various science and technology topics. He has authored and co-authored many books and research papers on varied subjects. He produced the first Hindi-language science serial for TV.

Jan Riise has a B.A. in urban and regional planning and works as a freelance science communication project manager and senior adviser. He is based in Gothenburg, Sweden, where he was the co-founder of the International Science Festival in the mid-1990s. Since then, he has been working closely with science festivals, science events and science centres in EU-funded projects. He was the president of the European Science Events Association from 2008 to 2010. His work has included project management of the PCST-10 conference in Malmö–Lund–Copenhagen in 2008, communication and web management of the 2WAYS project in 2009 and 2010, and project management at the Gothenburg Center for Public Learning and Understanding of Science and the Center for Culture and Health at the University of Gothenburg. He has participated as speaker and moderator at many international conferences and organized parts of the outreach programme of the European Commission during the EU–China Science Week in Shanghai in 2010.

In 2011, he works with Mistra Urban Futures, a transdisciplinary research centre focusing on urban development. He has a special interest in the spatial dimension of science communication.

Xavier Roigé is professor of social anthropology at the University of Barcelona and a specialist in museology and cultural heritage. He is Director of the Master's in Cultural Heritage Management as well as Vice-Dean for Research in the Faculty of Geography and History at the university. He is a member of a number of museum and ethnographic heritage committees. In the field of cultural heritage, he has undertaken studies on local, ethnological and national museums. He has also carried out a number of museological and exhibition projects and has published numerous books and articles in specialist journals.

Bernard Schiele is Head of the Graduate Program in Science, Technology and Society at Université du Québec à Montréal (UQAM), Researcher at the Interuniversity Research Centre on Science and Technology, and Professor of Communications at the Faculty of Communication at UQAM. He frequently teaches and lectures in North America, Europe and Asia. He has been working for a number of years on the socio-dissemination of S&T.

Bernard is a member of several national and international committees and is a regular consultant on scientific culture matters to governmental bodies and public organizations. He is also a founding and current member of the scientific committee of the PCST Network. Recently, he chaired the International Scientific Advisory Committee for the New China Science and Technology Museum (Beijing). At the moment, he chairs the scientific committee for the next Hubert Curien Conference (Nancy, France, 2012).

Vladimir de Semir born in 1948, has been a journalist since 1975 and has specialized in science journalism and science popularization since 1982. In 1994, he began teaching and training new journalists and popular science writers as Associate Professor of Science Journalism in the Faculty of Communication at Pompeu Fabra University, Barcelona, where he also directs the Science Communication Observatory, a research centre specializing in analysis of the transmission of scientific, medical and environmental knowledge to society. At Pompeu Fabra University, he is also director of the Master's in Scientific, Medical and Environmental Communication and of the Diploma in Scientific Communication in Buenos Aires (Argentina). He is a member of the international Public Communication of Science and Technology Network, a member of the expert group, Monitoring Policy and Research Activities on Science in Society (MASIS), of the European Commission, and a member of the European Science Communication Network (ESConet).

Shunke Shi is Head of the Division of Theoretical Studies on Science Popularization of the China Research Institute for Science Popularization, with which he has worked since 1988. His main research interest concerns the history of science popularization and the theoretical study of S&T communication development around the world.

He has published articles on science popularization, edited *The English–Chinese lexical dictionary*, collated *A Chinese–English dictionary of classic and current expressions*, published *Brief introduction to science popularization in USA*, and collaborated in publishing the books *At the human scale* as co-author and *Communicating science in social contexts* as co-editor.

He helped organize the Beijing PCST Working Symposium in 2005 and the International Forum on PCST Studies in Beijing in 2010.

Surjit Singh is a researcher and a member of the core team conducting research on public understanding of science at the National Institute of Science Communication and Information Resources in New Delhi. He has been involved in large-scale survey studies on the public understanding of science, especially in collecting time-series during *Kumbh* and *Ardh-Kumbh Melas* (mass Hindu pilgrimages to the Ganges). He is also interested in science popularization activities, especially among children, and is involved in HIV/AIDS awareness programmes. He participated in the *Bharat Jan Vigyan Jatha* ('Campaign for Taking Science to the Public') in 1987. He has 3 books and about 20 research papers and articles to his credit.

Brian Trench is a writer, trainer and researcher in science communication. He is a former senior lecturer and Head of School in the School of Communications, Dublin City University (retired 2010), where he directed or co-directed the Master's in Science Communication for 15 years. He was also the inaugural chair of the university's Research Ethics Committee (2004–2009) and co-organiser of annual Science//Society seminars (from 2006). Brian's research interests are focused on internet communication in science, models of science communication, practices in science journalism and public engagement with new sciences. He is currently taking part in three projects under the European Union's Science in Society programme.

He is chair of the conference committee of the international Public Communication of Science and Technology Network and a member of the local organising committee for the Euroscience Open Forum in Dublin in 2012.

Brian has lectured, given conference presentations or led workshops on science communication in many countries. He has contributed chapters on science communication to several collected volumes and is co-editor (with M. Bucchi) of the *Handbook of public communication of science and technology* (Routledge, 2008).

Before joining Dublin City University in 1992, Brian was a journalist for 20 years, working for magazines and newspapers and as a freelance contributor for many national and international broadcasters and publications.

Contributors

Martin W. Bauer Institute of Social Psychology and Methodology Institute, London School of Economics and Political Science, London, United Kingdom

Massimiano Bucchi Science and Technology in Society Programme, Università di Trento, Trento, Italy

Yurij Castelfranchi Federal University of Minas Gerais (UFMG), Belo Horizonte, Brazil

Sook-Kyoung Cho Korea Foundation for the Advancement of Science and Creativity (KOFAC), Seoul, Korea

Sung Kyum Cho Department of Communication, Chungnam National University, Daejeon, South Korea

Michel Claessens ITER Organization, St Paul-lez-Durance, France

Fujun Ren China Research Institute for Science Popularization (CRISP), Beijing, China

Toss Gascoigne Toss Gascoigne & Associates, Campbell ACT, Australia

Michèle Gellereau University of Lille, Lille, France

Honglin Li China Research Institute for Science Popularization (CRISP), Beijing, China

Maja Horst Department of Media, Cognition and Communication, Faculty of Humanities, University of Copenhagen, Copenhagen, Denmark

Huiliang Zhang China Research Institute for Science Popularization (CRISP), Beijing, China

Yves Jeanneret Paris Sorbonne University, Paris, France

Hak-Soo Kim College of Communication, Sogang University, Seoul, South Korea

Ock Tae Kim Institute of Communication Research at Seoul National University, Seoul, South Korea

P.V.S. Kumar SCM Division, National Institute of Science Communication and Information Resources, New Delhi, India

Anik Landry University of Quebec at Montreal (UQAM), Montreal, Canada

Markus Lehmkuhl Research Center Jülich, Freie Universität Berlin, Berlin, Germany

Lin Yin China Research Institute for Science Popularization (CRISP), Beijing, China

Joëlle Le Marec University Paris-Diderot, Paris, France

Jenni Metcalfe Econnect Communication, South Brisbane, Australia

Hester du Plessis Research Use and Impact Assessment, Human Sciences Research Council (HSRC), Pretoria, South Africa

Sustainable Energy Technology and Research (SeTAR) Centre, Faculty of Science, University of Johannesburg, Johannesburg, South Africa

Mapungubwe Institute for Strategic Reflection, Johannesburg, South Africa

Carmelo Polino Centro REDES, Associated to the National Council of Science and Technology, Buenos Aires, Argentina

Gauhar Raza Head, SCM Division, National Institute of Science Communication and Information Resources, New Delhi, India

Jan Riise Agadem AB, Onsala, Sweden

Xavier Roigé Department of Social Anthropology, Research Project CSO2011-29413, University of Barcelona, Barcelona, Spain

Bernard Schiele International Research Center on Science and Technology (IRCST/CIRST), University of Quebec at Montreal (UQAM), Montreal, Canada

Vladimir de Semir Science Communication Observatory, Pompeu Fabra University, Barcelona, Spain

Shunke Shi China Research Institute for Science Popularization (CRISP), Beijing, China

Surjit Singh SCM Division, National Institute of Science Communication and Information Resources, New Delhi, India

Brian Trench School of Communications, Dublin City University, Dublin, Ireland

Abbreviations and Acronyms

ACFAS	Association canadienne-française pour l'Avancement des Sciences (Canada)
ANC	African National Congress (South Africa)
ANU	Australian National University
ASC	Australian Science Communicators
CCSVI	Chronic cerebrospinal venous insufficiency
CPAS	Australian National Centre for the Public Awareness of Science (ANU)
CPC	Communist Party of China
CRISP	China Research Institute for Science Popularization
CSIRO	Commonwealth Scientific and Industrial Research Organization (Australia)
CSL	Civic scientific literacy
ENSCOT	European Network of Science Communication Teachers
EU	European Union
FP	Framework Programme (EU)
GDP	Gross domestic product
HASS	Humanities, arts and social sciences
HSRC	Human Sciences Research Council (South Africa)
IKSs	Indigenous knowledge systems
KOFAC	Korea Foundation for the Advancement of Science and Creativity
KSF	Korea Science Foundation
KSJA	Korea Science Journalists Association
KSSP	Kerala Sastra Sahitya Parishad (India)
MS	Multiple sclerosis
NGO	Non-government organization
NPO	Non-profit organization
OECD	Organization for Economic Co-operation and Development
PCST	Public communication of science and technology
PR	Public relations
PSL	Public scientific literacy
PUS	Public understanding of science

PUSH	Public Understanding of Science and Humanities (Germany)
R&D	Research and development
RD&I	Research, development and innovation
SAASTA	South African Agency for Science and Technology Advancement
SCI	Science Culture Index
SCR	Science communications research
SET	Science, engineering and technology
SISSA	Scuola Internazionale Superiore di Studi Avanzati (International School for Advanced Studies) (Italy)
SP	Science popularization
STAGE	Science, Technology and Governance in Europe
STC	Science and technical culture
STEM	Science, technology, engineering and mathematics
STS	Science, technology and society

List of Figures

Fig. 5.1 Science popularization policy documents, distribution over 5-year intervals, 1949–1955 to 2006–2010 ... 84

Fig. 11.1 'Science and Future' section in Korean newspapers ... 175

Fig. 17.1 Nobel articles, *Il Corriere della Sera*, 1901–1999, by newspaper section (Note: *Lines* represent the percentage of articles, per section, over each decade. Total number of articles=263) ... 263

Fig. 17.2 Articles on Guglielmo Marconi, *Il Corriere della Sera*, 1901–1911 (*n*=215) ... 266

Fig. 20.1 The Yin and Yang of science: scientific performance and mentality of science in society as mutually reinforcing processes ... 302

Fig. 20.2 Two hypothetical relationships between competence and attention and aspirations towards science. In an industrial context, this might be positive; in a post-industrial context, negative ... 307

List of Tables

Table 2.1 Where science communication practitioners get ideas to inform their practice ... 29

Table 4.1 Models and standpoints of public communication of science ... 74

Table 5.1 Science popularization policies, grouped by actor ... 85
Table 5.2 Programs supported by science popularization policies ... 86
Table 5.3 Outputs from science popularization investments between 2004 and 2008 ... 90

Table 8.1 Initiatives targeting science popularization in Germany ... 128

Table 11.1 Science Korea movement programs ... 173

Table 15.1 Some major initiatives supported by the EU's Framework Programmes to stimulate and promote science communication and education through transnational research projects (Source: CORDIS (www.cordis.lu)) ... 230

Table 17.1 Nobel laureates most often cited in articles, *Il Corriere della Sera*, 1901–1999 ... 264

Table 18.1 Correlations between P-Engagement and SPS-Engagement (PEP/IS-I Index) ... 276
Table 18.2 Correlations between P-Engagement and SPS-Engagement (PEP/IS-II Index) ... 276

Introduction

Simply put, the public communication of science and technology (PCST), beginning from the 1960s, underwent an expansion that would lead to the predominant place PCST now assumes in the public sphere. That development occurred along two axes: first, fewer but more diversified practices of public awareness, promotion and communication; and second, a growing number of major theoretical developments. This expansion of PCST practices and the formation of a specific theoretical field reflect the awareness that science and technology are today deeply embedded in our society and tremendous vectors of economic, social and cultural change.

For this reason, all countries to varying degrees now promote, support and value PCST actions. They draw up and implement national policies, master plans for development and action programs. These initiatives range from the introduction of major cultural installations such as science centres, as part of national projects, to hosting public awareness programs and activities aimed at providing and disseminating science and technology in cities, communities, schools and other associations to meet the particular needs of different publics. One need only consider the significance of the current construction program of science museums in China to glimpse what the valorization and diffusion of science represent for that country in particular, and for all other countries similarly engaged in a fierce race to maintain their scientific and economic competitiveness.

In this context, research on the communication and appropriation of science has become a strategic priority.

Up to the turn of the 1960s, theoretical reflection was largely the purview of science communicators. It was mainly the science journalists themselves who proposed the models to describe, explain and justify their practices. The 'deficit model' being debated today—first formulated by C.P. Snow in his famous thesis on the 'two cultures', describing the gap between science and culture in modern society—was often cited by science journalists. They have long held that science created its own mystique by forging a wide gap between scientists and the general public, since the reality constructed by modern science no longer corresponds to people's ordinary sensate perception of the world. So they assumed responsibility for repairing the broken link by demystifying science. The idea of the 'deficit model' became so popular and entrenched that much work conducted in the nascent field of PCST did not even question the presuppositions on which it was based.

However, subsequent work in the history and sociology of science, communications, linguistics and discourse analysis quickly showed the limits of such a conceptualization. For 40 years, this work has revolutionized our understanding of the processes of dissemination and propagation of knowledge, our understanding of the modalities of appropriation of knowledge by social actors, scientists and laypeople, and our understanding of the social, economic and political issues associated with any undertaking to valorize and disclose science and technology.

This research, conducted worldwide, has helped create and 'autonomize' a new research field. International conferences, held regularly since 1989, specialized publications, university programs in this area and the creation of teaching positions in science communication have firmly established the field. This book therefore reflects vigorous development in research and a growing professionalization of these activities. PCST research is an extremely dynamic field that brings together the whole spectrum of scientific approaches (theoretical description, experimental initiatives, pragmatic approaches, formalization, etc.) and is in the process of becoming a fully separate academic discipline, as evidenced not only by its intellectual productions but most of all by the institutionalization of university studies into three levels (bachelor's degree, master's degree, doctorate). Faculty members offer these courses and work on research in this field.

But despite the growing number of projects in the field, there has until now been no global overview of their development or theoretical scope. This was largely due to the nature of the research work, which was often polarized on national priorities, and due to the dispersion of the research teams, which formed around those priorities and focused on ad hoc demands. The precise goal of this volume is to overcome that deficiency by providing an overview of the development of the theoretical field of PCST beginning from the 1960s, when it was first forming, up to the present day.

We therefore wished to provide an overview in the context of globalization—that is, in a world where societies are interdependent, and where each one perceives its development in dynamic relation to the others. This means we have taken great care to clearly reflect the diversity and complementarity of scientific contributions in constructing the field. We sought the collaboration of authors whose work stands as the authority in their respective countries and who are also recognized by the international community of researchers in the field.

This volume is first and foremost a theoretical report. In recent years, the main focus has been on the excellence of practices. Considerable work has been devoted to this, aimed at drawing and sharing lessons from success stories. Research results in the PCST field have also been the subject of a growing number of publications on subjects ranging from the measurement of spin-offs triggered by 'consensus conferences' to the growth of theatrical performances in museums, by way of the evaluation of science boutiques and similar mechanisms. But no existing publication had sought to chronicle the theoretical development of the field for the scientific community at large. This is what we wished to achieve. Until now, the theoretical work has been confined to specialized journals, giving the appearance of closure to anyone

Part I
National Overviews

Chapter 1
The ‘Communicative Turn’ in Contemporary Techno-science: Latin American Approaches and Global Tendencies

Carmelo Polino and Yurij Castelfranchi

Abstract Between the Second World War and the end of the Cold War, dramatic changes have occurred both in the mode of production of scientific knowledge and in the relationships between knowledge, innovation, the economy and civil society. We hypothesize that the reconfigurations in science made possible by a closer relationship with capitalism stimulated a ‘communicative turn’ in science communication. Public communication of science and technology (S&T) has transformed into a structural value within the core axiological pluralism of contemporary techno-science: journalistic values, persuasion, publicity, opinion etc. converge within the axiological core of techno-science. Therefore, science communication is today not only a ‘right’ for the publics and a ‘moral duty’ for scientists, but a need for society and an unavoidable, intrinsic process in the metabolism of contemporary capitalist democracies. We map the present forms of techno-scientific practices and discourses, and show how public communication of S&T has a central role (within fluid, conflictive, global and mediatized scenarios) in understanding both the production of knowledge and the governance of contemporary techno-science. We describe three different aspects of communication: in research institutions, in global mass media, and in civil society and ‘participatory movements’. We show how new mediators and communicators (or no mediators at all), new stakeholders in science communication and new ‘sources’ of scientific knowledge emerge today

---

In the case of Yurij Castelfranchi, part of his research was funded through a grant of the Federal University of Minas Gerais: ‘Programa Institucional de Auxílio à Pesquisa de Doutores Recém- Contratados- PRPq/UFMG’.

C. Polino (✉)
Centro REDES, Associated to the National Council of Science and Technology,
Buenos Aires, Argentina
e-mail: cpolino@ricyt.org

Y. Castelfranchi
Federal University of Minas Gerais (UFMG), Belo Horizonte, Brazil
e-mail: yurij@fafich.ufmg.br

B. Schiele et al. (eds.), *Science Communication in the World: Practices, Theories and Trends*, DOI 10.1007/978-94-007-4279-6_1,

in this interesting scenario. Finally, in a preliminary analysis to be completed in further work, we present examples and possible case studies from Latin America. The region is exposed to global pressures and trends similar to those in the developed countries (S&T policies; intellectual property rights; relationships between government, university and market; public participation, inclusion and engagement in S&T etc.) but, at the same time, is subject to very diverse and very specific boundary conditions and historical trajectories.

Keywords Sociology of S&T • Contemporary techno-science • Science communication theory • Models and practices in public communication of S&T • Latin America

1.1 Introduction

Scholars investigating public communication of science and technology (S&T) today face an insidious interdisciplinary challenge: that of unraveling changing configurations and recombination in the co-evolving entanglement between the production, circulation, appropriation and governance of scientific knowledge and socio-technical systems. On the one hand, almost all the sociological models and diagnoses of the contemporary world ('risk society', 'network society', 'reflexive modernization' and so on) focus explicitly on the roles played today by S&T in all spheres of human activity: subjectivity and cultural changes; social stratification and public policies; social control and governance; market dynamics and social networks; and so on. At the same time, such theoretical approaches emphasize important changes in both the functioning and practices of science communication and its actors. On the other hand, S&T studies have also focused, in the past decades, on the hypothesis that scientific practices today may be quite different from those of the times of Newton, Darwin or even Einstein in their norms, organizational and epistemic structure, and relationships with the market and politics.

These circumstances came together with an increasing public exposition of S&T during recent decades. Different social movements since the 1960s began to denounce the industrialization model, as well as the serious environmental consequences of science-based technological applications. The present configuration of techno-science shows an engagement of different agents and institutions participating in global science communication fluxes that do not accept a more or less passive role as 'audience' or 'consumers', as in traditional communication models, and reclaim a role as participants and producers of information and knowledge. All this has important impacts on the dynamics of both the production and the diffusion of knowledge, and also changes communicative strategies for internal communications between scientists, communication in the public arena, relationships with mass media, marketing, political propaganda and expert consulting. Public communication of S&T has, therefore, a central role (within fluid, conflicting, global and mediatized scenarios) in our understanding of both the production of knowledge and the governance of contemporary techno-science.

1.2 Science, Technology and Capitalism: From 'Big Science' to 'Techno-science'

Between the Second World War and the end of the Cold War, dramatic changes occurred both in the mode of production of scientific knowledge and in the relationships between knowledge, innovation, the economy and civil society. The transformations in the regimen of capitalist accumulation meant a kind of reorganization in which the production and appropriation of scientific knowledge came to play an important and strategic role.

Several scholars studied those transformations, and some of them formulated a strong claim: that science changed radically during the past 40 years, having become today an activity epistemically and institutionally different from the 'Big Science' we were used to in western nations for much of the twentieth century. Some authors have stressed the economic aspects (reorganizations in the capitalist regime in which the production and appropriation of scientific and technological knowledge play novel, deeper strategic roles). Others have emphasized institutional, organizational or epistemological changes in science. 'Post-industrial' society, as proposed by Bell (1994) and others, Weinberg's 'trans-science' (1972), 'regulatory science' (Jasanoff 1995), 'post-normal' science (Funtowicz and Ravetz 1993), the 'Mode 2' of knowledge production (Gibbons et al. 1994; Nowotny et al. 2001), 'post-academic' science (Ziman 2000), and 'techno-science' (Latour 1992; Echeverría 2003) are some examples. More generally, recent sociological diagnostics of contemporary societies, such as 'risk society' (Beck 1999) and 'network society' (Castells 1997) analyses, also showed the crucial role of these changes in the relationships between science, technology, politics and the market.

Among several factors, one that most authors emphasize is the remarkable role played by private capital in contemporary techno-science in most developed and emergent countries. In the decades of the Cold War, science was strongly and mainly supported and funded by nation states, and rhetorically seen as a 'common good'. That configuration began to change in the 1980s with strong growth of private funding for R&D, and scientific knowledge began to be seen as something that could, or even should, be commercialized, sold and patented (Bauer 2008). Of course, we are not thinking of a sharp and rigid historiographical periodization, separating two 'eras' of science, nor claiming that a revolutionary paradigmatic shift has happened. Several elements of contemporary science were already functioning during Big Science, or were always present in modern science. Science studies already showed in the 1960s and 1970s that science is immersed in a field in which economic pressures, political strategies and interactions with the public are relevant. However, as we shall show, the contemporary modes in which such factors function, as well as their meanings, their relative strengths and their effects, are different.

Today, techno-science shares with business and industry several norms and practices. Economic rationality plays a role in the force field that shapes what science is and how it is done, so that S&T systems are being thought of, in most developed countries, as big companies with mixed capital, and many concepts of the business world (such as flexibility, mobility, venture capital, competitiveness, performance and productivity) are being applied to them. Synergy, efficiency, spin-offs, failure/success, marketing, proactivity and entrepreneurialism have entered the daily vocabulary and practices of researchers in many different areas of science. Narratives of techno-science tell us today a story in which the production and circulation of scientific and technical knowledge have to be managed in 'efficient', 'calculated' forms more directly linked to the 'national security', 'social demands' and 'economic performance' of the nation states. Both in Latin America and in the 'developed' countries, policymakers, managers and techno-scientific leaders repeat slogans that emphasize the need for a reconfiguration in the role of universities and research: they tell the story of the 'challenge' and of the 'urgent need' to create 'entrepreneurial universities' able to 'commercialize' and 'sell' research to society (Etzkowitz 2001).

If such reconfigurations in the relationship between scientific research and the markets are linked to the present shape of capitalism, the links between science and politics are also affected by the growing importance of risk and socio-environmental issues. Both in 'risk societies' (Beck 1999) and in 'reflexive modernization' (Giddens et al. 1997), the problem of the social consequences of S&T is central, intrinsically political and global. A crisis of legitimization in the Cold War 'social contract between science and society' (Nowotny et al. 2001) also emerged from the increasing visibility of conflicts of interests—biomedicine, genetically modified organisms, patents—as well as the publicizing of several recent cases of misconduct in science (Castelfranchi 2008).

Focusing on the axiological aspects, some authors also claimed that some relevant changes are occurring in contemporary techno-science. Echeverría (2003), for example, like other authors, argues that economic, political and military progress constitute the fundamental pragmatic principle that guides techno-science. From this, it follows that a fundamental principle in techno-science is not knowledge, the basic principle of modern science, but 'the capacity of action' (Echeverría 2003:267). Techno-science praxiology therefore assumes wider and more complex values than modern science. The classic epistemic values are maintained; however, techno-science incorporates new values or radically modifies the relative weight of previous values. Echeverría's axiology incorporates typical technical and technological values (efficiency, innovation, etc.); economic values (patents, resources optimization, benefits, management, competitiveness, profitability, etc.); military values (national autonomy and security, etc.); and ecological, human, political and social values in a broad sense.

1.3 New Values, Practices and Meanings for the Public Communication of S&T

In this (at least partially) new scenario, we can claim that public communication of S&T is today not only a moral duty for scientists, a necessity for the publics, or a tactical need of scientific institutions that try to politically legitimate their activity or to gain funds and sponsors, but also a spontaneous, necessary, physiological process in the functioning of techno-science. Science and technology are communicated not only through traditional channels (formal and informal, education, popularization, science journalism etc.) but by complex fluxes of communications that do not always have scientists, institutions or professional communicators as authors (mailing lists, patient groups, social movements, debates in the media etc.). Our hypothesis is that the reconfigurations in science made possible by the closer relationship with capitalism also stimulated a 'communicative turn' in science communication: journalistic values, persuasion, publicity and opinion converge within the axiological core of techno-science. Certainly, public communication has always been an important value since the period of science professionalization in Europe. But with the emergence of Big Science and, especially, techno-science, communication has acquired a new status: it has become a structural and structuring feature of contemporary techno-science.

The confluence between new values and contexts of practices stimulates quite remarkable research questions on science communication in global and mediatized societies. One is about the types of communication and the ways and channels where communication occurs. Several theoretical proposals have been made, such as that by Cloître and Shinn (1985) identifying four scenarios for communication processes, based on the type of product produced (*intra-specialist*, *inter-specialist*, *pedagogical* and *popular*). Other typologies look at the relationships between the publics involved (see Verón 1999) or the distinctions among 'endogenous' or 'exogenous' fields with respect to science. Such taxonomies suggest frontiers at different levels of science communication and are useful for analyzing traditional scientific popularization, where the key factor is to explain (to share) scientific knowledge or ideas with non-specialized audiences. However, communication practices in techno-science are more complex. Many authors acknowledge that the traditional frontiers between the communicative contexts have become more porous and fluid, that the separation between 'science' and the 'lay public' had become less radical, and that communication had started to look more like a communicative *continuum* (Bucchi 1998). Several case studies criticize models in which there is a 'science which is made' and, after that, another science that is 'popularized' or 'socially widespread'. Some examples are Clemens' study (1986) on the extinction of the dinosaurs at the K-T boundary; Nieman (2000) in relation to the popularization of physics; Lewenstein (1995) on the 'cold fusion' saga; Bucchi (2000) on COBE satellite discoveries; and Kiernan (2000) on NASA's information control during the announcement of the 'Martian meteorite'.

In what follows, we show some tendencies in global and local (Latin American) practices that are connected to the emergence of communication as a structural feature of science, the global mass media, and civil society and its ways of appropriation and participation in social debates.

1.3.1 Research Institutions and Communication

Public communication (in an institutional, political, mediated or marketing context) is today an essential need for many researchers and techno-scientific organizations.

The search for visibility, legitimization, funding and alliances and the need for negotiations and dialog with different stakeholders generate new impulses for science communication. In some cases, scientists metaphorically wear their white coats when they enter political debates and conflicts, as a symbol of a pure, neutral, universal knowledge. In other cases, they accept the rules of the mediated game. In order to sell books, market a company or become 'visible', they accept the use of 'hype'. Of course, 'selling science' to the public is not new: rhetorical strategies to popularize scientific claims to wider audiences are known since the seventeenth century, and the use of the media in order to gain prestige or political legitimization has been an important part of the work and tactics of 'academic scientists' since the nineteenth century. What is new is not that behavior, but the way the relationship with media is felt, encouraged, operationalized and institutionalized, for example by means of media offices and explicit incentives for science communication and engagement activities.

The media offices of many big scientific institutions today produce multimedia news and materials for journalists. In several cases, they use (in titles, images and metaphors) 'sex appeal' and the kind of hype that scientists used to see (and criticize) as a product of 'bad' journalists, not of their own media offices. 'God's machine', 'the face of God', 'God's particle', 'the pillars of Creation', 'the missing link between ape and man' and 'the Holy Grail' are just some examples of hyperbolic expressions that have their source not in journalists, but in scientific institutions or even in the official declarations of scientists. At the same time, journals (*Science* is a good example), forums and scientific institutions stimulate blogs, wikis and social networks to communicate science. The same occurs with 'open days' in the labs (for journalists, politicians, businessmen and the 'general public'). This type of activity is increasingly used by scientists and their institutions to improve the penetration and the impacts of their communication.

Science advocacy is another interesting aspect of these processes: marketing, lobbying and publicity are increasingly important in S&T (Castelfranchi 2002). What is remarkable is that an important part of such activities is directed not only at politicians or the business sector, but at civil society, too. The need of techno-science for accountability, lobbying and public legitimization today joins with the need of politics to legitimate itself through science. Techno-science needs politics as much as politics needs techno-science. A strong legitimacy argument in contemporary politics is to claim that some position is based 'not on ideology, but on facts'.[1] Efficiency and efficacy are used as strong arguments in favor of some public policy, as are 'justice', 'equality' or 'moral values', so that almost every NGO, party and coalition searches for some researcher saying that, for example, global warming exists (or does not exist), that it is not (or is) crucially linked to anthropogenic emissions, and so on.

[1] See Castelfranchi (2008) for an analysis and sources of this and following claims.

As a consequence, public communication of science today plays a complex role, not only in the public agenda but also in the governance of S&T and even inside the lab. It is not exogenous to science; nor does it happen diachronically 'after' the research to diffuse new knowledge. It is also an intrinsic, physiological process that happens synchronically with new research, shaping the research's forms and possibilities. Science communication, therefore, serves pedagogical goals, but also informative, strategic, marketing, and political ones. It is also an important battlefield for struggles between scientists for prominence, funding, and epistemic or political authority.

1.3.2 Science Communication in the Media: The Global and Latin American Contexts

The sociological map of science communication is not limited to the communicative strategies produced by agents and science institutions. Traditional and new mass media make possible science communication with a permanent presence and a differentiated audience on a global scale. The socio-institutional transformations in the dynamic of contemporary science and the global expansion of mass media are, in fact, social phenomena which experienced parallel evolutions (Bucchi 2008). Besides that, the mass media have come to play a decisive role in contemporary democratic societies, in which the media are now one of the fundamental social institutions. And, like many systems, the media have suffered the impact of capitalistic reconfigurations. Hallin (2008:43), for instance, says that 'the mass media are among the most important of those social institutions which have been subject to "enclosure" by the logic of the market in the Age of Neoliberalism.'

So, science communication is not comprehensible today outside the actions of mass media. In the past, most scientific controversies remained inside small circles of the 'experts' and 'colleagues'. Today, many of them are known by nearly everyone, amplified by the reverberations of the global mass media. Great empirical evidence exists to show how science follows logics of production, diffusion and battles typical of the 'cultural industries' (Schiele 2006, 2009; Cheng et al. 2008; Dierkes and von Grote 2000; Castelfranchi and Pitrelli 2007; Polino and Castelfranchi in press; Friedman et al. 1999; Bauer and Bucchi 2007).

The relation between journalists (mass media) and scientists (research institutions) is, however, controversial. Science journalists are culturally dependent on scientists because of scientists' capacity to produce 'autonomous information' (Bourdieu 2007:104). In contrast, journalism, fundamentally through the triumph of television and its effects on the social agenda, has imposed the logic of the audience on science.

While such a 'latent conflict' certainly exists, science–media interfaces have experienced transformations that could be interpreted as structural changes in the traditional relationship between scientific and media cultures. Empirical evidence shows that interactions between the media and science, present since the professionalization and institutionalization of science in the nineteenth century, have been increasing in recent years. Weingart (2001), quoted by Peters et al. (2008a), observed a close connection between scientific institutions and the media which was connected to the need for legitimacy and political influence, but which was also used to seek public support in cases of disputes within science itself. Scientists now have a deeper perception of the impact of media on their careers, activities and social purposes. A mail survey conducted by Peters et al. (2008a, 2008b) in the United States, Japan, Germany, the United Kingdom and France revealed that scientist–journalist interactions 'were more common than anticipated' (Peters et al. 2008a:204). The vast majority of the respondents acknowledged that 'increasing the public's appreciation of science was the most important benefit mentioned as an incentive to interact with the media' (Peters et al. 2008a:204). In all five countries, 'a plurality of scientists who had contact with the media in the past 3 years rated the impact of those contacts on their careers positively.' Another cross-national analysis of popular science publishing among university staff in 13 developed and developing countries counted the number of articles by scientists in newspapers and magazines over the 3-year period from 2005 to 2007. The data suggest that academic staff with popular publications also have higher levels of scientific publishing and academic rank—a fact that confirms previous findings in surveys by Kunth (1992).[2] So it is less the social reality of this correlation that is new than the fact that scientific communities now accept it as they previously did not. This finding, Bentley and Kyvik (2011) show, is consistent across all countries and academic fields. Other independent investigations have recently identified similar patterns in the relationship of science, popularization and the media in countries as different as Italy (Bucchi and Mazzolini 2003), Argentina (Kreimer et al. 2011), the United Kingdom (Royal Society 2006), France (Jensen 2011) and Spain (Torres Alberó et al. 2011).

The mass media also have changed science's work: for example, 'researchers are often among the most assiduous users of science coverage by the media, on which they draw to select among the enormous mass of publications and research studies in circulation' (Bucchi 2002:113). For example, famous work by Phillips et al. (1991) showed how a paper published in the *New England Journal of Medicine* is three times more likely to be cited in the scientific literature if it has first been mentioned by the *New York Times*. Kiernan (1997, 2003) demonstrated similar results. Journalism influences scientific controversies, too: the mass media shape discussion about the acceptance of scientific theories or technological developments (Brossard 2009; Clemens 1986; Epstein 1996; Wynne 1989). Moreover, the mass media may be used by scientists as platforms to assure their priority in discovery—a well-known phenomenon in the sociology of science (Collins and Pinch 1993). Live press conferences may be called before a formal paper is published, or even submitted. In addition, many scientific papers are published in specialized journals and simultaneously reported by the mass media, amplifying both the audiences and the contexts of evaluation.

[2] See also the analysis and theoretical explanation by Jacobi and Schiele (1988).

Other visible consequences are produced by the mediatization of science. The media have played a decisive role in the amplification of risks associated with the development of S&T, as has been apparent in the United States and Europe, and also in Latin America (see Polino 2009; Takahashi 2010; Polino et al. 2006; Ramalho et al. 2011; Da Silva Medeiros and Massarani 2010). In fact, the visualization of risks is probably associated with the emerging forms of professionalization and journalistic institutionalization, first in the industrialized countries and, lately, in the developing world. This is already evident in countries such as Argentina, Brazil and Colombia (despite local and regional disparities). As a consequence of the incipient processes of institutionalization, in many cases science journalists have begun to act like economic or political reporters in the past:

> [J]ournalists who had previously deferred to party and group leaders or to state officials increasingly began to assert their independence and their right to scrutinize elites and established institutions on behalf of their readers and of 'society' or 'the public'. (Hallin 2008:43).

A closer look at the media reveals that certain important social issues have undergone a 'shift' in media coverage, from a pattern centered on 'scientific discoveries' to a more balanced and complex treatment, taking into account risks, interests, connections with business, environmental and social impacts, science for policy and policies for science, and so on (see Polino 2009; Massarani and Polino 2008; Massarani et al. 2007; Vara 2007a).

There are other interesting tendencies in the institutionalization of science communication activities in Latin America. Countries where national S&T systems are more developed or growing faster (such as Brazil, Mexico and Argentina) are also countries that experienced, during the past two decades, a strong growth in science communication. In Argentina and, particularly, Brazil, science communication has been growing through both public and private activities. In the past year, the Brazilian ministries of Science and Technology and of Education have given a very strong push to science communication and to mechanisms of social inclusion and participation in S&T. Brazil's S&T Week is today a huge constellation of events, occurring from the Amazon rainforest to southern villages. Currently, there is research on the public perception of S&T financed by the public sector in both countries. In Argentina, a public TV channel (Encuentro) has for nearly 10 years produced a high-quality scientific program (*Científicos Industria Argentina)* with very good ratings. Other private sector channels are launching special programs in S&T. One of the most popular infotainment TV programs in Brazil, *Fantástico*, has a very strong S&T component and is broadcast by Globo Network to a huge audience. Other private enterprises also see an important niche market in popular science: increasingly, the main publishing houses in Argentina and Brazil produce popular science books covering a huge range of issues and publics, as do the publishing arms of public universities and other institutions. There are Argentine and Brazilian editions of *Scientific American* magazine, as well as several other very popular science magazines. Public research foundations have also developed magazines for the 'lay' public. Journals such as *Revista Pesquisa Fapesp* (of the São Paulo state foundation for the advancement of research), *Ciência Hoje* (Brazil) and *Ciencia Hoy* (Argentina) are today not only distributed to scientific institutions, libraries and universities, but also sold in kiosks. Some postgraduate or master's degree courses in science writing and public communication of S&T have been created in both public and private institutions. And some very famous scientists have decided to become science writers, editorialists or educators, or to take on active public roles (for example, in politics or as entrepreneurs).

From all these developments, we see that the growing importance of science communication seems to be strongly linked to the emergence of new configurations and needs in the relationships between science, politics and the market, but that the process is not identical to that happening in the North, since it is affected by important historical conditions linked to several factors. Those factors include a tradition of top-down policymaking, often not transparent and in some cases with an authoritarian flavor; a great cultural diversity across the country (mainly in Brazil) and an important role for social movements, especially in fights for indigenous rights, small farmers rights, the environment, etc.; the very recent re-democratization in the 1980s, after some 20 years of military dictatorship; and a strong presence of the private sector in basic and higher education alongside a weaker presence in R&D.

1.3.3 Civil Society, Participation and Techno-scientific Communication

The crisis in industrialization models and the environmental consequences of technological applications have affected the communicative field by opening it to the actions of other agents and social institutions. During recent decades, those agents and institutions have started to intervene directly in scientific and technological issues: a 'participatory turn' in science and technology, supporting the idea of governance and the democratization of science and technological decisions. (Lengwiler 2008; López Cerezo 2003).

'Lay' users want to participate in the construction of knowledge, or at least in its validation and governance. In the medical area, increasingly strong, organized and informed patient groups can help to guide the research agenda. In some cases, they even have a say in establishing what should be considered 'good science' or 'bad science', as shown, for example, in the classic case study by Epstein (1995). In fact, in contemporary medicine, NGOs not only defend the rights of patients, but in some cases can collect more money than governments and decide how to use it. In other cases, knowledge production occurs, at least in part, outside the universities and centers of traditional research. Local communities and pressure groups can order reports and experiments by 'independent scientists'. In some cases, grassroots organizations may orient research demands, produce data, facts, and truth effects, circulate discursive fragments and do research, having an influence on scientists' behavior or even over methodological decisions, and transforming themselves in an integral part of techno-science (Bucchi 2009; Castelfranchi 2008: Chap. 4).

Petras and Veltmeyer (2006), analysing the cases of Argentina, Brazil, Bolivia and Ecuador, identified three basic modalities of social change and political power in the region: electoral politics, social action in the direction of local development, and the construction of social movements. In Latin American countries, there have been some remarkable recent events that have put civil society into the center of the technological development discussion and, more widely, of debates on democracy and sustainable development. Examples include social resistance to open-pit mining in Argentina, Chile, Bolivia and Peru (Svampa and Antonelli 2009), social mobilization and organization against the installation of pulp mills on the margins of the River Uruguay (Vara 2007b), and the public discussion on lithium extraction in Bolivia. In Brazil during the past decade, governmental rhetoric has placed a stronger emphasis on e-democracy, social inclusion, participation and engagement. Several mechanisms for 'bottom-up' deliberation, such as participative budgeting, public consultations and plebiscites, have assumed stronger roles in governance processes. Science and technology were not immune to the process. 'Civil society' participated in the debate on the constitutionality of stem-cell research (Cesarino 2007), and participates, at least partially, in bioethics and bio-security committees (Leite 2007); an on-line public consultation was recently held to formulate a proposal for a new law on intellectual property rights to music; consensus conferences were organized on biotechnology; public engagement programs were planned to stimulate the debate on nanotechnology; and so on. Recently, Amazonian indigenous people participated actively, not as 'subjects of research' or 'informants' but as co-authors, in researches published in international journals (Heckenberger et al. 2003). Social movements and indigenous NGOs have also produced scientific data and funded scientific research useful to their campaigns (Castelfranchi 2008).

Some academic areas in Latin America are also traditionally, and strongly, linked to social movements, and a culture of civic duty, of a responsibility to the people and the nation, is quite strong among some intellectuals. Every university in Brazil, Argentina and other countries in the region has three duties: to research, to teach, and to engage in 'extension' (that is, to diffuse and transfer knowledge to generate social inclusion and transformation). In some senses, 'accountability', the social responsibility of science and 'engagement' are not novel here. Gramsci's theories, as well as Paulo Freire's critical 'pedagogy of the oppressed', greatly influenced some intellectuals and practices of knowledge communication in community radios, schools etc. As a consequence, on the one hand, communication theory and practices were developed in Latin America with peculiar inflections and political flavors, as in Martin Barbero's theory of mediations. On the other hand, neoliberal science policies and university management based on market demands, as well as globalized science communication practices, are not the only forces in action, and in some cases may not be the dominant ones.

One conclusion is that interest groups, pressure groups and social movements cannot be viewed as 'passive' or 'ignorant' 'lay' people. In some cases, they refuse to be considered simply as 'audiences' or 'consumers'. They use new information technologies (blogs, social networks, wikis, YouTube etc.) and validate themselves as activists and information producers. They know their actions can influence policies (and politics) and affect the dynamics of knowledge production.

1.4 Concluding Remarks

In this chapter, we argue that reconfigurations that occurred in the past 40 years in the relationships between science, politics and the market in developed and developing countries are deeply linked to the emergence of new practices, new modes and (partly) new roles for the public communication of S&T. Science communication is today an unavoidable and intrinsic process in the metabolism of contemporary societies. A trend is clear: techno-science is synchronically produced and communicated. The alliance between techno-science and capitalism has stimulated a 'communicative turn' in science communication, incorporating new agendas, problems, agents, institutions and social scenarios. Such a turn has two aspects. On the one hand, R&D, S&T policies and science communication practices are more and more linked to market demands, cultural industries and political interests: this is a globalized, neoliberal techno-science. On the other hand, that configuration also creates interesting possibilities for novel mutual feedbacks between techno-science and society, and for expanded social participation mechanisms in the production and governance of S&T, catalyzed and strengthened by new models and practices in public communication of S&T.

New actors emerge: science communication today does not only pass through the traditional channels of science journalism, popular science and formal or informal education. Several modes of diffusion are active. Sometimes, there is no mediator (for example, technical scientific information diffuses among mailing lists of patient pressure groups or environmental NGOs, or scientists communicate directly to their publics via blogs, wikis etc.). In some cases, the 'public' is also a producer of scientific information (for example, indigenous movements and environmental activists collect and produce data).

Latin America, where these global trends are immersed in specific situations and diverse boundary conditions, is a very interesting region to study such phenomena. In most countries of Latin America, a tradition of centralization, statism and authoritarianism has existed alongside strong and important social mass movements, cultural processes of resistance and assimilation, and an academic culture that eventually showed itself to be not so closed up in its ivory tower, but willing to listen to and cope with societal demands. In parts of Latin America, globalizing trends in techno-science and science communication are visible and accelerating strongly.

At the same time, the region shows that market is not the only force shaping science communication today. Latin America does not merely 'follow' global trends but, on the contrary, is immersed in trends and global forces with different conditions and follows diverse, peculiar and interesting trajectories. We think the examples and commentary we have provided give an idea of some peculiar and partly heterodox practices in Latin America. Nonetheless, we are aware that more research is needed to produce analytical comparative models. We hope to develop such studies soon.

Acknowledgement We would like to thank the referees for their very careful work on our text and for several important questions they raised, which generated insights both for this chapter and for future research.

References

Bauer, M. (2008). Survey research on public understanding of science. In M. Bucchi & B. Trench (Eds.), *Handbook of public communication of science and technology*. London/New York: Routledge.

Bauer, M., & Bucchi, M. (Eds.). (2007). *Journalism, science and society: Science communication between news and public relations*. New York: Routledge.

Beck, U. (1999). *World risk society*. London: Polity Press/Blackwell.

Bell, D. (1994). *El advenimiento de la sociedad post-industrial*. Madrid: Alianza.

Bentley, P., & Kyvik, S. (2011). Academia staff and public communication: A survey of popular science publishing across 13 countries. *Public Understanding of Science, 20*, 48–63.

Bourdieu, P. (2007). *Sobre la televisión*. Barcelona: Anagrama.

Brossard, D. (2009). Media, scientific journals and science communication. *Public Understanding of Science, 18*(3), 258–274.

Bucchi, M. (1998). *Science and the media: Alternative routes in scientific communication*. London: Routledge.

Bucchi, M. (2000). A public explosion: Big-bang theory in the UK daily press. In M. Dierkes & C. von Grote (Eds.), *Between understanding and trust: The public, science and technology*. London: Routledge.

Bucchi, M. (2002). *Science in society: An introduction to the social studies of science*. London/New York: Routledge.

Bucchi, M. (2008). Of deficits, deviations and dialogues: Theories of public. In M. Bucchi & B. Trench (Eds.), *Handbook of public communication of science and technology*. London/New York: Routledge.

Bucchi, M. (2009). *Beyond technocracy: Science, politics and citizens*. London/New York: Springer.

Bucchi, M., & Mazzolini, R. (2003). Big science, little news: Science coverage in the Italian daily press, 1946–1997. *Public Understanding of Science, 12*(1), 7–24.

Castelfranchi, Y. (2002). Scientists to the streets: Science, politics and the public moving towards new osmoses. *JCOM—Journal of Science Communication, 1*(2), 1–14.

Castelfranchi, Y. (2008). *As serpentes e o bastão*. Tecnociência, neoliberalismo e inexorabilidade, Ph.D. thesis, Campinas State University (UNICAMP).

Castelfranchi, Y., & Pitrelli, N. (2007). *Come si comunica la scienza?* Roma–Bari: Laterza.

Castells, M. (1997). *La era de la información: Economía, sociedad y cultura*. Madrid: Alianza.

Cesarino, L. N. (2007). Nas fronteiras do 'humano': Os debates britânico e brasileiro sobre a pesquisa com embriões. *Mana, 13*(2), 347–380.

Cheng, D., Claessens, M., Gascoigne, T., Metcalfe, J., Schiele, B., & Shi, S. (Eds.). (2008). *Communicating science in social contexts: New models, new practices*. New York: Springer.

Clemens, E. (1986). Of asteroids and dinosaurs: The role of the press in the shaping of scientific debate. *Social Studies of Science, 16*(3), 421–456.

Cloître, M., & Shinn, T. (1985). Expository practice: Social, cognitive and epistemological linkage. In T. Shinn & R. Withley (Eds.), *Expository science: Forms and functions of popularization* (Sociology of sciences yearbook, Vol. 4). Dordrecht/Boston: Reidel.

Collins, H., & Pinch, T. (1993). *The golem: What you should know about science*. Cambridge: Cambridge University Press.

Da Silva Medeiros, F., & Massarani, L. (2010). Pandemic on the air: A case study on the coverage of new influenza A/H1N1 by Brazilian prime time TV news. *Jcom, 9*(3), 1–9.
Dierkes, M., & von Grote, C. (2000). *Between understanding and trust: The public, science and technology*. London: Routledge.
Echeverría, J. (2003). *La revolución tecnocientífica*. Madrid: Fondo de Cultura Económica.
Epstein, S. (1995). The construction of lay expertise: AIDS activism and the forging of credibility in the reform of clinical trials. *Science, Technology & Human Values, 20*(4), 408–437.
Epstein, S. (1996). *Impure science: AIDS, activism and the politics of knowledge*. Berkeley: University of California Press.
Etzkowitz, H. (2001). The second academic revolution and the rise of entrepreneurial science. *Science, Technology and Society Magazine, IEEE, 20*(2), 18–29.
Friedman, S., Dunwoody, S., & Rogers, C. (Eds.). (1999). *Communicating uncertainty: Media coverage of new and controversial science*. New Jersey/London: Lawrence Erlbaum.
Funtowicz, S., & Ravetz, J. (1993). Science for the post normal age. *Futures, 25*, 739–755.
Gibbons, M., Limoges, C., Nowotny, H., Schwartzman, S., Scott, P., & Trow, M. (1994). *The new production of knowledge: The dynamics of science and research in contemporary societies*. London/Thousand Oaks: Sage.
Giddens, A., Beck, U., & Lash, S. (1997). *Modernización reflexiva. Política, tradición y estética en el orden social moderno*. Madrid: Alianza.
Hallin, D. (2008). Neoliberalism, social movements and change in media systems in the late twentieth century. In D. Hesmondhalgh & J. Toynbee (Eds.), *The media and social theory*. London/New York: Routledge.
Heckenberger, M. J., et al. (2003). Amazonia 1492: Pristine forest or cultural parkland? *Science, 301*(5640), 1710–1714.
Jacobi, D., & Schiele, B. (Eds.). (1988). *Vulgariser la science: Le proces de l'ignorance*. Seyssel: Champ Vallon.
Jasanoff, S. (1995). Procedural choices in regulatory science. *Technology in Society, 17*(3), 279–293.
Jensen, P. (2011). A statistical picture of popularization activities and their evolutions in France. *Public Understanding of Science, 20*, 26–36.
Kiernan, V. (1997). Ingelfinger, embargoes, and other controls on the dissemination of science news. *Science Communication, 18*(4), 297–319.
Kiernan, V. (2000). The Mars meteorite: A case study in controls on dissemination of science news. *Public Understanding of Science, 9*, 15–41.
Kiernan, V. (2003). Diffusion of news about research. *Science Communication, 25*(1), 3–13.
Kreimer, P., Levin, L., & Jensen, P. (2011). Popularizations by Argentine researchers: The activities and motivations of CONICET scientists. *Public Understanding of Science, 20*, 37–47.
Kunth, D. (1992). La place du chercheur dans la vulgarisation scientifique. *Rapport CNRS*. Retrieved from http://sciences-medias.ens-lsh.fr/scs/article.php3?id_article=276.
Latour, B. (1992). *Ciencia en acción*. Barcelona: Labor.
Leite, M. (2007). Arautos da razão: A paralisia no debate sobre transgênicos e meio ambiente. *Novos estud.—CEBRAP, 78*, 41–47.
Lengwiler, M. (2008). Participatory approaches in science and technology: Historical origins and current practices in critical perspective. *Science, Technology & Human Values, 33*, 186–200.
Lewenstein, B. (1995). From fax to facts: Communication in the cold fusion saga. *Social Studies of Science, 25*, 403–436.
López Cerezo, J. A. (Ed.). (2003). *La democratización de la ciencia*. San Sebastián: Erein.
Massarani, L., & Polino, C. (organisers) (2008). *Los desafíos del periodismo científico en Iberoamérica. Jornadas Iberoamericanas sobre la ciencia en los medios masivos*. Río de Janeiro, RICYT, Scidev.net, CYTED, AECI, OEA.
Massarani, L., Buys, B., Amorim, L. H., & Veneu, F. (2007). Growing, but foreign source dependent: Science coverage in Latin America. In M. Bauer & M. Bucchi (Eds.), *Journalism, science and society: Science communication between news and public relations*. New York/London: Routledge.

Nieman, A. (2000). *The popularization of physics: Boundaries of authority and the visual culture of science*. Doctoral thesis, Faculty of Applied Sciences and Faculty of Humanities, University of the West of England, Bristol.
Nowotny, H., Scott, P., & Gibbons, M. (2001). *Re-thinking science: Knowledge and the public in an age of uncertainty*. Oxford/New York: Wiley–Blackwell.
Peters, H. P., Brossard, D., de Cheveigné, S., Dunwoody, S., Kallfass, M., Miller, S., & Tsuchida, S. (2008a). Interactions with the mass media. *Science, AAAS, 321*, 204–205.
Peters, H. P., Brossard, D., de Cheveigné, S., Dunwoody, S., Kallfass, M., Miller, S., & Tsuchida, S. (2008b). Science–media interface: It's time to reconsider. *Science Communication, 30*, 266–276.
Petras, J., & Veltmeyer, H. (2006). Social movements and the state: Political power dynamics in Latin America. *Critical Sociology, 32*, 83–104.
Phillips, D. P., Kanter, E. J., Bednarczyk, B., & Tastad, P. L. (1991). Importance of the lay press in the transmission of medical knowledge to the scientific community. *New England Journal of Medicine, 325*, 1180–1183.
Polino, C. (2009). La ciencia en las noticias de América Latina: Una aproximación empírica para el caso de la salud. In J. A. López Cerezo & J. Gómez (Eds.), *Apropiación social de la ciencia*. Madrid: Biblioteca Nueva.
Polino, C., & Castelfranchi, Y. (in press). Comunicación pública de la ciencia: Historia, prácticas y modelos. In E. Aibar, & M. A. Quintanilla (Eds.), *Ciencia, Tecnología y Sociedad*. Enciclopedia Iberoamericana de Filosofía. Madrid: Trotta.
Polino, C., Fazio, M. E. & Chiappe, D. (2006). Análisis de la oferta informativa sobre ciencia y tecnología en los principales diarios argentinos. Observatorio de Ciencia, Tecnología e Innovación Productiva, SECYT. www.observatorio.secyt.gov.ar.
Ramalho, M., Polino, C., Cevallos, M. C., Cruz, P., Intriago, M., Terán, A., Chiappe, D., Neffa, G. & Massarani, L. (2011). La ciencia en los telediarios de Brasil y Ecuador. *XII Reunión Bienal RedPop*, Campinas.
Royal Society (2006). Factors affecting science communication: A survey of scientists and engineers. Royal Society. www.royalsociety.org.
Schiele, B. (2006). Communicating science in the real context of society. In C. Donghong, J. Metcalfe, & B. Schiele (Eds.), *At the human scale: International practices in science communication*. Beijing: Science Press.
Schiele, B. (2009). Journalisme scientifique: Le rôle des journalistes comme médiateurs doit être repensé. In A. Char & R. Côté (Eds.), *La révolution internet*. Québec: Presses de l'Université du Québec.
Svampa, M., & Antonelli, M. (2009). *Minería transnacional, narrativas del desarrollo y resistencias sociales*. Buenos Aires: Biblos.
Takahashi, B. (2010). Framing and sources: A study of mass media coverage of climate change in Peru during the V ALCUE. *Public Understanding of Science, 20*(3), 1–15.
Torres Albero, A., Fernández Esquinas, M., Rey Rocha, J., & Martín Sempere, M. (2011). Dissemination practices in the Spanish research system: Scientists trapped in a golden cage. *Public Understanding of Science, 20*, 12–25.
Vara, A. M. (2007a). Periodismo científico: ¿Preparado para enfrentar los conflictos de interés? *Revista CTS, 9*(3), 189–209.
Vara, A. M. (2007b). Sí a la vida, no a las papeleras: En torno a una controversia ambiental inédita en América Latina. *Redes, 12*(025), 15–49.
Verón, E. (1999). *Entre la epistemología y la comunicación*. CIC, no. 4, Servicio de Publicaciones UCM, 149–155.
Weinberg, A. (1972). Science and trans-science. *Minerva, 10*, 209–222.
Weingart, P. (2001). *Die Stunde der Wahrheit? Zum Verhältnis der Wissenschaft zu Politik, Wirtschaft und Medien in der Wissensgesellschaft*. Germany: Velbrück.
Wynne, B. (1989). Sheep farming after Chernobyl: A case study in communicating scientific information. *Environment Magazine, 31*(2), 10–39.
Ziman, J. (2000). *Real science: What it is, and what it means*. Cambridge: Cambridge University Press.

Chapter 2
The Evolution of Science Communication Research in Australia

Jenni Metcalfe and Toss Gascoigne

Abstract The strength of science communication in Australia has until now been in practice rather than theory, driven by a demand for practical solutions to problems. Science communicators are resourceful in devising solutions, either adapting international experience to suit local circumstances or inventing their own. The theoretical study of science communication in Australia has been slower to develop. Only recently has Australia recognized that many science-based issues require a more considered approach, in which practical actions are governed by a deeper theoretical understanding. Prior to 1990, the limited number of university departments researching science communication-related issues worked principally from a social sciences perspective. Based in units with names such as 'History and Philosophy of Science', they had little to do with practitioners in science communication. The practitioners usually worked for research organizations, science centers and museums, and came from a wide variety of disciplinary backgrounds. They performed a range of tasks and the titles of their positions varied widely. There were no established training programs, and the role of 'science communicator' was only beginning to be defined. Dialog between theoreticians and practitioners was virtually non-existent. Since the 1990s, the practice of science communication has become more professional through the development of Australian Science Communicators, the consolidation of three centers for training science communicators, and an increase in academic research into science communication. Academic research into science communication currently takes a

J. Metcalfe (✉)
Econnect Communication, South Brisbane, Australia
e-mail: jenni@econnect.com.au

T. Gascoigne
Toss Gascoigne & Associates, Campbell ACT, Australia
e-mail: director@tossgascoigne.com.au

B. Schiele et al. (eds.), *Science Communication in the World: Practices, Theories and Trends*, DOI 10.1007/978-94-007-4279-6_2,

mostly multidisciplinary approach and has moved away from a deficit model focus to one that is more participatory. This may at least partly explain the closer links emerging between researchers and practitioners.

Keywords Science communication • Engagement • Research • Public communication of science and technology • Australia • Australian Science Communicators • Science communication courses.

2.1 Introduction

This chapter traces the growth of science communication research in Australia. It examines the emergence of three strands of science communication, practice, teaching and research, and the way those strands have interacted with each other. In particular, it looks at how science communication practice has influenced the nature of science communication research.

We outline the emergence of science communication practice in Australia in the 1980s and early 1990s, how that practice linked with and influenced the growth of related university courses, and a recent increase in research. In Australia, research on science communication with direct implications for practice was rare until the 1990s, but now most research is driven by the perceived needs of practitioners.

When we talk about ‘science communication’ in this chapter, we are referring to all the associated terms, including science popularization, public communication of science and technology, public understanding of science and technology, science literacy, and social appropriation of science and technology. When we refer to ‘science communication research’, we are talking about academically based research into science communication theories or the rigorous evaluation of practice.

In researching this chapter, we reviewed the available literature, interviewed conveners of university science communication courses, reviewed publications relevant to Australian science communication over the past 60 years, and conducted a survey of researchers and practitioners.

In examining the evolution of research, we reviewed 22 journals that potentially contained articles on Australian science communication research (see Appendix A). An online search was conducted using the key phrases ‘science communication’, ‘public communication of science’, ‘media and science’ and ‘science museums’. The search extended over a 60-year period, from 1951 (*Journal of Communication*) to contemporary publications such the Marquette Communication Journals, which have only recently been published online.

To discover the extent and nature of current research in Australia, in 2011 we surveyed the activities of science communication researchers and practitioners. The survey was promoted through the discussion list of Australian Science

Communicators (ASC), which has 800 subscribers and a good reach into the sector. It aimed to find out:

- Who was doing science communication research
- What research projects they had completed or were working on
- What tools and disciplines are being brought to research
- How the research is being applied
- How practitioners are using research to inform their practice.

There were 65 responses to the survey: 30 respondents said they were involved in both science communication research and practice; 30 were involved only in science communication practice; and five were involved only in research. Most researchers are associated with universities, and the five who were involved only in research were all from the university sector.

2.2 Early Science Communication Was Driven by Practical Needs

Every expedition that explored first the coast and then the interior of Australia in the seventeenth, eighteenth and nineteenth centuries included a person with scientific interests. The scientists documented and charted the coastline, the land, the geology, and plant and animal life in the period leading up to the first settlement by Europeans in 1788 and then beyond that as the continent was explored and opened up by the new settlers.

Those traditions of scientific inquiry were extended into urban life in the nineteenth century by the formation of mechanics institutes,[1] botanic gardens, learned societies, museums, public libraries and universities. The formation of such institutions accelerated after the 1850s, based on new wealth from the Australian gold rushes and generous government support:

> By the 1870s it was clear that the program that had unfolded in these [learned] societies was one largely committed to the collection, description and classification of Australian natural history, phenomena and resources, combined with a discussion of practical matters involved in colonial development. This reflected the mood of the times, which had little patience with abstract theorizing. (Home 1989)

From the very beginning, communication about science was rooted in practicalities. The demands of establishing a settlement in an environment often hostile to European approaches to farming and management of the environment shaped the

[1] The objective of the typical mechanics institute was 'the diffusion of scientific, literary and other useful knowledge among its members and the community generally and particularly among the young as well as the operative classes.' R.W.E. Wilmot, quoted in Home (1989).

discourse of science. It was not an enterprise conducted in a rarefied and scholarly atmosphere: in 1870, 60% of the membership of the Royal Society of New South Wales had no scientific background or involvement (Home 1989).

The first formal studies on topics most closely related to science communication began at the University of Melbourne with the establishment of the Department of General Science and Scientific Method in 1946 (coincidentally, the year that the first Ph.D. programs were offered in Australia) (University of Melbourne 2011).

The University of Sydney (1945) was another early entrant; in fact, with the University of Melbourne, it was among the first universities in the world to establish departments of that type. In describing its role as to 'mediate change and help us understand the world and our place in it', the University of Sydney pre-empted many of the questions dealt with by researchers now working under the broad banner of 'science communication' internationally.

2.3 The 'Profession' of Science Communication Emerges

In 1994, ASC was formed. This was an important milestone in the formalization of the term 'science communication' and the emergence of a new profession. At the time, people involved in science communication had a wide variety of titles and came from different educational and disciplinary backgrounds. They tended to operate in a professional vacuum because there was no place where they could seek advice or discuss problems with professional colleagues, and science communicators felt that lack of collegiality.

In the first 2 months after the intention to form ASC was announced, 375 people from across Australia demonstrated their interest in science communication (and their support for the new body) by joining as foundation members. This was a strong showing of support: the personal benefits were negligible, but the funding helped the fledgling body to become established in September that year.

Science communicators came from a variety of training backgrounds and former occupations, and included people who described themselves as media officers, scientific editors, public relations officers, librarians, scientists and science journalists. Their commonality was in their shared interest in communicating science to a diverse range of audiences.

Australia's leading national research organization, the Commonwealth Scientific and Industrial Research Organisation (CSIRO), played a powerful informal part in shaping the new 'science communication' role in the early 1990s. Many of the people working in science communication roles were staff members of CSIRO, with its 37 divisions at 100 sites across Australia. It provided the most comprehensive and cohesive approach to science communication and offered a framework where like-minded professionals could discuss issues and share experiences. The CSIRO experience was in sharp contrast to the sense of isolation other 'science communicators' felt in their individual jobs in research organizations, museums and libraries, and as writers and editors.

2.4 University Courses Seek to Make Scientists Better Communicators

The same groundswell of interest that led to the formation of ASC also encouraged the emergence of new courses in science communication at Australian universities. The first, in 1996, was at the Australian National University (ANU), which initially offered graduate diplomas (in conjunction with the National Science Museum) and bachelors' degrees. A survey in 1996 showed that 16 Australian universities were offering or planning to offer courses or subjects in science communication, mostly at postgraduate levels.

These courses were focused initially on improving the communication skills of scientists through specific subjects (for example, the Queensland University of Technology offered its science students 'Scientific and technical writing' as a subject); joint degrees (for example, the University of New South Wales offered Bachelor of Science degrees that included arts, media or communication subjects); or postgraduate diplomas open to science graduates.

Thus, the initial focus of university science communication courses was on training scientists rather than catering to the newly emerging profession of 'science communicator', and many of the teachers of science communication had scientific rather than communication qualifications and experience.

Many early courses and subjects in science communication failed to survive more than a few years because scientists saw communication as a lower priority than other subjects, and without support the courses lost funding. But some courses targeting scientists rather than science communicators survive. For example, Monash University offers a joint Bachelor of Arts (Professional Communication) and Bachelor of Science. The announcement on the Monash University website is typical:

> There is an increasing need for scientists to be able to communicate their work and its importance to colleagues in other rapidly diverging fields and to grant-awarding bodies, as well as to industry and the community in general. (Monash University 2011)

2.5 Science Communication University Courses Shift Direction

The closures of some of the early science communication subjects and courses led to the consolidation of courses at three universities: the ANU, the University of Western Australia (UWA) and the University of Queensland. Those universities recognized the increasing professionalization of science communicators, and took advantage of the opportunity to provide courses with a vocational emphasis so that graduates had the skills to take up positions in this emerging area.

The focus was still on training those with a science background, rather than those from the arts or humanities, but the training aimed to turn scientists into professional science communicators rather than teaching them communication skills to be used in the course of their scientific careers.

Science communication was now most commonly undertaken in a natural sciences context rather than (as happens in other countries) through departments affiliated with English, media studies or social sciences (University of Wisconsin–Madison 2010). The exception is the University of Queensland, where science communication courses are offered through the School of English, Media Studies and Art History.

The normal prerequisite to enter a science communication course is a degree in science. For example, the ANU's Centre for the Public Awareness of Science (CPAS) advertises that entry to its graduate research programs requires an honors degree in science. Its website promotes the course with the promise:

> We train a new generation of highly qualified scientists to become skilled communicators who can engage people with the science, technology, or medical information that is most relevant to them.[2]

The University of Queensland's science communication courses have similar requirements:

> This field is intended for science graduates, or those with strong science backgrounds, who wish to communicate effectively with scientists and professionals in business, industry, government, and the media. (University of Queensland 2011)

At the UWA, science communication is offered through the Faculty of Life and Physical Sciences. A bachelor's degree requires a strong maths background, and a master's requires a Bachelor of Science. The UWA Master of Science Communication and Education is described thus:

> Students learn principles of effective science communication, develop practical skills and design strategies that address the communication needs of groups such as government organizations, informal museums, science centers and research centers. (UWA 2011)

The stronger science communication courses now include disciplines other than natural sciences in their programs to cater better to the needs of science communicators. The CPAS courses at ANU include social science disciplines:

> In this degree, not only will your science writing and presentation abilities be honed, you will also study how people think about risk and ethics, consider what types of communication techniques are most persuasive, and discuss the underlying social and cultural influences on science as it exists today. (ANU 2011)

Research may be a compulsory unit in undergraduate courses in science communication. The University of Queensland, for instance, states on the course website:

> Courses are designed to increase understanding of the application of communication theory and research to health promotion, business, public relations, policy and politics, intercultural relations and globalization. (University of Queensland 2011)

The Australian situation reflects an analysis of worldwide science communication courses and their intellectual foundation conducted by Mulder et al. (2008), which

---

[2] http://cpas.anu.edu.au/.

concluded: 'Science communication programs at universities appear to combine aspects from the four key cognate areas of science, education studies, social studies of science, and communication studies.'

2.6 Science Communication Research Driven by Practitioners

As science communication courses grew and matured and became concentrated in a handful of universities, interest also increased in conducting science communication research at the postgraduate level.

The number of students has increased steadily. In 2011, the University of Queensland had enrolled 15 masters and five Ph.D. students in research projects. The number of research students has expanded at ANU's CPAS, from three Ph.D. students in 1997 to 20 in 2010. The first postgraduate research students at the UWA began their research in 2005 and graduated in 2009. In 2010, UWA enrolled five Ph.D. students and two master's students in science communication research. A review of recent research projects shows the breadth of topics that the students tackled and also the strong influence of the natural sciences.[3]

Twenty-two journals are referred to above as being relevant to science communication. A search of those journals discovered 73 articles describing research in science communication in Australia or written by Australian researchers. Two were published prior to 1990, 23 in the period from 1990 to 1999, and 48 since 2000. These figures demonstrate the marked increase in science communication research over the past 20 years, particularly in the past decade. The increase has been driven by the increasing numbers and professionalization of science communicators, and the more intensive activities of university departments in this area.

These views are supported by our survey results, which show that most science communication researchers (86% of respondents) in Australia are also practitioners. The researchers are mostly based in universities, but they are involved in science communication practice through consultancies and training. Others work for the CSIRO, which employs a number of people to carry out the dual role of research and practice in science communication.

2.7 Science Communication Research Evolves into a More Multidisciplinary Approach

The beginnings of research in science communication were in departments with names such as 'History and Philosophy of Science', or 'Science and Technology Studies' at the University of Melbourne (began in 1946) and the University of

[3]N. Longnecker, Science Communication Program Coordinator, Faculty of Life and Physical Sciences, University of Western Australia, pers. comm., 2010.

Sydney (began in 1945). Work in those departments continues today (although the departments may have diversified and reorganized), but it continues largely separately from the new departments of science communication described above.

The effect had been to create two separate strands in Australia. Those training and teaching in the new 'science communication' areas worked from a natural sciences disciplinary background and often dealt with more mechanical and descriptive topics; while those dealing with 'science and society' issues worked from a social sciences perspective and looked at a different set of issues. Until recently, there was not much contact between the two sides.

Those boundaries appear to be breaking down. For example, the Ph.D. students at ANU's CPAS are now studying more varied topics than previously:

> Topics include everything from mental illness and illicit drugs to climate change and data visualization as well as science and public policy, science communication capacity building and evaluation in multi-national NGOs … [There is] also [research into] what makes a science hero, science theater and performance … The change really has been in the sheer volume of work happening, and also in the increased interest in new (particularly social) media.[4]

The convenor of science communication courses at UQ, Dr. Joan Leach, says that science communication research has increased in both quantity and sophistication:

> Things have gotten a bit more sophisticated than 'how many images of scientists do you see on mainstream TV' or the content analysis that was dominating about eight years ago. [Researchers] are theoretically more savvy and much more interdisciplinary. It seems that fewer scientists are trying to move from scientific research to research in science communication.[5]

These observations are backed up by our review of the Australian science communication papers published in relevant journals. The topics have changed in the past 20 years, from looking at the impact of communication on natural sciences to a more multidisciplinary focus. From 1990 to 1999, the most favored topics of research could be categorized under media and science (22% of papers), public attitudes to science (13%), and museums and exhibitions (13%). In the past decade, a greater diversity of research topics is apparent. While research on public attitudes to science (19%) and media and science (17%) have retained their popularity, the newer areas include policy or politics related to science communication (16%), science and culture (13%) and risk communication (12%).

There has also been a shift in the disciplines represented by the research papers' authors, which suggests a move to more interdisciplinary research. Prior to the mid-1990s, many authors were writing from the disciplines of education (26%),

---

[4] R. Lamberts, Deputy Director, Centre for the Public Awareness of Science, Australian National University, pers. comm., 2010.

[5] J. Leach, Senior Lecturer in Rhetoric and Science Communication, University of Queensland, pers. comm., 2011.

communication (26%), science and technology studies (22%) or technical/scientific fields (22%). Since the early 2000s, other disciplines have emerged in the papers, especially the social sciences, political science and psychology.

Overall, papers researched from a humanities, arts and social sciences (HASS) point of view have increased from 57% in the 1990–1999 period to 71% in the decade from 2000. Authors with a natural sciences background are also making more use of HASS themes (from 22% of papers to 33%). This suggests an increasing interest in multidisciplinary research—an observation borne out in an examination of the authors' affiliations. For example, 'Something to talk about: Affective responses to public health mass media campaigns and behavior change' (Dunlop et al. 2009) was written by joint authors from the School of Public Health at the University of Sydney, the Cancer Council of Victoria and the Department of Psychology at Melbourne University.

Our survey of science communication researchers revealed a mix of disciplines and university departments supporting a multidisciplinary approach to current research. Among the 30 respondents involved in both research and practice, double degrees and multiple qualifications were common. Most (57%) had a Ph.D., and 87% had at least one science degree. One-third had an arts, communication or media degree, and 27% had had some involvement in social science, psychology or sociology. Twenty-three percent said they had a specific science communication qualification.

Four of the five researchers who did not practice science communication had Ph.D.s, and the other was a Ph.D. student. Four had at least one science degree, and four of the five had a degree in arts, communication, media, social science, psychology or sociology.

The multidisciplinary nature of science communication research is reflected in the mix of qualitative and quantitative tools researchers use. The most common research tools are surveys (66%), interviews (60%) and focus groups (34%). To a lesser extent, researchers are also using document/media analysis, literature reviews, action research and content analysis.

2.8 Science Communication Research Shifts Its Focus: From Deficit to Engagement Models

Interest in research into models of science communication has risen, more than doubling over the past decade from 22% of all papers to 48%. There has been an increase in interest in looking at science communication from an engagement perspective (23% of papers), rather than a 'deficit' one-way communication perspective, as illustrated by the titles of recent research papers:

- Research partnerships with local communities: Two case studies from Papua New Guinea and Australia (Almany et al. 2010)
- The public production and sharing of medical information: An Australian perspective (Ko 2010)

- Identifying and testing engagement and public literacy: Indicators for river health (Metcalfe and Riedlinger 2009).

Such papers contrast with the first science communication papers from the 1970s, all of which used the deficit model, rather than a participatory one, as a theoretical base.

This increasing focus on engagement is illustrated by responses to the survey. We asked researchers to list up to three significant research projects they had completed recently or were working on. In total, they described 45 projects. The most common topics dealt with engagement in science (20%) and formal or informal science education (20%). The next most common topics were skills development (13%) and audience attitude analysis (13%). Examples of some of the engagement topics were:

- Developing guides and evaluating engagement using social network analysis
- Seeking to discover how public engagement on new technologies influences attitudes, policy directions, behaviors etc.
- How to best engage with the public/s in online environments.

We asked researchers about their use or evaluation of science communication models. The types they mentioned included participatory, behavioral, decision-making and learning theory. This emphasizes the point that researchers are focusing more on participatory models of science communication.

2.9 Evaluation: An Unresolved Issue

Three-quarters of the researchers nominated areas where the research they had conducted was being applied by practitioners. The most common applications were in the design and management of events or public programs (41%); informal public communication (33%); developing policy (26%); professional development workshops or training (26%); media (26%); and developing communication strategies (19%).

However, a concern expressed by both researchers and practitioners is that research is needed to provide for the accurate targeting, management and monitoring of programs that aim to influence public attitudes to science and technology. Respondents said:

> There is an extreme lack of data on the long term effectiveness of any specific science communication activity. Basically we do not know how to construct programs very much beyond just good ideas. I hope my work, and that of my students, will contribute to filling that gap.

> Unfortunately it is limited to investigating the interest level or measured in terms of how much exposure one gets. I would love to have an effective method of measuring the true effectiveness of my communication efforts (in their many forms).

> Evaluation has depended on the project. Some evaluation has been simple due to ready determinant of baseline as effectively "zero", so easy to measure change in practice against this. In other instances, can be difficult to determine the baseline against which to measure effectiveness depending on when I get brought into the project.

Evaluation is an unresolved issue in science communication. While activities to measure the 'success' of an event are quite common (usually using rudimentary tools such as counting numbers, exit surveys or media monitoring), measuring attitude changes over time is more challenging and much rarer. Most practitioners (92%) said that they had evaluated their practice in some way, usually through a survey (47%), feedback forms (30%), interviews or discussions (19%), or analysis of the number of website hits, visitors, media coverage or requests for information (17%).

They reported that they used the results of such evaluations to tailor or refine their practices (32%). One person noted that it led to further resource allocation for their science communication activities.

2.10 Where Do Practitioners Seek Ideas to Improve the Way They Do Their Job?

The 65 survey respondents involved in science communication practice were asked about their sources for ideas to inform and improve their work. More than half actively seek out research results, attend workshops or do their own research. They nominated informal conversations, conferences or seminars, and books and papers as the most common sources of ideas to inform their activities. Their responses are detailed in Table 2.1.

Practitioners had a mixed response when asked about the value of reading journal articles about science communication. They are more likely to read the journals if they also do their own research (60%), and are also more likely to find them useful or very useful. Their reservations centered on the issues of quality, relevance, rigor and accessibility. They commented:

> [I am] frustrated by papers that do not have the same rigor as applied to science papers.

> I find sometimes there is a disconnect between practicing science communication and the research that is published. I particularly struggle with academic presentations that contain a lot of jargon—the exact kind of thing we try to avoid!

Table 2.1 Where science communication practitioners get ideas to inform their practice

Source	Those involved in practice and research (30 respondents)	Those involved only in practice (30 respondents)	Total (60)
Informal conversations	22	27	49 (82%)
Conferences/seminars	21	22	43 (72%)
Books/papers	21	20	41 (68%)
Finding out about research	20	16	36 (60%)
Workshops	17	15	32 (53%)

There was a strong preference for practitioners to use their colleagues both inside and outside science communication as sources of advice, and most found such advice very useful. Seventy-eight percent reported personal advice as their preferred source of information, but they were aware of variations in quality:

> Some people are wedded to a single theory direction that aligns with their research area or their area of interest, and attempt to make it fit any other area.

> Science communication research is still emerging and there are many techniques from other disciplines which are yet to be tested/applied to science communication research. It is also a very messy problem to investigate the understanding of science.

> I have good relationships with particular researchers and science communication academics and I find my discussions with them useful—they are also very approachable.

2.11 Discussion

Science communication is now a recognized occupation in Australia. The term is commonly used to describe a set of activities, and the title is well established in the employment market at research organizations, museums and universities. A number of vibrant small companies have emerged over the past 18 years, each providing a range of services to research organizations and government departments. Their services include writing, event management, conducting surveys, designing signage, developing strategies, media relations and training.

Universities in Australia offered courses in science communication prior to 1990, but the courses were fragmented and often short-lived. 'Science communication' was defined and shaped in the late 1980s and early 1990s, and the emergence of the new profession gave energy and purpose to a number of universities. New units were formed, new courses written, and a training framework established.

The need for research followed. Postgraduate qualifications in science communication by coursework and research are well established at three universities. Other universities contribute to the research and training effort, at times using a different approach or a different disciplinary basis.

The growth in training and research programs has also marked a shift in the ideological approach. During the 1990s, science communicators focused on one-way communication via formal education, the media, publications, lectures and museum displays. That focus probably reflected their professional backgrounds as editors, journalists, teachers and librarians. Much of the science communication research reflected this deficit-style communication, and educational, attitude and media studies dominated the literature.

Science communication research leading up to the 1990s was largely from a natural sciences perspective or associated with the history and philosophy of science. There was little connection between practitioners and science communication researchers during this period.

However, increasing numbers of tertiary courses or subjects in science communication led subsequently to a corresponding increase in research specifically focused on science communication. Much of that research today appears to be carried out by practitioners in universities or research organizations. It is becoming more collaborative across organizations and disciplines, and includes an increased humanities and social sciences component. This is reflected in the main tools used for research: surveys, focus groups and interviews. More research is evaluating or using science communication models as a framework for investigation.

Today, there appears to be a much stronger connection between practitioners and researchers. This may be partly due to the fact that many practitioners also do research, but it also reflects the professionalization of the science communicator. Those practicing science communication are actively seeking advice from others, and a significant number are finding new ideas from published papers and articles as well as conferences and seminars. There is a move away from the deficit mode of communication to one that seeks more actively to engage people in science and to understand better the concerns and needs of those with whom the science communicator is trying to communicate.

These developments bode well for an egalitarian model of science communication that can more effectively engage in the big environmental, economic and social issues currently facing Australian society.

Appendix A

Journals searched for articles on science communication

Journal	Searched online back to
Australasian science	2005
Communicating astronomy with the public	2007
Communication theory	1991
Critical studies in media communication	2000
European journal of communication	1986
Human communication research	1974
Journal of communication	1951
Journal of science communication	2002
Marquette communication journals	2008
Nature	1986
Public understanding of science	1992
Science and society	1989
Science and technology studies	1986–1987
Science as culture	1997
Science communication	1979
Science in context	1987
Science, technology and human values	1978
Science, technology and society	1996
Social studies of science	1971
The journal of science education and technology	1992
Web journal of mass communication research	1997
Written communication	1984

References

Almany, G. R., Hamilton, R. J., Williamson, D. H., Evans, R. D., Jones, G. P., Matawai, M., Potoku, T., Rhodes, K. L., Russ, G. R., & Sawynok, B. (2010). Research partnerships with local communities: Two case studies from Papua New Guinea and Australia. *Coral Reefs, 29*, 567–576.

ANU (Australian National University). (2011). *Bachelor of Science (Science Communication).* Retrieved April 2011, from http://studyat.anu.edu.au/programs/3602XBSCM;overview.html

Dunlop, M., Wakefield, M., & Kashima, Y. (2009). Something to talk about: Affective responses to public health mass media campaigns and behaviour change. *Journal of Health & Mass Communication, 1*(3/4), 211.

Home, R. W. (Ed.). (1989). *Australian science in the making*. Cambridge: Cambridge University Press.

Ko, H. C. H. (2010). The public production and sharing of medical information: An Australian perspective. *Journal of Science Communication, 09*(01), 1–10.

Metcalfe, J., & Riedlinger, M. (2009). Identifying and testing engagement and public literacy: Indicators for river health. *Science, Technology & Society, 14*(2), 241–267.

Monash University. Retrieved April 2011, from http://monash.edu/pubs/handbooks/courses/2130.html

Mulder, H. A., Longnecker, N., & Davis, L. S. (2008). The state of science communication programs at universities around the world. *Science Communication, 30*, 277–287.

University of Melbourne. (2011). *School of historical and philosophical studies, history and philosophy of science, about us*. Retrieved May 11, 2011, from http://hps.unimelb.edu.au/about/

University of Queensland. (2011). *Master of communication.* Retrieved April 2011, from http://www.uq.edu.au/study/program.html?acad_prog=5334

University of Wisconsin–Madison. *The directory of science communication courses and programs.* Retrieved September 5, 2010, from http://dsc.journalism.wisc.edu/allEntries.php

UWA (University of Western Australia). (2011). *Bachelor of science (Science communication).* Retrieved from http://courses.handbooks.uwa.edu.au/courses/c5/5011.27

Chapter 3
The Development of Science Communication Studies in Canada

Bernard Schiele and Anik Landry

Abstract This chapter is in three parts. The first discusses the level of achievement in structuring the science communications research field in Canada and elsewhere. The second examines the historical development of the research. We recall the role of the state, underscoring the influence of the OECD in formulating national policies. In the third part, we describe the major orientations of the research by distinguishing two phases: an initial one, in which the work devolved mostly from government concerns or was conducted directly at the behest of various ministries; and a maturation characterized by the development of university research. The chapter concludes with the observation that research still remains underdeveloped in Canada, although it has diversified and grown significantly in recent years.

Keywords Policies • OECD • Public • Health • User • Media • Canada • Research

Research has been conducted on science popularization, science vulgarization and science communication for over 40 years. But if public communication of science and technology (PCST) researchers agree on one thing, it is that 'science communications' research has not yet emerged as a discipline; nor do they expect that it will

---

B. Schiele (✉)
International Research Center on Science and Technology (IRCST/CIRST), University of Quebec at Montreal (UQAM), Station Centre-Ville, P.O. Box 8888, H3C 3P8 Montreal, QC, Canada
e-mail: schiele.bernard@uqam.ca

A. Landry
University of Quebec at Montreal (UQAM), Station Centre-Ville, P.O. Box 8888, H3C 3P8 Montreal, QC, Canada
e-mail: landry.anik@uqam.ca

B. Schiele et al. (eds.), *Science Communication in the World: Practices, Theories and Trends*, DOI 10.1007/978-94-007-4279-6_3,

any time soon (Gascoigne et al. 2010). However, this research already fulfills certain conditions that define it as such. It is:

> a bounded field of studies [with] significant presence in teaching and research in the higher education sector, international research, specialist scholarly publishing, organized communities of networks of scholars, and a body of theoretical work that underpins empirical study. (Trench and Bucchi 2010)

3.1 A Field of Research Whose Boundaries Remain Blurred in Canada as Elsewhere

A point of clarification before we look at the difficulties hampering science communications research (SCR) in gaining the status of a discipline: we distinguish it from 'science mediation', denoted by the acronym PCST (public communication of science and technology).[1] PCST includes all science mediation, interpretation, dissemination and explanation activities—the range of efforts, among others, to inform, sensitize and mobilize the public. This could be a radio program on a particular discovery and its potential spin-offs, hosted by media professionals; a discussion on a topical controversial scientific issue, initiated by a science club, science café, public awareness group, etc.; or science and technical information made publicly available by researchers on the internet. The actors, means, channels and goals of science mediation are many and varied. We prefer to use the term 'science mediation' rather than any synonymous term currently in use ('popularization', 'disclosure', etc.). In the interests of clarity, we distinguish mediation activities from their outcomes, real or anticipated, for which we prefer the term 'science and technical culture' (STC) to denote a state of knowledge, opinions and representations at a particular moment and the conditions contributing to it.

So, what difficulties do SCR researchers point out?

The first difficulty raised is in delimiting the field. A priori, SCR 'concerns the communication between communities of scientists, interest groups, policy makers and various publics' (Trench and Bucchi 2010). But should the exchanges between scientists themselves also be included, especially when addressing specialists in a different field? (see Jacobi 1986, 1999). And what about risk communication, environmental communication, health communication? Are they part of the science communications field, or are they subdisciplines of risk science, environmental science (environmental studies) or health sciences? One could also signal other fields where the boundaries are blurry, such as research in science journalism, science museology, or again the whole stream of research on informal learning in informal environments (Bell et al. 2009). How, then, not to encroach on the neighbors' territory and vice versa? Such reciprocal overlapping is quite normal, as a field forms progressively through a process of dissociation, speciation and autonomization.

---

[1] The acronym PUS (public understanding of science) is also frequently used in Anglo-Saxon countries to denote the actions of dissemination (that is, the results thereof).

But the fact that a field is being formed also means that a group of social actors is engaged in that process. And that group is also a social organization, a sub-community with its own interests within the scientific community. It is a new sub-community existing de facto amid other sub-communities of divergent interests. And those other groups either seek to preserve the status they enjoy (their interests) or they also try to gain recognition to promote their interests. So, on the one hand, delimiting a field involves a search for internal coherence by participants mutually engaged in the same process who recognize themselves within the definition of what they do (Fourez 2002:93–102, passim). On the other hand, it is an attempt to impose a certain perspective on other sub-communities, to be recognized within the scientific field ('internal recognition') and to impose it outside the scientific field ('external recognition'). In short, it is to demand a legitimacy in order to affirm an authority (see Bourdieu 1980, 2001; Latour 1984).

In this perspective, the question of the boundaries of the SCR field also implies its conditions for emergence, since it is made up of pre-existing knowledge (theories, models, facts, etc.) mobilized by researchers seeking a new research program. Another consideration is its connection to those related fields from which it is striving to distance itself, although they gravitate around the same core concepts. The determining factors, *in fine*, are the conditions that brought the researchers together in a sub-community, around a qualified research program in science communications research, the knowledge they have mobilized, and the strategies they use to gain recognition of their uniqueness by the scientific community and society.

The second difficulty mentioned by researchers is in defining the purpose of SCR. How, they ask, does SCR differ or distinguish itself from those disciplines from which it often borrows models and methods, such as sociology, psychology, communications, linguistics, discourse analysis, etc.? The terms used to formulate this difficulty suggest that disciplines are defined by their subjects. But, on the contrary, disciplines are defined in the process of their development. Because 'science attains solely what its own mode of representation preconceives as its potential purpose' (Heidegger 1958:199), it follows that delimiting the field and defining the purpose happen jointly, so the purpose is never irrevocably fixed. Field and purpose interact with each other. They mutually reconfigure as knowledge progresses and as the contexts in which they are produced evolve.

In this vein, one must consider the trends of contemporary research. The contexts in which knowledge is developing and how it is produced are rapidly changing. Research results less and less from questions specific to academic disciplines and more and more in an application context, with specialized teams focused on target questions.[2] This research is called 'object-oriented':

> This type of research concerns ... particular objects which demand the expertise necessary to answer the questions posed. This necessarily involves the development of an interdisciplinary approach. (Gingras 2004:20)

[2] The cultural sector, to take an example outside science communications, participates like others in this movement. See Schiele (in press).

Should SCR then be conceived as a 'hybrid', rather than trying to limit its object to a disciplinary framework? Should it be seen as an 'interdisciplinary' field when contributions from various specialties are integrated in order to tackle a problem? Or as a 'multidisciplinary' field (Gingras 2004: passim) if examining the implications of answers to a problem, such as, for example, choosing a policy to promote science once certain factors are determined—let's say, the lack of young people's interest in science careers—by bringing in science mediators, teachers, sociologists and so on? For Priest (2010), the science communications field 'is both interdisciplinary and multi-disciplinary—and likely to remain so', both in terms of research in science communications and science mediation practices.

By insisting on science communications research as being both interdisciplinary and multidisciplinary, Priest (2010) reintroduces into the questioning of the status of SCR what the discussion is actually seeking to dispose of: the practice of science mediation and the interplay of social actors involved. This is also the position that Bell (2010) adopts, faced with the current impossibility of making SCR exist based on a single conceptual framework: 'science communication is less a community of researchers'—and we add ideas, concepts and theories since it is essentially in and by this symbolic mediation that a community of researchers is formed—'but more a space where communities coexist and the work of a science communication worker (be they academic, practitioner or a bit of both) is one of constant negotiation' (Bell 2010). Thus amalgamated, SCR and science mediation therefore signify, in Bell's mind, a constellation of heterogeneous sub-communities with distinct objectives, practices and discourses, aggregated in the same space and interacting with each other. This is somewhat like occupants claiming the same territory, while their habits and customs make them virtual strangers to each other: they co-exist, without really cohabiting, yet they interact nonetheless. The field would therefore be structurally splintered into groups of diverse interests but which interact constantly. In short, SCR understood in this sense would mean heterogeneous subspaces ranging from theoretical coherence to ideological conviction.

A third difficulty is the absence of theories and models (Trench and Bucchi 2010) that can structure the field by focusing questions around a group of systematically organized ideas and notions forming a unique part of the science communications field.[3] On this topic, the *deficit model*, though widespread, is at best common sense transposed into the science communications field to characterize a link to knowledge, which may be deemed ignorance or knowledge, but definitely not theory (Schiele 2008a). Other attempts to develop models have so far remained schematic as well (see Bucchi 2008).

Time will tell whether SCR is to remain relegated, albeit with its unique characteristics, on the periphery of communication studies as Priest anticipates (2010), is sufficiently autonomized to be considered as a genuine discipline, succeeds in

[3] Let us recall that a 'theory is a set of interrelated constructs (concepts), definitions, and propositions that present a systematic view of phenomena by specifying relations among variables, with the purpose of explaining and predicting the phenomena' (Kerlinger 1973:9).

developing theories and models, develops around a specific purpose or 'object-oriented' research by each time mobilizing the required competencies, or remains a heterogeneous field in which all manner of researchers and mediators continue to rub shoulders.

This long discussion was necessary to demonstrate that, after nearly a half century of SCR and nearly 30 years of international conferences and seminars, the debates sparked by the early work continue unabated and the same problems persist. Ultimately the real difficulty to overcome is as much in defining the field and its purpose as in reducing the tension between a tendency to be closed off, in an act of autonomization accompanying the attempts to define the boundaries, and the effects of the pressure that orients its research. This palpable tension dates back to the initial work and still operates today.

In this respect, Canada is an exemplary laboratory because of the heterogeneity of the work conducted, and because of the specific form the tension takes between the questionings emerging from the field and those from the social environment on which the field's development depends. Moreover, exogenous factors—historical and political—that we shall recall very briefly are also at play.

3.2 Issues of Science Literacy Development in Canada

To be fully understood, the development of science communications research in Canada must be reconstructed in the context of transformations in Canadian society.

Beginning in the 1970s, the development of a science and technical culture, long relegated to an adjunct cultural role, became a political priority. This evolution was not unique to Canada. STC elicits an interest in nearly all industrialized countries. To understand that interest in Canada, and ultimately its impact on SCR, certain features of the Canadian federation must be pointed out and the general stages in the development of science mediation must be described.

3.2.1 Exogenous Factors in the Canadian Context

First of all, Canada is located at the crossroads of several major intellectual traditions. On the one hand, French and British cultures have historically marked its development. Québec, which is very largely French-speaking, developed two parallel networks of universities and research centers: one in which research and teaching activities take place solely in French, and a second that is English-speaking. The other Canadian provinces and territories are majority English-speaking and higher education is therefore transmitted solely in English,[4] but with several bilingual universities in

[4] One exception is the University College of Saint Boniface in Manitoba.

areas where the French presence prevailed up until the conquest of Canada by England—ratified in 1760—when the French colonies were ceded to the British crown.[5] Thus, intellectual reflection in Canada draws on those two traditions, but without their necessarily communicating, collaborating or being mutually enriched by each other. These *Two solitudes*[6] partly explain, as we will see later, why certain SCR questions are dealt with in perspectives that have little or no counterpart in the other cultural community, and for which these questions either simply do not arise, or are viewed differently. However, the geographical and cultural proximity of the United States has made of these a privileged point of contact. American culture, ubiquitous via media, permeates Canadian society. Hence the following paradox: although French-Canadian and English-Canadian researchers virtually ignore each other, they nonetheless share a kinship due to the American work in their respective fields. So, exploring the development of SCR in Canada means considering these three traditions simultaneously, along with their mutual interactions.

Second, one must also consider the fact that Canada is a confederation in which the Canadian provinces have precedence over the federal government in cultural matters, and hence on all questions pertaining to teaching, research and the dissemination of science and technology (S&T). It follows that each province has its own objectives in promoting science culture, and each one develops its own research programs while the federal government in turn supports its own initiatives. This creates a major disparity in programs from province to province, and between the provinces and the federal sphere. To some extent, this Canadian heterogeneity resembles that in the United States and intrinsically leads to a proliferation of initiatives in science mediation and science communications,[7] but it differs at the political level since 'the central power in the United States is able to profoundly influence the decisions taken by the states in their field of competency by means of conditional subsidies' (Massicotte and Vaillancourt 2008:76), while in Canada the provinces are more resistant to any attempted encroachment.

[5] The two official languages of Canada (population 34.5 million) are English and French, the English-speakers totalling 66.3% and the French speakers 21.4%. Visible minorities amount to 16.2% of the Canadian population, South Asian and Chinese being the most important (4% and 3.7% of the total population, respectively). Furthermore, 19% of the Canadian population speaks neither English nor French as their mother tongue. The province of Quebec is a major exception: French speakers make up 81.8%, compared to 10.6% for English speakers and 7.6% for those who speak neither as a mother tongue. The province with the second highest number of French speakers is New Brunswick (29.7%); in no other province do French speakers account for more than 2.5%. Yet, the English and French peoples were not the original settlers of Canada: aboriginals amount to 3.75% of the total population.

[6] In 1963, Hugh MacLennan published *Two solitudes*, a novel in which he presented the complex relationships between English Canadians and French Canadians. The novel became a symbol of the Canadian reality: two cultures that prosper side by side, but most of the time superbly ignore each other.

[7] For an idea of this diversity, and the context in which it is possible, even if the data is becoming outdated, see Lewenstein (1994).

Third, in Canada, as in many other countries, science mediation practices substantially preceded the theoretical work. Moreover, after taking care to share and democratize knowledge, the federal and provincial governments increasingly mobilized around questions of promoting science careers and supporting economic development. They have supported the actions of developing and disseminating S&T, and are continuing to do so, especially in those two perspectives. This trend, which is not unique to Canada, has been reinforced as the development of S&T has pursued a course of recomposition in a logic of innovation (Castells 1996). This is seen as the very motor of economic development—a transformation of the role of S&T which has reconfigured the relationship with knowledge, redefined the motives and objectives of disseminating knowledge, and thereby marked reflection on SCR in Canada.

Fourth, everything seemingly indicates that theoretical reflection on science mediation in Canada was spurred by the political will to establish and justify programs promoting science literacy. That was where the heteronomy of the field was most in evidence and the interplay of cultural influences was most influential. Their combined effect makes it possible to understand the perspectives in which these mediation and research programs were conceived and developed. In a parallel movement, the theoretical reflection sought its autonomization. In other words, in Canada the theorization of science communications research apparently first emerged in the context of an auxiliary role to which it was subjected, and more recently as a specific preoccupation.

Before examining SCR in Canada, it is appropriate to briefly recall the historical development of science mediation activities for the simple reason that the theoretical development work did not arise *ex nihilo*, as we showed in Sect. 3.1. The theoretical reflection is part of a given context of the development of science mediation activities, the advancement of knowledge and the organization of research. The particular features of the Canadian federation—keeping in mind the factors mentioned above—overdetermined the development of PCST and influenced the orientation of research in this field.

3.2.2 Historic Milestones

Science disclosure, to recapitulate briefly, developed in parallel with the institutionalization of science and with early systematic training in science in a progressively renewed teaching system. In fact, they were three facets of the same movement affirming the role of science in society. While the eighteenth century was marked by the advent of the spirit of the Enlightenment, in the nineteenth century this became materialized as the Industrial Revolution, producing impacts on the evolution of society first through the development of technology, and then through science. Within this overall movement, popularization—not yet signifying public PCST—helped ‘make scientific progress a culture’. Popularization also propagated ‘the scientific spirit’ (Raichvarg and Jacques 1991:20, passim). In France, this spirit was notably spread by scientists themselves, among other things, via public lectures (quite in vogue at the time) and by popularizers, although relations between those two groups were not always cordial (Raichvarg and Jacques 1991). Heedful of the need for transparency and served by a dynamic press in full expansion, journalists quickly joined in this mission of propagation by publishing the reports of the Académie (Colin 1990). In short, 'mediatized and adapted to various publics, science [gained] ground in the culture of the century' (Béguet 1994:20).

This was also the case in Canada,[8] but by different routes: here, the learned societies played an important role with newspapers, scientific leaflets, public lectures, museums and demonstrations. In the United States, however, dissemination was mostly via conference speakers and itinerant science demonstrators constantly criss-crossing the country, which, unlike Paris, London or Montreal, did not have cultural centers (Lewenstein 1994). In Canada, the Canadian learned societies emerged from a rising group of merchant bourgeoisie and elites who prided themselves on their culture and who put science on an equal footing with the fine arts and music. The societies were a 'powerful instrument for advancing and disseminating knowledge', and 'their success bears witness to the place of science in the encyclopedic culture of fashionable society of the time' (Chartrand et al. 2008:78). The birth of the popularization movement in Canada in the nineteenth century was in large part due to them.

Although the institutionalization of science took a leap forward in Canada at the turn of the twentieth century, it was only with World War I that the impact of research on industrial development became evident. The Canadian Government then acknowledged the role of science and relieved the learned societies of their responsibilities by creating the National Research Council of Canada. The council helped create a host of structures that would stabilize and stimulate the development of research in the country (Duchesne 1978; Gingras 1991; Chartrand et al. 2008). However, unlike their English-speaking colleagues, French-Canadians in the province of Québec were still lagging behind. Among other factors at play in Québec society, the heavy influence of the Catholic Church in education curbed the teaching of science and the development of research in French Canada up until the 'Quiet Revolution' of the late 1950s and early 1960s, when the Canadian Government began emphasizing science more and more. The creation of the Association canadienne-française pour l'Avancement des Sciences (ACFAS) in 1923 sought to pull Québec forward by federating numerous French-speaking associations that were promoting and popularizing scientific research, advocating science teaching in high school, and soliciting government intervention. The relentless efforts by ACFAS, which were as much nationalist affirmation (Fournier and Maheu 1975) as promotion of science literacy on all fronts (Gingras 1994), did not really bear fruit until the Quiet Revolution.

[8] At the beginning of the nineteenth century, Canada was made up of only two British provinces created by the *Constitutional Act* of 1791: Upper Canada, part of what is today Ontario, and Lower Canada, which included the south and east of Québec and what is now Labrador.

During the 1930s in Canada and Québec, as elsewhere in the world, popularization became increasingly a part of cultural activities, particularly with leading spokesmen such as Brother Marie-Victorin in Québec—the archetypal image of popularizer–scholar. He combined a life of research with a mission of public disclosure of science. In founding the Cercles des Jeunes Naturalistes, he created a true mass movement promoting the natural sciences. At the same time, Marie-Victorin worked to sensitize young people, encouraging scientific vocations and a greater awareness by government to develop research. In France, to present another example of this dual scientist–protagonist role, the presence on the public scene of physicist Jean Perrin—to whom we owe the Palais de la Découverte, inaugurated in 1937—illustrates, just as in Canada and Québec, how the movement asserted the social necessity of science (Eidelman 1988). This accompanied the formation of a social group (the professionalization of physicians between the wars), which also crystallized in forms of dissemination. This is why Marie-Victorin was pleading for science as well as its popularization. Thus emerged an 'autonomous cultural sphere' (Eidelman 1988:1) of production and dissemination of knowledge (from teaching to popularization), in which the actors were called upon to play several roles: researcher, teacher, popularizer. The 'demonstrators' of the Palais de la Découverte were researchers in action as well.

This sphere then splintered into separate professional fields (science on one side and popularization on the other), as shown by the push for science journalism at the end of World War II (Lewenstein 1992). This trend to professionalism occurred rapidly. It was already evident among popularizers such as Camille Flammarion and Jules Verne in the nineteenth century (Raichvarg and Jacques 1991) and gained strength as the press developed. And the scientists who had been the principal purveyors of popularization before World War II, while still visible on the popularization scene, had to contend during the ensuing peacetime years with new professional popularizers in newspapers, publishing and radio. Professional popularization gained ground as mass means of communication developed and new media actors emerged during the 1950s, especially as television played an increasingly pervasive role in social and cultural life. In Québec, Fernand Seguin during the 1960s (Carpentier and Ouellet 1994), and in Canada more widely, David Suzuki from the early 1970s epitomized the social figure of the researcher turned professional popularizer.[9] In other words, in Canada as everywhere else, 'the general field of production that constitutes popularization became autonomized compared to the limited production field that constitutes scientific information in the true sense of the term' (Carle and Guédon 1988). However, while popularization was already playing a major role in society, distinct from that of science information and formal training, science literacy had not yet emerged as a specific project; nor, *a fortiori*, had research in this field. The turning point would be at the end of the 1960s.

---

[9] David Suzuki is an exception, as he held a position as genetics professor at the University of British Columbia until his retirement in 2001. Carl Sagan in the United States, to name him only, shared the same profile.

The 1960s were years of profound social and cultural transformations worldwide (Hobsbawn 1994), and Canada and its provinces did not escape. In Québec, these changes were termed the 'Quiet Revolution', which was the decade from 1959 to 1968—historians don't agree on the length—during which the Québec Government pursued fast-track modernization on the welfare state model and a policy of national affirmation. The Quiet Revolution asserted the French-speaking collectivity as the 'main vector of transformations of Québec society' (Linteau 2000:21). Canada was keeping pace. With Man and His World, the international exhibition of 1967 that celebrated the centenary of the Canadian federation, Canada gained recognition on the international stage. With this movement, popularization, accelerating its autonomization, mutated into science communications, as it was not only relaying information from the science community but also questioning the development and spin-offs of science. The term 'popularizer' became outmoded, and popularizers were now seen somewhat derogatively as hanging on to the coat-tails of the scientific field. In the rest of Canada, as in Québec, the preferred term became 'science communicator'—communication was in vogue. It was a term, attuned to its time, that allowed a distancing from the scientific community. So science communicators were now demanding an exclusive mediation with the general public, maintaining that scientists were incapable of addressing that public. Moles and Oulif (1967) demanded a role for journalists which they felt was imperative: to intercede between the scientists enclosed in their 'abstract language' and a public which 'with information' should 'have the right to scrutinize the resulting decisions' (Moles and Oulif 1967:33, passim; see also Schiele and Jacobi 1988). In 1971, to clearly distinguish themselves from the scientific community, science journalists founded the Canadian Science Writers' Association at the national level, and in 1977, in Québec, the Association des communicateurs scientifiques du Québec.

But more profoundly again, the 1970s marked the entry on the scene of a new actor: the state. If, until then, science mediation had been the business of scientists, followed by popularizers, and then communicators only, now the valorization of a science culture—'science literacy' in English-speaking countries—became a societal issue. During that decade, the development of S&T became intertwined with economic development. From then on, in the minds of its leaders, the nation's competitive capacity would be maintained by producing new knowledge and inserting it into industrial cycles to sustain and develop domestic and international markets. As well, shaken by the structural crisis disrupting their economies, western countries intensified their capacity for research and innovation, seeing in the progress of S&T the conditions for economic recovery. Thus, science literacy, until then essentially left on its own, was now perceived as an imperative priority for the state. Science had to be more closely aligned with society and also integrated into the culture. The school became a focus of attention to ensure that each and every child received a basic science education. But that wasn't enough: it was also necessary to 'put science into culture' (Lévy-Leblond 1994), which meant the valorization and public disclosure of science throughout all of society (Schiele 2005a).

Government commitment to STC had been preceded by various cries of alarm, such as W. Weaver's conference paper presented to members of the

American Association for the Advancement of Science, which since 1951 had insisted on the need 'to increase public understanding and appreciation of the importance and promise of the method of science in human progress' (Weaver 1951:471); that of C.P. Snow, who in 1959 decried the growing hiatus between the culture of scientists and that of the humanities; and the repeated demands of popularizers. However, it was the positions taken by the Organisation for Economic Co-operation and Development (OECD) and the repercussions of the 1985 Bodmer Report in England that for all practical purposes would crystallize the interest in promoting science literacy in Canada (Bodmer 1985). In Québec, apart from the admonitions of the OECD, it was rather the valorization measures adopted in France, notably with the creation of the Centres de Culture Scientifique, Technique et Industrielle, spurred by a conference convened by the Ministère de la Recherche et de la Technologie in 1982, that would open up the barriers of the scientific world to society and generate initiatives that would inspire the provincial government.

It was in this context that research work on science and the public developed in Canada and elsewhere in the 1970s. Researchers' interest in questions of science disclosure seemed to have been sparked at the same time as interactions between science and economic development became an issue for governments, inducing them to adopt policies to rationalize the relations between research and industry and measures to interest the public in science and make scientific information available. These concerns marked the work in Canada, much of it directly commissioned by the federal and provincial governments. In other words, the SCR researchers were not necessarily the initiators of the questions they were asking, even if those questions would contribute to structuring the field. Only progressively would the research autonomize, or, shall we say, become less overdetermined by government policies.

To sum up, science popularization in Canada developed much like elsewhere but a little out of sync. First, the roles were a greater combination of researcher and popularizer. Then those roles diverged, the split coinciding as much with the professionalization of the research as with the vocations of science mediation. The mass media drive sped up the autonomization of science journalism and opened the way for science communicators. Finally, governments recognized the potential for S&T innovations, with their economic impact, which led to a bevy of policies created to support research, stimulate science careers and encourage the disclosure of knowledge to sensitize and inform the public. It was when the state itself became an actor that the science communications research field crystallized.

3.3 Orientations of Science Communications Research in Canada

The role of the state in the development of science communications research cannot be discounted. This is particularly true in Canada, where education and culture are under provincial jurisdiction. However, before examining the themes of SCR in Canada, we must look more closely at the relationship between science policy and science literacy.

3.3.1 The State and STC

Despite jurisdictional factors, the federal and provincial governments both quickly committed to science communication. As briefly noted above, that commitment is understood only in the context of a recomposition of the relation to S&T which generated a series of coordinated measures ultimately referred to as 'science policies'. There were generally two major and complementary components. The first concerned everything dealing with producing and using knowledge; the second involved both the acquisition of scientific knowledge and the promotion of science.[10] Support for the cause of science literacy in Canada flowed directly from science policies, as a major outcome of reflection on science literacy.

Science historians quickly noted this correlation. Duchesne wrote in 1981:

> For several years we have seen the Government of Québec and certain organizations reporting more or less directly to it give increasing importance to the notion of science literacy. Variously cited and referred to by ministers and elected members of government alike, the 'thinkers' associated with the regime have been seen responding to contexts as wide-ranging as the debate on economic development, the project to reform the educational system, the museums creation policy, intergovernmental relations or the definition of sovereignty-association! (translated from Duchesne 1981:109)

It suffices to recall that a federal reflection begun in the early 1960s accompanied a political will for science research that was seen as expanding in a piecemeal, inadequately coordinated fashion (Canada 1962–1963). This led to the creation of the Science Secretariat (1964) and the Science Council of Canada (1966). This reflection continued throughout the decade, culminating in the establishment of the Ministry of State for Science and Technology (1971). Contrary to the 'anything goes' image criticized by the Glassco Commission in the early 1960s, Canadian governments going back to the nineteenth century have always been present in science development, first with the Geological Survey, the Dominion Experimental Farms and the Fisheries Research Board, and later with the National Research Council of Canada, which was established after World War I and the role of which increased significantly during World War II. From the early 1970s, in keeping with the global trend, the emphasis on fundamental research since the end of World War II (see Bush 1945) was displaced in favor of applied research and closer ties with industry (see Canada 1970, 1972, 1973). But it was not until the 1980s that Canada, in accord with the OECD, made S&T a priority and made efforts to promote and develop science literacy in Canadian society by launching the Science and Culture Canada grants program. This public valuing of science was part of the dual pursuit of economic development and national unity.

---

[10] We will not deal here with questions of innovation, industrial development or education in the schools.

We have already mentioned the OECD. It is necessary to recall its influence on decisions by governments to develop policies highlighting science literacy.[11] Because, as we've noted, government decisions hold sway as determining factors over the demands of scientists or communication professionals.

Aware of new economic factors and the necessity to redefine the role of science, the OECD had since its founding in 1961 encouraged member countries to establish national science policies and specific ministries apart from those of education and culture, which until then naturally tended to include science because, in the spirit of the Enlightenment, science continued to be linked with the advancement of thinking and knowledge. Now it had to integrate with economic and social policies, not just its potential spin-offs but in its very orientations. This meant going beyond previous research policy to create a true science policy in the service of economic and social development.

From the 1960s on, the OECD encouraged its members to adopt science policies while at the same time investing in research to boost their scientific potential in order to confront new S&T issues and compete better economically. In 1963, the OECD issued *Science and government policy*, a report that advocated the development of a genuine science policy: 'every nation' it stated, should 'have a reasoned and structured science policy, like those in the more traditional sectors of national activity, an economic or foreign policy' (OECD 1963:59). In 1971, a second report, *Science, growth and society*, noting that most countries had established measures to coordinate national scientific efforts, insisted on the fact that 'science and technology are an integral part of economic and social development' and that this 'involves a much closer liaison than in the past between science and technology policies, between all areas of socio-economic concern and government responsibility' (OECD 1971:107). In short, the OECD took into account the increasing integration of science and society and the structuring effect of that integration—an evolution that researchers had long observed. It thus concluded that current society does not develop apart from S&T and likewise that the idea of economic and social progress is now so intimately associated with the development of S&T that the two tend to merge.

But the 1970s were also a period of collective awareness of the rapid social and cultural changes that had been experienced by societies in the postwar years. This was accompanied by a crisis in values—to which we have already alluded—reflecting a huge social effervescence with a strong democratic thrust. The crisis was also marked by an expression of disenchantment towards the development of S&T based on observations of the damage referred to in the 1972 report of the Club of Rome, *The limits to growth*—observations to which MIT researchers lent their support. The OECD reacted to these social transformations and the rise of the environmental question by reasserting the imperative to pursue the progress of S&T, made tangible 'in the form of physical equipment and professional capabilities'. However, this

[11] We are indebted to Richard Pitre (1996) for this analysis of the role of the OECD.

pursuit of progress, while deemed 'imperative', was 'not at all sufficient' of itself to 'resolve the global problems' (OECD 1981:98) as perceived during those years. At least two conditions must combine. On the one hand, the 'indispensable progress' must not stymie 'growing aspirations to greater democracy and greater equity' (OECD 1981:99). Rather, it had to recognize and incorporate them. However, since the economic context had evolved and a structural crisis followed the rapid growth that characterized the postwar period, the need to adapt the workforce to technological changes became an imperative for governments. The OECD stated that:

> A high level of science and mathematical literacy of the population as a whole might be a prerequisite if the nation is to have an active population able to meet the very high level of professional qualifications required to rapidly implement the new technologies in the national economy. (OECD 1981:100)

Thus, for the OECD, the pursuit of social and economic development occurs through greater democracy and equity, but most of all by a valorization and dissemination of S&T literacy and by a general rise in the level of scientific competency among the population. In the OECD's thinking, an increase in collective innovative potential and greater adaptability of the population to the evolving sociotechnical environment are necessary in the pursuit of 'progress'.

It is useful, therefore, to distinguish the structuring of the field of science communications practices, like all significant practices purveyed in society—in particular those of science journalism—from the state's assertion of the necessity for a science and technical culture, accompanied of course by a retinue of means to promote its development.

Indeed, things were often presented as if state involvement was either the natural extension of the development of science communications practices or the result of a long-awaited response to the successive demands made on the state in the preceding years by those from the science communications community. Even if those promoting science literacy, including journalists, often sought government backing and demanded a policy of valorization and support, and although they were pleased by the adoption of such a policy, the reasons that drove government were, as we have just shown, completely different. In other words, the pursuit of individual and collective emancipation that impels the notion of 'knowledge sharing' so often demanded by science communicators dissipates into a radical reversal of perspective: science literacy henceforth manifested in the emergence of new productive forces. The economic development potential inherent in S&T can only be fully realized in a society that can perpetually sustain it by producing new knowledge and can perpetually adapt to the relentless transformations being generated as this knowledge is implemented in the form of innovations. So a new mode of socialization is sought. Science literacy is expected to yield active, individual and collective conduits, adapted to a high-productivity economic system. Michel Amyot, a former deputy minister in the Québec Government, wrote:

> In 1987, recognizing the need to promote the development of science and technology literacy throughout society, the Ministries of Science and Technology of OECD member countries lent a certain genuine credibility to this aspect of national policies. (Amyot 1994)

Following this example, Canada created the Science and Culture Canada grants program and embarked on a public awareness campaign. Federal and provincial ministers of science and technology agreed to make S&T a national priority (Amyot 1994). Thus, the OECD exercised considerable influence on the establishment of science policies on the valorization of science literacy in Canada and among OECD member countries.

Until then in its infancy, SCR began to develop on several fronts, generally focused on usefulness. The research conducted for government ministries and departments, for organizations dependent on the states (such as the Conseil de la science et la technologie du Québec) and for industry and the business community was pertinent and goal-oriented, coinciding pretty much voluntarily with the prevailing consensus or issues of the day, no matter what margin for maneuvre the researchers were given. A critical outlook arose in tandem, spurred by the idea of science literacy itself as advanced in the political discourse. Though seen as vaguely defined, its ideological dynamic captured the attention of researchers.

3.3.2 Measuring the Public's Science Literacy

Once the development and valorization of science literacy were on the political agenda, one of the first useful things governments wanted to know was the level of science literacy of the population, their sources of information and their attitudes towards science and scientists.

Therefore, one of the first surveys carried out in Québec (Tremblay and Roy 1985) examined the representations and attitudes of the public. Four main themes were covered:

- The level of information of the population and the amount of exposure to science news
- The perception of the positioning of science research compared to other research sectors and priorities
- The population's image of the researcher and science
- The perception of the consequences of scientific and technical development.

These four themes, as advocated by the OECD and already expressed in work consulted to prepare the survey, continuing from one study to another, would progressively shape the hard core of the federal and provincial governments' concerns. Work already done in the United States, France and the European Economic Community (only in 1993 did it become the European Union) set the basis for Canadian work. And it was around these themes that the questions of measurement would first crystallize.

Very quickly, in fact virtually immediately, governments began comparing their nations with others. In addition to assessing Quebecers' perceptions, the survey by Tremblay and Roy (1985) was intended to be 'a comparison with other surveys carried out in recent years on the same questions, notably in Canada, Great Britain, France and the United States' (Tremblay and Roy 1985:8). The authors concurred that 'The themes selected for the questionnaire were chosen only after perusing the available surveys. We were careful to adopt a formulation as similar as possible to questions already set forth in the surveys for which detailed results were available' (Tremblay and Roy 1985:153). Logically, this would keep the elements of comparison intact in subsequent work. For instance, Gagnon and Morin (1986:65) noted that 81.9% of Quebecers considered themselves capable in science matters if they were well explained, while in France that proportion was only 62%.

Five years later, Filiatrault and Ducharme (1990) used all the same categories in their investigation of perceptions of science by the population, although their investigation was more fully developed than the one by Tremblay and Roy. This was also true for the study by Albert et al. (2002), although they focused more on interest in science, sources of information and judgments on the effects of scientific development. They did, however, introduce the category of recreational science, a theme rarely dealt with.

Following Decima (1988), Einsiedel (1990a) for her part embarked on a study of the level of science literacy among Canadian adults. Unlike the francophone work, it drew directly on research by Miller (1983a, 1989; see also Miller et al. 1993, 1997; Miller 1999, 2000) conducted in the United States and that of Durant et al. (1989) in England. Since Einsiedel's study dealt with the Canadian population, this made possible comparisons between the various Canadian regions (West, Ontario, Québec, Atlantic),[12] which Québec studies were not doing. Moreover, rather than restricting itself to a simple factual presentation of results, such as the proportion of correct answers to questions of science knowledge categorized according to level of education or age, it sought to construct a Science Literacy Index and an explanatory model. In the proposed science model, science literacy would result from two groups of factors combined: antecedent factors (education, exposure to science courses, gender and age) and consequent factors (attitudes, attentiveness) (Einsiedel 1990a:29–30). What we call 'science literacy' will therefore be the product of interaction between the antecedent and consequent factors. In a more encompassing approach, Godin et al. (1998) also proposed a model. They considered it systemic and pluridimensional in that it sought to integrate the individual dimension, the collective dimension and the relationships between the two by focusing on three modes of appropriation they saw as determinants: the learning modes, the social organization modes and the social involvement modes. They professed their model to be founded on a 'complete reversal of the perspective on science and culture relationships' (Godin et al. 1998:25). For them, neither the opposition between science and culture (Snow 1959; Lévy-Leblond 1984), nor even the conception that science culture derives from science as a reduced knowledge, is satisfactory. Their model asserts that 'science culture is primary' (Godin et al. 1998:25). 'Science is

[12] The results of Einsiedel's (1990a) study were used by Miller et al. (1997) in a comparative study of perceptions of science and technology in the European Union, the United States, Japan and Canada.

not an "other" culture, but constitutes the heart of the current culture' (Godin 1999:5). Consequently, for them the indicators developed up to now do not measure science literacy (Godin 1993a).

Except for the attempts at formalization, the first thing that is striking in these efforts largely conducted at the request of government agencies[13] is of course their homogeneity: they all copy each other. Imitation served to consolidate the bodies of notions and the indicators used in these studies, which make up the bulk of SCR efforts today. Only progressively do other research perspectives develop, in the process of measuring the full complexity of the notion of science literacy, but those approaches are still marginal compared with the dominant paradigm. The second thing that is striking in Canada is the sparse amount of work devoted to creating a portrait of the status of science literacy, especially if we compare the amount of Canadian work with that done in the United States, in England, in China or in Korea, for example (see Bauer et al. 2011). This is all the more surprising in that a plethora of reports and consultations have emphasized the need to promote and valorize science literacy (see, for Québec, Santerre 2008).[14]

In other words, despite the often repeated assertion of the importance of a scientific culture, governments have made little attempt to learn the effects of their valorization measures. Nor have Canadian researchers shown much greater enthusiasm. An analysis of 185 articles published in the journals *Public Understanding of Science* and *Science Communication* from 2003 to 2009 showed that the surveys, with 49 articles tabulated, made up 26.5% of the total, or one topic out of four (Borchelt and Carollo 2011). Canada, with just a few surveys, was far short of the norm.

The idea of a science culture quickly became a topic of critical work, which first of all criticized the vagueness of the notion. 'The current success of the science culture theme should not be viewed as indicating it has gained a clear meaning, nor even that it has gained a general meaning by those using it' (Duchesne 1981:109). For Duchesne, who carefully scrutinized the political discourse, showing the common threads from one report to another, it was essentially an ideological operation, which in the guise of 'everyone's participation in science literacy' aimed to 'transform the debate on the control of scientific institutions and the appropriation of knowledge' (Duchesne 1981:110). It was less a matter of inviting the public to become 'learned'

---

[13] Einsiedel's survey was mixed: it was at once funded by the Social Sciences and Humanities Research Council, as part of research in several areas, and by Industry, Science and Technology, a Canadian government department.

[14] We should recall that the valorization of science culture also translates into a reform of science teaching. We may recall that the studies by Orpwood and Souque (1984) and by Olson and Russel (1984) led to a reform of science teaching programs, especially in Ontario. This question frequently returns in the forefront of debates in Canada, but without this necessarily being the case in Québec (Godin 1993b:314). However, a reform of science teaching had begun during the Quiet Revolution. Unfortunately, it didn't yield the expected results (Désautels 1980). That said, and while much research on science teaching broaches questions akin to SCR, lack of space prevents us elaborating further here.

than a matter or creating 'good consumers' (Duchesne 1981:121). Ten years later, in the same vein, Godin (1993b) observed that the notion had yet to be clarified and must still be considered as a 'political discourse', meaning that 'various attempts to operationalize the concept by the public authorities inevitably [suffer] the effects of historical contingency' (Godin 1993b:306). Schiele (2005a) agreed with that assessment: the notion remains vague and malleable amid hope that successive reformulations may yield an enduring core of meanings organized around four key ideas: 'science and technology … are at the center of the productive system'; 'maintaining economic capacity … demands rapid adaptation to change'; the advances in science transform our understanding of the world and how we 'perceive ourselves and how we think'; and 'the democratic resolution of environmental, societal and philosophical debates provoked by these disruptions demands the enlightened commitment of everyone' (Schiele 2005a:18). But, in the spirit of the work of Bauer (1998), who saw the recurring cyclical debate on science culture as important, since science culture fully plays its role at those moments when its social necessity is being reasserted, it would fulfill a dual function during those moments: 'on the one hand, a destabilization of the knowledge and skills demanded till then … and, on the other hand, a valorization of those emerging' (Schiele 2005a:38). Thus, the moments of revival of science literacy operated around 'lengthy economic cycles and structural adjustments when emerging from crisis, when the innovative potential of science and technology is instrumental in the recovery' (Schiele 2005a:38). The important thing for the idea of science literacy per se is the cyclical reassertion of its necessity.

3.3.3 *The Media and STC*

Even in a quick overview like this, one must acknowledge the Canadian work on the media. The capacity of the media to contribute significantly to the development of science literacy has long been debated. There have been a significant number of works on this topic, as shown by the above-mentioned survey of 185 articles in *Public Understanding of Science* and *Science Communication*: 49 dealing with science coverage in newspapers, 21 with science journalism, 17 with science coverage in mass media, and 13 with science coverage on television—101 articles out of the total, or 54.6% of topics presented. One out of two articles therefore dealt with the role of the media.

This ongoing investigation is fully justified because, as mentioned above, the media have become increasingly important in social and cultural spheres since World War II, imposing their own dynamic on culture as catalyzers and instigators of cultural change. Major happenings are transmitted via the media, especially television, which from the 1960s on has been the prime and pervasive purveyor of information and entertainment (Schiele 1978:5–21). 'Communication'—extended to all social relationships—was being constructed in a utopia of substitution like that of the Enlightenment, just when the Enlightenment was fading from the social consciousness (Breton 1997). It was natural that governments and researchers would focus on the question of the impact of the media, since the media were embodying modernity and were perceived as one of the vectors of change.

The first Canadian study on science information in the media appeared in 1975 (Dubas and Martel 1975), produced at the request of the Ministry of State for Science and Technology. Such enquiries into the media were not new. The Canadian Senate had already commissioned a major study on the influence of the media on the public, which showed that Canadians turned mostly to TV and newspapers as primary sources of information: newspapers were preferred to TV 'news, in-depth information and interpretation of events' (Special Senate Committee 1970:19).[15] What was new, however, was the emphasis on science information. Here again, the media were the primary source of information, but with TV and magazines this time coming before newspapers. Furthermore, the public felt there was insufficient science information in the media: they demanded more, and wished for better. This research also showed that interest in science increased with the level of schooling, reinforced by science education—a recurring observation from study to study (Miller 1983b; Miller et al. 1997; Banchet and Schiele 2003).

So it was natural that media coverage of science attracted the attention of actors in the science literacy milieu, already alert to the issue.[16] Note, moreover, that Dubas and Martel (1975) had also shown the media as granting little space to science. Ten years later, Boucher et al. (1985) reported that the situation had not improved. The place reserved for science news was only 3.1% of new publishings. Making the same observation, Einsiedel (1989, 1990b, 1992), from an analysis of content in major Canadian daily newspapers (seven in English and one in French), reported that Canadian newspapers gave little importance to science information; that this information was often relegated to secondary sections; and that, when there was some, it was presented as event-oriented, time-bound reports. In her study, Einsiedel also looked at the sources of information, the journalistic tone adopted, and the mode of presenting the information. For her:

> The typical story appears to be one on medicine and health, occurring in an area outside of the reader's community, reliant almost exclusively on scientific expertise, and generally about some positive event: an innovation, a medical advance, a cure. Although risks or negative consequences are covered, these are generally not highlighted in the way positive consequences are. (Einsiedel 1990b:20)

Each of these three research projects sought to produce an overall vision. Other efforts were more specific: Jacobi and Schiele (1993) did a semiolinguistic analysis of information in a newspaper; Einsiedel and Thorne (1998) examined public reactions to situations of uncertainty presented in newspapers, such as the risk factors in the transmission of Creutzfeldt–Jakob disease (mad cow disease) associated with eating contaminated meat; Maillé et al. (2010) examined the relationships between

[15] A fact to note: according to this study, francophone citizens of Québec were, after those of Saskatchewan, the least likely to read a newspaper daily, and they read them much less than Québec anglophones.

[16] *Québec Science*, nos. 12, 17 and 25, 1974.

scientists and journalists; Gauthier (2011) looked at how foodborne microbial risks were presented in newspapers. This list could go on, but would add little to the fact that the question of science information in the media had not elicited much interest either by the governments or among researchers in Canada. Literature reviews by Logan (2001) and Weigold (2001) provided an idea of the importance accorded to it elsewhere.

With a few exceptions, Canadian research largely adhered to the conception of the role of media presented in work developed in the United States. 'Essentially,' maintains Logan (2001: passim) 'most science communication research has revolved around (1) the sources of science news; (2) how news is reported, edited and written; (3) the appropriate media channel to communicate science; and (4) the audience for science.' However, the contribution of semiotics, linguistics and discourse analysis proved particularly fruitful (see Schiele and Larocque 1981; Schiele 1986; Jacobi 1987, 1999; Provençal 2011).[17] But those approaches, with few exceptions, did not grab the attention of Canadian researchers. Nor was there much recurrence of the critical perspective, except in a few works such as Schiele (1983), which maintained that, contrary to claims, the media contribute very little in disseminating science information. Having opted for entertainment when presenting science, their work consisted of producing 'knowledge as simulacrum', and producing 'the semblance of knowledge as merchandise offered for consumption' (Schiele 1983:174).

[17] The rarely discussed conceptual model that nourishes most of this work on the media is the so-called 'classic' communication model. This linear model, proposed by Lasswell (1948) to describe the structure and function of communication in society, describes a flow of information from source to receptor by means of messages. The Lasswell formula describing this flow remains famous: Who (says) What (to) Whom (in) What Channel (with) What Effect. Also sometimes called S-R (stimulus–response), this formula was enriched by the notion of 'retroaction' (Westley and MacLean 1957) from one source of information to another. There is, one might say, collusion between the recourse to the 'deficit model', which assumes a flow of information from those who know towards those who do not know, and the Lasswellian diagram of communication, which poses an action from the source to the receptor. And this without examining the conditions and modalities of the appropriation of messages, how the public uses them, the conditions of production of the messages, or even the very idea of the form and content of the messages. The debate usually proceeds towards the relationships between the journalists and the scientists (Nelkin 1995; Maillé et al. 2010), or the traditional relationship of accuracy and truth of the information (Bell 1994). The idea is that the quality of the information guarantees that of the reception. These efforts are not devoid of interest (far from it). However, even if the questioning has evolved through the years, going from the idea of 'science literacy' (from the 1960s onwards, focused on knowledge) to that of 'public understanding' (after 1985, focused on attitudes and education), to arrive today at the more complex relationship of 'science and society' (the 1990s to the present) (Bauer et al. 2007:80), the perspective in which the work on media was conducted perpetuates in the SCR field a conception of the role of media that no longer has currency in studies on communication. It is the conception inherent to communication that marks the boundaries of reflection in SCR. Moreover, the realist approach—in the limited sense of amassing facts and items noted by Bachelard (1966:5)—generally adopted rubs out as the social and political factors that influence the discourses and practices of the media. It impels a necessary decentralizing to envisage other models. The hesitant attempts to propose other perspectives, such as work on social representations (Farr 1993; Locke 2005; Wagner 2007; Goodwin et al. 2011), have not yet received the attention they deserve.

3.3.4 New Questions, New Issues

For several years, the number of Canadian efforts increased. This work also diversified, in keeping with a general evolution observed elsewhere, but most of all it explored areas that did not yet include SCR. Space limitations prevent the listing of all these works here, and one work is scarcely enough, but at the risk of simplifying one can characterize some trends using selected examples.

3.3.4.1 Research on Science Museology

Science museology is one field to consider. Essentially, it is for all useful purposes ignored in SCR, while science museums and science centers play a primary role in publicizing science literacy.[18] Two research trends garner attention. The first is the evolution of the institution itself, and the effects of changes on the exhibiting of S&T and consequently on the appropriation of the content by visitors. Some work has examined the role played by science museums and science centers at a time of an assertion of political will to promote the development of science literacy. Paquette (2011) showed that while governments, stimulated by the OECD, initiated policies to enhance science literacy, they discovered in the movement to establish science centers, then in full expansion, powerful and active allies that exemplified their objective. So did science journalists and those in the science culture milieu who were benefiting from the interest in science literacy (Schiele 2005a).

The SCR work on the evolution of journalistic practices showed that they were affected by structural changes: on the one hand, the rise of cyberculture (Trench 2007; Schiele 2009), and, on the other, the development of public relations (Göpfert 2007). Similarly, Paquette showed that museums, too, are subject to structural changes, especially in imposing a 'merchant logic' to the detriment of an 'institutional logic geared to public service' (Paquette 2009:64). This evolution profoundly altered the mission of science museums and science centers and hence the nature of the relationship to science that they created with their publics. Ultimately, the very image of science was changed. 'Managerialism,' wrote Paquette, 'had the effect of transposing the public mission into a quasi-commercial mission, of reducing the public gesture to a delivery of services and transforming the citizen into a consumer' (Paquette 2009:64).

Another research trend examines the nature of the communication established with the visitors in a context where they are now the central concern of museum institutions, which design their exhibitions and develop their activities to meet that

[18] Just as for science journalism, research in this area is greatly autonomized, to the point of forming a totally separate field of research. However, a fair amount of this work is akin to SCR, dealing as it does with the same questions and examining the same issues, so it is worth approaching SCR work in this perspective.

priority (Schiele 2000, 2005b, 2008b; Schiele and Koster 2000). These studies examine the interactions between the offer of information and the modalities of appropriation. Thus, it is the relationship to knowledge induced by the form and manner of exhibiting science—in both senses of the term 'exhibition': that of presenting science to the public, and that of simultaneously holding a discourse on science—which is the subject of enquiry (Montpetit 2000; Schiele 2011).

3.3.4.2 Research on the Usages

As briefly mentioned (see Note 17), most of the work conducted in SCR does not question the canonic model of communication, or does so only superficially. For this reason, it is mostly a matter of the effects of media on the audience and the measurement of them. Nonetheless, some Canadian efforts that questioned this conception adopted another approach: they explored the appropriation of science literacy according to how its users use it. This meant breaking from the naive vision of knowledge, which was widely prevalent: that of transposing school knowledge into daily life. Such knowledge is rarely operating because it is apprehended independently of the contexts in which it is mobilized. Hence, the measurement of science literacy based on indicators thus far developed seemed for some a futile exercise because it did not measure how that knowledge is actually implemented in social situations. The notion of usage, however, harks back to the idea of a knowledge developed and shared in and by a network of collaborators. As a consequence, we no longer think in terms of an essentially isolated actor invited to memorize information and recall it as needed. Conversely, the notion of usage involves actors mutually committed to each other. This approach developed from research on the dissemination of technical innovations and the ways they were adopted, initiated in the 1950s in the United States (Rogers 1962) and from the work of Wynne (1989, 1991) on interactions between actors mobilizing different levels of knowledge and forms of learning.

As for science museology, some work on the usages compels a revision of the current conception of producing and acquiring knowledge, obliging us to reconsider the definition of science literacy, to rethink it in a perspective of co-construction of knowledge since developed and acquired in contexts in which they are operating. Heaton et al. (2010, 2011) studied *Tela Botanica*, a worldwide communication and exchange network between French-language botanists.[19] Developed like Wikipedia, free and open, its special feature is its exclusive reliance on communication technologies. It seeks the free circulation and availability of botanical learning. Through its online notebook, *Tela Botanica* 'makes it possible to submit observations of plants, … sort them, and research them. These floral reports of terrains can be retrieved online' (Heaton et al. 2010:63). But the most interesting aspect affecting

[19] *Tela Botanica* is a French project, but the research is carried out by Canadian researchers.

the development of a genuine science literacy is the collaboration between scientists and laypeople with varying levels of knowledge, since *Tela Botanica* enables not only the 'dissemination of botanical learning'—a traditional function considerably broadened by the new communication supports—but also an 'active construction of knowledge within the network via the device' (Heaton et al. 2010:63). This lends weight to the notion of lay expertise—the idea that the lay public can also produce useful science knowledge—and argues 'for a conceptual reconsideration of the frontier between formal expertise and expertise flowing from the common sense of ordinary people, favoring an outlook that encourages permeability and fluidity between the two categories' (Heaton et al. 2010:64). We have developed this example because it singularly illustrates the potential of communication technologies. By now permitting distance interactions and reciprocal communication (Jauréguiberry and Proulx 2011), communication technologies help new organizational forms of production, dissemination and appropriation of knowledge to emerge. They therefore impel a rethinking of the spirit in which most SCR is still currently conceived. Other work, less focused on the role of media and more on the dynamics of interactions between the members of a network, where experts and non-experts rub shoulders, also examines the co-production of knowledge. This is the case among a community in British Columbia, studied by Boyer et al. (2009), composed of scientists and laypeople who are collaborating on a common project to draw up a map of the habitat of coastal eelgrass. One readily imagines that communication technologies are playing a role at least as important as direct relationships are in these new forms of organization.

3.3.4.3 Research on Participation

One final area merits a brief look: participation, or what is now more generally called 'citizen participation' or 'public engagement'. Einsiedel (2008) talked about the 'participatory explosion' to stress that the mobilization of the public had become a major social phenomenon. And it had diversified, since the term embraces forms of action that differ from each other.[20] This movement was not unique to Canada, but it was more recent here than elsewhere. Phillips and Orsini, in the synopsis of their study on citizen involvement, noted that:

> Canada's primary institutions are not assuming as effective a part in citizen involvement as they might. In particular, political parties, Members of Parliament and parliamentary committees play an unnecessarily weak role in involving citizens. (Phillips and Orsini 2002:iii)

In the same vein, Turnbull and Aucoin (2006) pointed out that in Canada the public is not very involved in developing policies, and that the levels of government

[20] Included are 'citizen involvement, stakeholder engagement, participatory technology assessment, indigenous people's rights, local community consultation, NGO intervention, multi-stakeholder dialog, access to information, access to justice' (Einsiedel 2008:173).

hesitated to move forward. But things have evolved: the Romanow Commission on the Future of Health (2001–2002) diverged from the usual mechanisms and listened to the public by organizing, in addition to the usual consultations, 'televised forums … held on university campuses, and an online consultation workbook' (Medlock 2011:105). Furthermore:

> the predominant model of communication with citizens has been a one-way flow of information from policy and scientific elites to the mass public. More recently, though, there has been a broad shift towards experimentation with more participatory forms of public engagement that encourage a two-way model of communication, one that provides a forum for dialog and mutual learning among citizens, experts, and policymakers. (Medlock 2011:103)

The push for public participation spurred researchers' interest in the impact of that participation and the means for evaluating it even before the movement was very evident in Canada. Thus, Einsiedel et al. (2001) compared three consensus conferences that took place in Denmark, Canada and Australia in March 1999. Held in western Canada, with no federal government support, the Canadian example was a national first. Like the other two, it looked at the question of food safety.

And here is where participation becomes an interesting issue for research: it is frequently associated with health or risks to health. So everything proceeds as if the actions of mobilization (those chosen by the researchers, of course) are associated with health questions, or as if the health questions (still those chosen by the researchers) include potential mobilizations. During the citizens' conference it was a question of 'deliberative democracy' (Einsiedel and Eastlick 2000), but closely associated with health and health risks. We can understand this interest on the part of researchers: the development of biotechnologies affects the food chain, and to varying degrees everyone is alert to the potential impacts of genetically engineered products. In Canada, that interest developed sufficiently to raise questions about participation. Other issues involve xenotransplantation (transplants from animal donors), along with inherent questions on the social and moral aspects of science (Bickford et al. 2005), genetically modified food (Blue 2010), the creation of biobanks (Longstaff and Burgess 2010; MacLean and Burgess 2010; Godard et al. 2010; Walmsley 2010) and assisted human reproduction (Jones and Salter 2010). This interest in health topics is also reflected in studies on media coverage of health and health-related questions (Ungar and Bray 2005; Hivon et al. 2010; Maillé et al. 2010; Gauthier 2011). In short, 'life' questions take precedence regardless of the purpose of the research.

* * *

The federal and provincial governments' infatuation with science has declined. The only federal program, Science and Culture Canada, was abolished in 1999, and the Société pour la promotion de la science et de la technologie, an organization supported by the government of Québec, was dissolved in March 2011 without a mention in the press. Science literacy is no longer a political priority, even for the government of Québec, which nonetheless had been very active (see Santerre 2008). However, the science literacy milieu is particularly dynamic, with programs in all the Canadian provinces and territories offered to all categories of the public (see Schiele et al. 2011).

SCR has undeniably developed in Canada in recent years. It has also, so to speak, been freed from governmental pressure. Still fragmented, however, it is far from constituting a field: the work comes from a range of horizons. But three poles can be defined where research activities are concentrated: the Montreal (Quebec) region, which has five universities[21]; the region of Calgary (Alberta); and to a lesser degree the region of Vancouver (British Columbia). The three poles do not denote schools of thought, but rather are focal points for researchers and teams.

The university programs are concentrated in Québec and Ontario. In Québec, Laval University (in Quebec City) offers a graduate program in public communication with a major in science journalism, the Université du Québec à Montréal (in Montreal) offers a graduate certificate, the University of Sherbrooke (in Sherbrooke) runs several courses, and the University of Montreal manages just one. In Ontario, Laurentian University (Sudbury) awards a Graduate Diploma in Science Communication; for all useful purposes, it is the only complete training in science communication in Canada. There is also a science journalism course offered at the University of British Columbia (in Vancouver). At present, these programs do not have researchers in the SCR field.

To return to research, we can distinguish two phases. The first coincided with the drive for science literacy at the turn of the 1980s. There were two kinds of research: evaluations to draw a picture of science literacy among the population, and assessments of the impact of the media. These efforts were mostly conducted at the behest of government ministries and departments and continued to be specific, ad hoc projects. Unlike other countries, Canada until now has never undertaken such studies in any regular and systematic way to compile comparative data.

A second phase ensued with a diversification of research, predominantly through university projects. While the research on media remained important, other topics were undertaken to broaden the scope of reflection in SCR. While university research was developing over time and tending to become structured around certain issues (and increasingly self-referencing), the preferred topics of researchers in eastern Canada remained significantly different from those in the western provinces. Both of the traditions we noted at the beginning of this chapter continue to operate. Generally speaking, the work done in western Canada seems more closely related to immediate political and social issues, while that in the east is more embedded in the long term, tending to be more fundamental and critical.

References

Albert, M., Marchal, C., & Robitaille, J.-P. (2002). *Enquête sur la culture scientifique et technique des Québécois et Québécoise*. Sainte-Foy: Conseil de la science et de la technologie.

Amyot, M. (1994). Bilan des actions québécoises en culture scientifique et technique. In B. Schiele (Ed.), *Quand la science se fait culture, Communications, Actes II*. Saint-Foy: Editions MultiMondes, Montréal: Université du Québec à Montréal. (Edition électronique.)

[21] Including the campus of the University of Sherbrooke (Sherbrooke, Québec), which now has a location in the Montreal region.

Bachelard, G. (1966). *Le rationalisme appliqué*. Paris: Presses Universitaires de France.

Banchet, J., & Schiele, B. (2003). Comparaison de quelques enquêtes nationals et internationals sur la compréhension et la perception de la science par le public. In B. Schiele & R. Jantzen (Eds.), *Les territories de la culture scientifique* (pp. 95–114). Montréal: Les Presses de l'Université de Montréal.

Bauer, W. M. (1998). 'La longue durée' of popular science, 1830 – present. In D. Devèze-Berthet (Ed.), *La promotion de la culture scientifique et technique: ses acteurs et leurs logiques* (pp. 75–92). Paris: Université Paris 7—Denis Diderot.

Bauer, M. W., Allum, N., & Miller, S. (2007). What can we learn from 25 years of PUS survey research? Liberating and expanding the agenda. *Public Understanding of Science, 16*(1), 79–95.

Bauer, W. M., Shukla, R., & Allum, N. (2011). Towards cultural indicators of science with global validity. In W. M. Bauer, R. Shukla, & N. Allum (Eds.), *The culture of science: How the public relates to science across the world* (pp. 1–17). New York/London: Routledge.

Béguet, B. (1994). La vulgarisation scientifique au XIXe siècle. In B. Béguet, M. Cantor, & S. Le Men (Eds.), *La science pour tous* (pp. 5–48). Paris: Editions de la Réunion des musées nationaux.

Bell, A. (1994). Media (mis)communication on the science of climate change. *Public Understanding of Science, 3*(3), 259–275.

Bell, A. R. (2010). Notes from some spaces in-between. *Journal of Science Communication, 9*(3) C02.

Bell, P., Lewenstein, B., Shouse, A. W., & Feder, M. A. (2009). *Learning science in informal environments: People, places, and pursuits*. Washington, DC: The National Academies Press.

Bickford, J., Mather, C., & Fleising, U. (2005). Time and the structuring of ritual performance in the xenotransplantation debate. *Public Understanding of Science, 14*(3), 235–247.

Blue, G. (2010). Food, publics, science. *Public Understanding of Science, 19*(2), 147–154.

Bodmer, W. (1985). *The public understanding of science*. London: Royal Society.

Borchelt, R. & Carollo, B. (2011). *Science communication literature review and analysis: Preliminary research report, 5–9 December*. Office of Public Affairs and Communication, United States National Cancer Institute.

Boucher, L., Dupuis, D. & Schiele, B. (1985). *La nouvelle scientifique dans la presse québécoise*. Département des communications, Université du Québec à Montréal.

Bourdieu, P. (1980). *Questions de sociologie*. Paris: Les Editions de Minuit.

Bourdieu, P. (2001). *Science de la science et réflexivité*. Paris: Raisons d'agir Editions.

Boyer, L., Roth, W.-M., & Wright, N. (2009). The emergence of a community mapping network: Coastal eelgrass mapping in British Columbia. *Public Understanding of Science, 18*(2), 130–148.

Breton, P. (1997). *L'utopie de la communication*. Paris: La Découverte.

Bucchi, M. (2008). Of deficits, deviations and dialogues: Theories of public communication of science. In M. Bucchi & B. Trench (Eds.), *Handbook of public communication of science and technology* (pp. 57–76). London/New York: Routledge.

Bush, V. (1945). *Science the endless frontier: A report to the President on a program for postwar scientific research*. Washington, DC.

Canada (1962–1963). *The Royal Commission on government organization—dite Commission Glassco du nom de son président*. Ottawa: Queen's Printer.

Canada (1970). *A science policy for Canada: Report of the Senate Special Committee—dit comité Lamontagne du nom de son président*, vol. 1, Ottawa: Supply and Services Canada.

Canada (1972). *A science policy for Canada: Report of the Senate Special Committee* (Vol. 2). Ottawa: Supply and Services Canada.

Canada (1973). *A science policy for Canada: Report of the Senate Special Committee* (Vol. 3). Ottawa: Supply and Services Canada.

Carle, P., & Guédon, J.-C. (1988). Vulgarisation et développement des sciences et des techniques: Le cas du Québec (1950–1950). In D. Jacobi & B. Schiele (Eds.), *Vulgariser la science* (pp. 192–219). Seyssel: Champ Vallon.

Carpentier, J.-M., & Ouellet, D. (1994). *Fernand Seguin, le savant imaginaire*. Montréal: Libre Expression.
Castells, M. (1996). *The rise of the network society*. Cambridge, MA/Oxford: Blackwell.
Chartrand, L., Duchesne, R., & Gingras, Y. (2008). *Histoire des sciences au Québec*. Montréal: Boréal.
Colin, F. (1990). Les revues de vulgarisation. In B. Béguet (Ed.), *La science pour tous—1850–1914* (pp. 71–91). Paris: Bibliothèque du Conservatoire National des Arts et Métiers.
Decima (1988). *Report to the Ministry of State for Science and Technology on public attitudes toward science and technology*. Unpublished report, Toronto.
Désautels, J. (1980). *Ecole + science = échec*. Québec: Québec Science Editeur.
Dubas, O. & Martel, L. (1975). *Sciences et media* (Vol. 2). Ministère d'Etat, Sciences et Technologie, Ottawa: Information Canada.
Duchesne, R. (1978). *La science et le pouvoir au Québec*. Québec: Editeur officiel du Québec.
Duchesne, R. (1981). L'Etat et la culture scientifique des Québécois. *Les Cahiers du Socialisme, 8*, 107–138.
Durant, J. R., Geoffrey, A. E., & Geoffrey, P. T. (1989). The public understanding of science. *Nature, 340*(3628), 11–14.
Eidelman, J. (1988). *La création du Palais de la Découverte : Professionnalisation de la recherche et culture scientifique dans l'entre-deux guerre*. Thesis, Université Paris V—René Descartes.
Einsiedel, E. (1989). *Portrayals of science and technology in the Canadian press*. Paper presented to the American Association for the Advancement of Science, January, San Francisco.
Einsiedel, E. (1990a). *Scientific literacy: A survey of adult Canadians*. Calgary: University of Calgary.
Einsiedel, E. (1990b). *Framing science and technology in the Canadian press: A preliminary report*. Calgary: University of Calgary.
Einsiedel, E. (1992). Framing science and technology in the Canadian press. *Public Understanding of Science, 1*(1), 89–101.
Einsiedel, E. (2008). Public participation and dialogue. In M. Bucchi & B. Trench (Eds.), *Handbook of public communication of science and technology* (pp. 173–184). London/New York: Routledge.
Einsiedel, E., & Eastlick, D. (2000). Consensus conferences as deliberative democracy. *Science Communication, 21*(4), 323–343.
Einsiedel, E., & Thorne, B. (1998). Public response to uncertainty. In S. M. Friedman, S. Dunwoody, & C. L. Rogers (Eds.), *Communicating uncertainty: Media coverage of new and controversial science* (pp. 43–57). Mahwah: Lawrence Eribaulm.
Einsiedel, E., Jelsøe, E., & Breck, T. (2001). Publics at the technology table: The consensus conference in Denmark, Canada and Australia. *Public Understanding of Science, 10*(1), 83–98.
Farr, M. F. (1993). Common sense, science and social representation. *Public Understanding of Science, 2*(3), 189–204.
Filiatrault, P. & Ducharme, J. (1990). *Le développement des sciences et de la technologie au Québec: perceptions de la population*. Montréal: Association canadienne-française pour l'avancement de sciences.
Fourez, G. (2002). *La construction des sciences*. Bruxelles: De Boeck Université.
Fournier, M., & Maheu, L. (1975). Nationalisme et champ scientifique québécois. *Sociologie et Societes, 7*(2), 89–114.
Gagnon, J.-M. & Morin, L. (1986). *La diffusion de la culture scientifique et technique au Québec*. Sainte-Fooy: Conseil de la science et de la technologie.
Gascoigne, T., Cheng, D., Claessens, M., Metcalfe, J., Schiele, B. & Shi, S. (2010). Is science communication its own field? *Journal of Science Communication, 9*(3)C04.
Gauthier, E. (2011). Foodborne microbial risks in the press: The framing of listeriosis in Canadian newspapers. *Public Understanding of Science, 20*(2), 270–286.
Gingras, Y. (1991). *Les origines de la recherche scientifique au Canada*. Montréal: Boréal.
Gingras, Y. (1994). *Pour l'avancement des sciences: Histoire de l'ACFAS, 1923–1993*. Montréal: Boréal.

Gingras, Y. (2004). L'université en mouvement. *Egalité, 50*, 13–28.

Godard, B., Ozdemir, V., Fortin, M., & Egalité, N. (2010). Ethnocultural community leaders' views and perceptions on biobanks and population specific genomic research: A qualitative research study. *Public Understanding of Science, 19*(4), 469–485.

Godin, B. (1993a). Les indicateurs de culture scientifique et technique: quand dire, c'est ne rien faire. In D. Côté, G. Paquet, & J.-P. Souque (Eds.), *La prospérité en péril* (pp. 49–60). Ottawa: ACFAS Outaouais.

Godin, B. (1993b). La politique scientifique et la notion de culture scientifique et technique: les aléas politiques d'une idée floue. *Recherches sociographiques, XXXIV*(2), 305–327.

Godin, B. (1999). *Les usages sociaux de la culture scientifique*. Québec: Les Presses de l'Université Laval.

Godin, B., Gingras, Y., & Bourneuf, E. (1998). *Les indicateurs de culture scientifique et technique*. Sainte-Foy: Conseil de la science et de la technologie.

Goodwin, R., Haque, S., Binti, S., Hassan, S., & Dhanoa, A. (2011). Representations of swine flu: Perspectives from a Malaysian pig farm. *Public Understanding of Science, 20*(4), 477–490.

Göpfert, W. (2007). The strength of PR and the weakness of science journalism. In M. W. Bauer & M. Bucchi (Eds.), *Journalism, science and society* (pp. 215–226). New York/London: Routledge.

Heaton, L., Millerand, F., & Proulx, S. (2010). 'Tela botanica': une fertilisation croisée des amateurs et des experts. *Hermes, 57*, 61–68.

Heaton, L., Millerand, F., Crespel, E., & Proulx, S. (2011). La réactualisation de la contribution des amateurs à la botanique: Le collectif en ligne 'Tela botanica'. *Terrains & Travaux, 18*, 155–173.

Heidegger, M. (1958). *Essais et conférence*. Paris: Gallimard.

Hivon, M., Lehoux, P., Denis, J.-L., & Rock, M. (2010). Marginal voices in the media coverage of controversial health interventions: How do they contribute to the public understanding of science? *Public Understanding of Science, 19*(1), 34–51.

Hobsbawn, E. (1994). *The age of extremes: 1914–1991*. London: Abacus.

Jacobi, D. (1986). *Diffusion et vulgarisation: Itinéraires du texte scientifique*. Paris: Les belles lettres.

Jacobi, D. (1987). *Textes et images de la vulgarisation scientifique*. Berne: Peter Lang.

Jacobi, D. (1999). *La communication scientifique: Discours, figures, modèles*. Grenoble: Presses Universitaires de Grenoble.

Jacobi, D., & Schiele, B. (1993). Science in magazines, and its readers. *Public Understanding of Science, 2*(1), 3–20.

Jauréguiberry, F., & Proulx, S. (2011). *Usages et enjeux des technologies de communication*. Toulouse: Editions érès.

Jones, M., & Salter, B. (2010). Proceeding carefully: Assisted human reproduction policy in Canada. *Public Understanding of Science, 19*(4), 420–434.

Kerlinger, F. N. (1973[1964]). *Foundations of behavioral research*. Toronto: Holt Rinehart and Winston.

Lasswell, H. L. (1948). Structure and function of communication in society. In L. Bryson (Ed.), *The communication of ideas: A book of readings*. New York: Harper.

Latour, B. (1984). *Les microbes: Guerre et Paix, suivi de Irréductions*. Paris: Métailié et Pandore.

Lévy-Leblond, J.-M. (1984). *L'esprit de sel: science, culture, politique*. Paris: Seuil.

Lévy-Leblond, J.-M. (1994). La vulgarisation: Mission impossible? *Interface, 15*(2), 36–41.

Lewenstein, B. (1992). Public understanding of science in the United States after World War II. *Public Understanding of Science, 1*(1), 45–68.

Lewenstein, B. (1994). A survey of activities in public communication of science and technology in the United States. In B. Schiele (Ed.), *When science becomes culture* (pp. 119–178). Sainte-Foy/Ottawa: Editions MultiMondes/University of Ottawa Press.

Linteau, P.-A. (2000). Un débat historique: l'entrée du Québec dans la modernité et la signification de la révolution tranquille. In Y. Bélanger, R. Comeau, & C. Métivier (Eds.), *La révolution tranquille 40 ans plus tard: un bilan* (pp. 21–41). Montréal: VLB.

Locke, S. (2005). Fantastically reasonable: Ambivalence in the representation of science and technology in super-hero comics. *Public Understanding of Science, 14*(1), 25–46.
Logan, R. A. (2001). Science mass communication: Its conceptual history. *Science Communication, 23*(2), 135–163.
Longstaff, H., & Burgess, M. M. (2010). Recruiting for representation in public deliberation of the ethics of biobanks. *Public Understanding of Science, 19*(2), 212–224.
MacLean, S., & Burgess, M. M. (2010). In the public interest: Assessing expert and stakeholder influence in public deliberation about biobanks. *Public Understanding of Science, 19*(4), 486–496.
Maillé, M.-E., Saint-Charles, J., & Lucotte, M. (2010). The gap between scientists and journalists: The case of mercury science in Québec's press. *Public Understanding of Science, 19*(1), 70–79.
Massicotte, L., & Vaillancourt, F. (2008). Le fédéralisme américain: évolution historique, institutions politiques et flux financiers. In M. Fortmann & P. Martin (Eds.), *Le système américain*. Montréal: Les Presses de l'Université de Montréal.
Medlock, J. (2011). Rubrique: 'Canada, science communication in'. In S. H. Priest (Ed.), *Encyclopedia of science and technology communication* (Vol. 1, pp. 103–107). Thousand Oaks: Sage.
Miller, J. D. (1983a). *The American people and science policy*. New York: Pergamon.
Miller, J. D. (1983b). Scientific literacy: A conceptual and empirical review. *Daedalus, 112*(2), 29–48.
Miller, J. D. (1989). *Scientific literacy*. Paper presented to the American Association of the Advancement of Science, January 1989, San Franscico.
Miller, J. D. (1999). *Public attitudes toward science and technology, 1979–1999: Integrated codebook*. Chicago: The Chicago Academy of Science, International Center for the Advancement of Scientific Literacy, mimeo.
Miller, J. D. (2000). Scientific literacy and citizenship in the 21st century. In B. Schiele & E. Koster (Eds.), *Science centers for this century* (pp. 370–413). Sainte-Foy: MultiMondes.
Miller, J. D., Pifer, L. K., & Ressmeyer, T. J. (1993). *Public attitudes toward science and technology, 1979–1992: Integrated codebook*. Chicago: The Chicago Academy of Science, International Center for the Advancement of Scientific Literacy, mimeo.
Miller, J. D., Pardo, R., & Niwa, F. (1997). *Public perceptions of science and technology: A comparative study of the European Union, the United States, Japan, and Canada*. Bilbao: Fundación BBV.
Moles, A. A., & Oulif, J.-M. (1967). Le troisième homme, vulgarisation scientifique et radio. *Diogène, 58*, 29–40.
Montpetit, R. (2000). Beyond the 'science center', toward a social interpretation of science and technology. In B. Schiele & E. H. Koster (Eds.), *Science centers for this century* (pp. 209–229). Sainte-Foy: Editions MultiMondes.
Nelkin, D. (1995). *Selling science*. New York: W.H. Freeman.
OECD (Organisation for Economic Co-operation and Development) (1963). *La science et la politique des gouvernements: L'influence de la science et de la technique sur la politique nationale et internationale*. Rapport de la Commission consultative spéciale pour la politique scientifique auprès du secrétaire général. OECD.
OECD (Organisation for Economic Co-operation and Development) (1971). *Science, croissance et société: Une perspective nouvelle*. Rapport du groupe spécial du Secrétaire général sur les nouveaux concepts des politiques de la science. OECD.
OECD (Organisation for Economic Co-operation and Development) (1981). *La politique scientifique et technologique pour les années 80*. OECD.
Olson, J., & Russel, T. (Eds.). (1984). *L'enseignement des sciences dans les écoles canadiennes* (Vol. III). Ottawa: Conseil de sciences du Canada.
Orpwood, G., & Souque, J.-P. (1984). *L'enseignement des sciences dans les écoles canadiennes* (Vol. I, II). Ottawa: Conseil de sciences du Canada.
Paquette, J. (2009). Communiquer la science: métier conflit de normes et harcèlement social. *Ethique publique, 11*(2), 61–71.

Paquette, J. (2011). Science as culture and leisure: Cultural policy, industry and scientific culture in the Canadian context. *International Journal of Cultural Policy, 17*(1), 39–51.
Phillips, S. & Orsini, M. (2002). *Mapping the links: Citizen involvement in policy processes.* Discussion paper F 21, prepared for the Canadian Policy Research Networks, Ottawa.
Pitre, R. (1996). *L'usage de la notion de culture scientifique et technique dans les discours politiques au Québec.* Thesis, Université du Québec à Montréal.
Priest, S. H. (2010). Coming of age in the academy? The status of our emerging field. *Journal of Science Communication, 9*(3), 1–9.
Provençal, J. (2011). Extending the reach of research as a public good: Moving beyond the paradox of 'zero-sum language games'. *Public Understanding of Science, 20*(1), 101–116.
Raichvarg, D., & Jacques, J. (1991). *Savants et ignorants: Une histoire de la vulgarisation des sciences.* Paris: Seuil.
Rogers, E. (1962). *Diffusion of innovations.* New York: Free Press.
Santerre, L. (2008). From democratization of knowledge to bridge building between science, technology and society. In D. Cheng, M. Claessens, T. Gascoigne, J. Metcalfe, B. Schiele, & S. Shi (Eds.), *Communicating science in social contexts: New models, new practices* (pp. 289–300). New York: Springer.
Schiele, B. (1978). *Incidence télévisuelle sur la diffusion des connaissances scientifiques—Science-Réalité: un cas particulier de vulgarisation scientifique.* Thesis, Université de Montréal.
Schiele, B. (1983). Les enjeux cachés de la vulgarisation scientifique. *Communication Information, 5*(2–3), 157–186.
Schiele, B. (1986). Vulgarisation et télévision. *Social Science Information, 25*(1), 189–206.
Schiele, B. (2000). The silences of scientific museology. In B. Schiele & E. H. Koster (Eds.), *Science centers for this century* (pp. 449–491). Sainte-Foy: Editions MultiMondes.
Schiele, B. (2005a). Publiciser la science! Pour quoi faire? In I. Paillart (Ed.), *La publicisation de la science* (pp. 11–51). Grenoble: Presses Universitaires de Grenoble.
Schiele, B. (2005b). La médiation: quand le public va à la rencontre de son musée. In J. Davallon & B. Schiele (Eds.), *Logiques du Patrimoine, Logiques de la culture* (pp. 99–127). Sainte Foy: Editions MultiMondes.
Schiele, B. (2008a). On and about the deficit model in an age of free flow. In D. Cheng, M. Claessens, T. Gascoigne, J. Metcalfe, B. Schiele, & S. Shi (Eds.), *Communicating science in social contexts: New models, new practices* (pp. 93–117). New York: Springer.
Schiele, B. (2008b). Science museums and science centers. In M. Bucchi & B. Trench (Eds.), *Public communication of science and technology* (pp. 27–39). London/New York: Routledge.
Schiele, B. (2009). Le rôle des journalistes comme médiateurs doit être repensé. In A. Char & R. Côté (Eds.), *La révolution internet* (pp. 109–119). Québec: Les Presses de l'Université du Québec.
Schiele, B. (2011). Suggesting cultural indicators derived from exhibitory science. In W. M. Bauer, R. Shukla, & N. Allum (Eds.), *The culture of science: How the public relates to science across the world* (pp. 385–399). New York/London: Routledge.
Schiele, B. (in press). La muséologie un domaine de recherches. In A. Meunier, *L'Avenir de la muséologie.* Québec: Presses de l'Université du Québec.
Schiele, B., & Jacobi, D. (1988). La vulgarisation scientifique: Thèmes de recherche. In B. Schiele & D. Jacobi (Eds.), *Vulgariser la science: Le procès de l'ignorance* (pp. 12–46). Seyssel: Champ Vallon.
Schiele, B., & Koster, E. H. (2000). *Science centers for this century.* Sainte-Foy: Editions MultiMondes.
Schiele, B., & Larocque, G. (1981). Le message vulgarisateur. *Communications, 33*, 165–183.
Schiele, B., Landry, A. & Schiele, A. (2011). *Science communication in Canada: An inventory for the major PCST initiatives carried out in Canada.* Research report to the Korean Foundation for the Advancement of Science & Creativity. Montreal: CIRST-UQAM, University of Quebec at Montreal.
Snow, C. P. (1959). *The two cultures.* Cambridge: Cambridge University Press.

Special Senate Committee. (1970). *Bons, mauvais, ou simplement inévitables ? Rapport du Comité spécial du Sénat sur les moyens de communication de masse* (Vol. III). Ottawa: Imprimeur de la Reine pour le Canada.
Tremblay, V., & Roy, J. (1985). *Sondage d'opinion en matière de science et technologie: Faits saillants et rapport d'analyse*. Québec: Gouvernement du Québec, Ministère de l'Enseignement supérieur, de Science et de la Technologie, mimeo.
Trench, B. (2007). How the internet changed science journalism. In M. W. Bauer & M. Bucchi (Eds.), *Journalism, science and society* (pp. 133–141). New York/ London: Routledge.
Trench, B. & Bucchi, M. (2010). Science communication, an emerging discipline. *Journal of Science Communication, 9*(3)C03.
Turnbull, L. & Aucoin, P. (2006). *Fostering Canadians' role in public policy: A strategy of institutionalizing public involvement in policy*. Research report P07, Canadian Policy Research Networks, Ottawa.
Ungar, S., & Bray, D. (2005). Silencing science: Partisanship and the career of a publication disputing the danger of secondhand smoke. *Public Understanding of Science, 14*(5), 5–23.
Wagner, W. (2007). Vernacular science knowledge: Its role in everyday life communication. *Public Understanding of Science, 16*(1), 7–22.
Walmsley, H. (2010). Biobanking, public consultation, and the discursive logics of deliberation: Five lessons from British Columbia. *Public Understanding of Science, 19*(4), 452–468.
Weaver, W. (1951). AAAS policy. *Science, 114*, 471–472.
Weigold, M. F. (2001). Communicating science: A review of the literature. *Science Communication, 23*(2), 164–193.
Westley, B. H., & MacLean, M., Jr. (1957). A conceptual model for communication research. *Journalism Quarterly, 34*(1), 31–38.
Wynne, B. (1989). Sheepfarming after Chernobyl. *Environment, 31*(2), 10–15, and 33–37.
Wynne, B. (1991). Knowledges in context. *Science, Technology & Human Values, 16*(1), 111–121.

Chapter 4
Science Popularization Studies in China

Fujun Ren, Lin Yin, and Honglin Li

Abstract Science popularization, or science communication, has held its ground as a favorable cultural device since the coming out of new China, but it was not until 1980s that science popularization studies in China moved into the stage of theoretical integration. This paper briefly reviews the scenarios of public science popularization in China over 60 years in different cultural contexts, traces the development of science popularization studies at the theoretical level by discussing the main issues in the period of theoretical integration, and summarizes the basic characteristics of science popularization studies in China.

Keywords Science popularization • Science communication • Theoretical study • Public scientific literacy

4.1 Introduction

In China, the term 'science popularization' refers to a prevailing cluster of new concepts such as science communication, public understanding of science, scientific culture and so on. It is used to cover all kinds of activities helping individuals to understand and learn science and technology (S&T) knowledge and to make use of it for a better social life. The beginning of science popularization for the public in China can be traced back to the sixteenth century,[1] but studies of science popularization at the theoretical level started rather late compared to the practice.

[1] At the end of sixteenth century, the first wave of 'science communication' from the west to China began. It mainly took the form of translations of S&T works into Chinese by Chinese intellectuals and western missionaries.

F. Ren (✉) • L. Yin • H. Li
China Research Institute for Science Popularization (CRISP), 86 Xueyuan Nanlu, Haidian District, Beijing 100081, China
e-mail: hljrenfujun@126.com; yinlin213@126.com; lihl04@gmail.com

B. Schiele et al. (eds.), *Science Communication in the World: Practices, Theories and Trends*, DOI 10.1007/978-94-007-4279-6_4,

The first attempts to build science popularization as a separate research domain can be found at the end of the 1970s. Nevertheless, after 30 years of development, the depth and breadth of theoretical studies still need to be improved. Some research has narrated case studies in an effort to find regularities, while other research has involved careful, rational, theoretical analyses. Although Chinese researchers still have a long way to go, science popularization studies in China have moved from the stage of spontaneous development into self-conscious development, which implies that more and more people have come to realize the necessity and importance of this research field and its significance for informing science popularization practices. From 2002 to 2007, 1,795 papers on the subject of science popularization were published, which indicates the growing body of work of researchers in this field.[2]

The main purpose of this chapter is to describe the development of science popularization studies in China based on analyses of material written by Chinese science popularization researchers. The analysis is based on monographs, proceedings, compilations, periodicals, theses for master's degrees and doctorates, and newspaper coverage.

4.2 Stages in the Historical Development of Science Popularization Studies

The evolution of science popularization studies in China can be roughly divided into three stages: 'awareness of science popularization' from the end of sixteenth century to 1949; 'cognition of science popularization' from 1949 to 1978; and 'theoretical integration' from 1978 onwards. Although the first two stages left us little that is comparable with today's research, they were indispensable preparation for the boost to science popularization theoretical studies when the sediments and reactions of long-time practice were cleared away in a new perspective. Before looking into the 'theoretical integration' period, we will first look briefly at the evolutionary path that science popularization studies took in China.

During the first period, advanced western science, technology and scientific thinking spread eastward and was absorbed gradually by Chinese culture. This was

---

[2] This number is obtained from the 'List of published literature on science popularization studies' in the *Report on the development of science popularization studies in China (2002–2007)* (Ren Fujun et al. 2009). The list includes of all the papers collected in the China Knowledge Resource Integrated Database (www.cnki.net) run by Tsinghua University. The database was searched using keywords such as 'public understanding of science', 'education of science and technology', 'science popularization', 'science communication', 'scientific literacy' and 'scientific culture'. The numbers of papers in each year were 2002 (125 papers), 2003 (280), 2004 (244), 2005 (246), 2006 (558), 2007 (342).

also a substantial modernization of Chinese society and a localization of scientific thought into Chinese culture. It started from the idea of 'Chinese culture the body, western science the limbs'[3] (Chang Chih-tung 1900), developed into 'saving the country with science', and finally went further to consider 'scientism'.[4] The main thrust of science popularization was through associations for S&T built up by the end of the nineteenth century, and science popularization was no longer an isolated practice of a few vanguard intellectuals. The abolition of the imperial examination system and reforms in education pushed the dissemination of scientific knowledge up to the level of science popularization to the broader public, which implies that the target audience for science popularization had expanded from the nobility and intellectuals to laypeople.

However, the inadequate development of science and the great power it brings led to a unilateral perception of nature and its functions. The practicability and functionality of science were overemphasized, while the significance and value of scientific thought were not fully understood. Nonetheless, the door was open for science popularization in China.

In the second period, science popularization began to develop in an institutionalized pattern at the national level. The government of new China provided beneficial conditions and powerful support for the development of S&T, of which science popularization is an auxiliary but irreplaceable part. The main task of science popularization at that time was to give to the public as much practical technology and basic scientific knowledge and theory as possible. Science popularization studies focused on how science popularization could serve the Communist Party of China, the government, social production and the life of the people. Items concerning

[3] This was the principle guiding the westernization movement after 1860 in the late Ching Dynasty. The westernization movement was a reform initiated and organized by the Yang Wu group, most of whom were imperial officials influenced by western culture who aimed to make the feudalist Ching Dynasty strong and prosperous. The idea of 'Chinese culture the body, western science the limbs' was well expounded in Chang Chih-tung's *Quan Xue Pian* (English edition: *China's only hope*). The work consisted of two parts with 24 separate papers. The first part persuaded people to stick to Confucian ideas and be loyal to the feudalist emperors, while the second part called for learning science and imitating the western institution without shaking the governance of traditional Chinese culture. Those ideas were incorporated into the guiding principle of the westernization movement.

[4] Chinese-American researcher D.W. Kwok made a sound comment on scientism in China in the first half of the twentieth century in his book *Scientism in Chinese thought, 1900–1950*: 'Scientism, in general, assumes that all aspects of the universe are knowable through the methods of science. Proponents of the scientific outlook in China were not always scientists or even philosophers of science. They were intellectuals interested in using science, and the values and assumptions to which it had given rise, to discredit and eventually to replace the traditional body of values. Scientism can thus be considered as the tendency to use the respectability of science in areas having little bearing on science itself. In China, the desire for national growth was accentuated by the weakness in technology, and it is thus not surprising to find among her Western-educated intellectuals great enthusiasm for science' Kwok (1965:3).

science popularization were included in S&T policies, and discussions about science popularization work included coverage of some theoretical factors.

The third stage was a consciously developing period of science popularization study. Influenced by the effectiveness of western S&T communication theories, some Chinese scholars and researchers began to translate and introduce them to China. Science popularization practices in China were first studied under the frame of communication science theory. After the National Science Conference in March 1978,[5] some researchers put forward the idea of constructing a 'science of science popularization'—that is, making science popularization an independent research domain.

Around the beginning of 1990s, the first proceedings of science communication studies were published and the theories of public understanding of science were introduced into China. Studies of S&T communication entered an active period. Some scholars in universities began to look critically at traditional science popularization ideas and models, expecting more scientific and rational ones to emerge.

At the turn of the twenty-first century, another hot spot appeared with the issue of the *Law of the People's Republic of China on Popularization of Science and Technology* (2002) and the *Outline of the National Scheme for Scientific Literacy (2006–2010–2020)* (2006). Studies on how to interpret and promote 'public scientific literacy' become the focus of science popularization studies.

4.3 Focus Issues in the Stage of Theoretical Integration of Science Popularization Studies

Science popularization in China entered into a new period after 1978. Deng Xiaoping pointed out in his speech at the opening ceremony of the National Science Conference that 'the key of the Four Modernizations is the modernization of science and technology' and 'science and technology is the primary productive force'. He also stated that 'science popularization must be strengthened forcefully' (Deng 1978). Fang Yi, who was then Vice Prime Minister and Director of the National Scientific Committee, in reporting at the conference, appealed for 'pushing forward the modernization of Science & Technology' (Xinhua News Agency 1978) and gave specific requirements for patterns and targets of science popularization. Science popularization in China developed speedily in a relatively relaxed and open atmosphere after the Cultural Revolution. Much science popularization practice urgently needed guidance and support at the theoretical level, which built science popularization studies into an increasingly developing field. However, the targets, content and approaches of science popularization studies are very complicated because there is a very tight

[5] The National Science Conference was a milestone that established an unshakable place for S&T from then on. This was the springtime for science in China.

connection between science popularization and communication science, education, sociology and anthropology.

Some scholars have tried to delineate tendencies in the development of science popularization studies in their research. Zhai Jiequan from the Beijing Institute of Technology classifies science communication studies (those concerning traditional ideas of science popularization were not included) in terms of research perspectives: the first is from the angle of social development, the second from the angle of scientific culture, and the third from the angle of technology communication (Zhai 2007). In the *Report on the development of science popularization studies in China*, edited by Ren Fujun, published research papers from 2002 to 2007 are divided into eight categories according to themes: research on theory, policy, scientific literacy, education in S&T, natural science museums, science popularization in urban communities, science popularization in rural areas, and science popularization by mass media (Ren et al. 2009).

Below, we follow the development path by elucidating the representative issues emerging in the theoretical integration stage.

4.3.1 Initial Attempt to Make Science Popularization an Independent Discipline

Drawing upon successful experiences and learning lessons from failures in more than 20 years of science popularization practice, practitioners proposed making science popularization a separate discipline when popularization began its redevelopment after 1978.[6] It was not until 1989 that a monograph, *Introduction to science of science popularization* (Yuan 1989) was published, although *Probing into science of science popularization* (Zhou and Zeng 1981) had been published 10 years before.

Zhou and Zeng wrote that 'There are special rules in realizing speedy, correct and most effective popularization of different scientific knowledge and techniques to different target objects, which has become a particular science.' This was the initial definition of the science of science popularization. Comparing the differences between science popularization and other sciences, the article concluded that 'The science of science popularization is a multidisciplinary domain studying the popularizing phenomena that exist in all sciences' (Zhou and Zeng 1981). Two kinds of research were suggested: 'theoretical science popularization' and 'applied science popularization'. The former concerns rules covering the history, position, function, motivation and perception of science popularization, while the latter concerns the different patterns, rules and approaches of science popularization practices, such as science popularization propaganda, education, exhibition and creation.

[6] The proposal was first initiated on the National Symposium on Science Writing in May 1978 and was reaffirmed at the Conference of Association of Chinese Science Writers in 1979 and the Second National Congress of CAST (the China Association for Science and Technology) in 1980.

As a result of Zhou and Zeng's paper, books on the subject were published in the following years: *Introduction to science of science popularization* (Yuan 1989), *Conspectus on science of science popularization* (Yuan 2002) and *Science of science popularization* (Zhou and Song 2007).

The *Introduction to science of science popularization* was the first monograph specifically discussing what science popularization is in China, and was the first to 'preliminarily construct a relatively complete theoretical system of science popularization'. The book begins with the history of science popularization, which it presents as three stages: science propaganda, science education and science service, the last of which is regarded as the final purpose and nature of science popularization. The three stages are not only divisions but also three functions of science popularization, which the book defines as:

> scientific activities to popularize scientific and technological knowledge as well as scientific thoughts, methods and spirits to self-educated people and receptors in the whole society by special carriers and diversified means, so as to reach the expected social, economical, educational and cultural effects. (Yuan 1989)

Scienology (the science of sciences), the science of education and the science of communication are considered to be the basic theoretical resources for the science of science popularization. The processes and steps of science popularization are analyzed under the theoretical framework of communication science. For example, based on the Braddock Model (the 7W Model), the author creates seven basic elements of the science of science popularization, including the environment, object, main body, content, method, audience and effect.

The *Conspectus on science of science popularization* was written by the same author 3 years later in 2002, with no essential changes compared to the earlier work.

In 2007, *Science of science popularization* was compiled by Zhou Mengpu (one of the authors of *Probing into science of science popularization*) and Song Ying. The authors put forward three basic principles of the science of science popularization:

- Science popularization is an activity to popularize scientific and technological knowledge as well as scientific thought, methods and spirit.
- Science popularization is an understandable, acceptable and interactive mutual exchange on the basis of equality of the main body and the audience.
- Science popularization is a social activity guided by the government and supported by the whole society.

Books such as *Conspectus on science and technology popularization* (Yuan 2002), *Introduction to modern science popularization* (Yang and Wu 2004a) and *Course on modern science popularization* (Yang and Wu 2004b) showed the intention of the authors to construct the theoretical system and discover working patterns of practices for modern science popularization based on sociology, education and communication science. Science popularization in modern China was seen as a systematic process consisting of components including the environment, purpose, undertakers, target, content, carriers, effects and so on. Basic problems encountered in theoretical construction and the social practices of science popularization were analyzed, and the composition and role of public scientific literacy in the processes of personal and social development were discussed.

With few fundamental differences in primary viewpoints, these books tried to find regularities in science popularization practices and put them together using a certain kind of logic, showing that the researchers understood science popularization as a systematic course of action with its own rules. However, this was newly cultivated land: there was not enough analysis of many aspects, such as the relationship between science popularization, the nature of science and the mechanisms of communication.

Influenced by perceptions of science popularization in the second period, the study of science popularization in these books unavoidably had a macroscopic and top-down viewpoint. The attempt to fit science popularization into a systematic framework continued from 1978 into the twenty-first century, mainly from the perspective of the subjects (government and organizations). Scholars considered the science of science popularization to be 'a science to study the rules of popularization and pass-on of knowledge and techniques' (Rao 1981).

Based on the analytical approach of communication science, the vision of these books did not go far beyond the basic theory of communication models. For example, the main chapters in the *Introduction to science of science popularization* corresponded to seven elements in the science of communication: the science popularization environment (in what situation), the objectives of science popularization (for what), the main body of science popularization (who), the contents of science popularization (what), the audience for science popularization (to whom), the media of science popularization (in which channel), and the evaluation of science popularization (with what effect). But the analysis in each chapter did not take the characteristics of scientific knowledge into consideration, so the communication of science differed little from the communication of other information. Although these studies made great contributions, they were constrained in their vision and depth. Science communication studies in the form of a 'simplified combination of science communication concepts with communication theory' are reported in many research papers (Liu and Hou 2007).

4.3.2 *Reflections on Traditional Science Popularization Ideas*

From the 1990s, criticism and review of traditional science popularization were influenced by the concepts of public understanding of science. *Study on science and technology communication* (Sun 1996), which was published after the first Science & Technology Communication Symposium at Tsinghua University, discussed such matters as the definition of science communication, the training of science communication professionals and the potentially significant influence of the internet on science communication.

Soon afterwards, *A guidebook on science and technology communication* (Sun 1997) was published by Tsinghua University Press. The book expounded on the nature of scientific and technological information; systems and functions of S&T communication; S&T news, publications, writings and translations; and information on communicating techniques and facilities.

Both these books considered the communication of S&T from the perspective of communication and science journalism.

Science popularization studies using the communication approach have so far not developed further because there are few academic connections between them and communication science (Tian 2004). With the introduction of public understanding of science concepts into China and the involvement of researchers with backgrounds in scientific philosophy, science history and scientific sociology, science popularization studies in China turned to exploring the relationships among science, society and individuals.

4.3.2.1 Connotations of the Concept of 'Science Popularization'

After the ideas of public understanding of science were introduced into China, there was debate about the connotations of 'science popularization' and 'science communication' (Wu 2000). It was considered that the term 'science communication' should be used instead of 'science popularization' because the former had a broader vision and more profound content compared to the latter, and better fitted the nature of science popularization in new times. Another view was that it was not necessary to completely replace 'science popularization' with 'science communication' because the preference for either term is not important provided that advanced ideas and thoughts could be included under the old term. The concept of 'science popularization' could no longer guide science popularization practices, so new ideas should be added to it. Three points summarize the new ideas:

- Science popularization should develop from one-way communication to two-way interaction. In the first, scientists deliver scientific knowledge to non-specialists; in the second, the public participates together with scientists in the creation of scientific knowledge, the formulation of science policies, the construction of scientific structure, and the interpretation of the role of science in society.
- Science communication is neither a device applied by the scientific community to achieve its own purposes, nor a unilateral one-way dissemination of scientific knowledge by the government, but an activity in the formation of culture.
- Science communication is a process of integration of science and the humanities.

The proposal to replace 'science popularization' with 'science communication' produced different views among researchers. Some took 'science communication' as a concept with extensive content, including science popularization and science journalism, but insisted that it could not substitute for either of them. Others thought that 'science popularization' put emphasis on the result while 'science communication' focused on the process, meaning that they could not replace each other because they were not equal pragmatically.

These debates showed different understandings of the content, purpose and role of science popularization, and brought new connotations to the concept of 'science popularization' as time passed. Comparing understandings of 'science popularization' in the 1980s and the twenty-first century, we can easily notice the difference. In the 1980s, 'scientific knowledge and techniques, scientific thoughts and methods' were considered to be what should be delivered to the masses. Science popularization was looked on as 'social activities and cognitive process', which emphasized the process of science learning. The purpose of science popularization was to 'improve one's knowledge and ability [and to] promote material and spiritual civilization', which implied the will of the government more than that of the people themselves (Zhang et al. 1983). In the *Outline of the National Scheme for Scientific Literacy (2006–2010–2020)*, a very important national policy at the state level now governing practices in this field, the task of science popularization is to promote the citizens' scientific literacy, which is expressed as:

> knowing some necessary knowledge of science and technology, mastering basic methods of science, building up science thoughts, advocating science ethos and having the ability to apply them to resolve practical problems and participate in public affairs. (State Council 2006).

The policy calls for public attention to scientific thinking and science ethos instead of merely scientific knowledge, and shows the requirements of the time for the citizens to understand the interaction between science and society, as well as to improve their ability to participate in public affairs. From these differences, we can sense that the relationships among science, society and individuals have gradually become important subjects in science popularization studies.

4.3.2.2 Three Stages of Science Popularization and Their Models and Standpoints

The debate on 'science popularization' and 'science communication' later led to a discussion about stages of science popularization and their models and standpoints. Liu Huajie from Peking University pointed out two traditions in science communication in China (Liu 2009). The first is science popularization, and the second is science journalism. Following the first tradition, there are three stages in science popularization: traditional science popularization, public understanding of science and science communication. Their models and standpoints are shown in Table 4.1.

According to Liu Huajie, the first model was used in times of the planned economy to meet the needs of the nation and government. It emphasized academic authorities and scientific beliefs, paying more attention to knowledge and technologies but less to scientific methods and processes and saying nothing of social operations, the limitations of science or the faults of scientists. Science popularization ideas in such a model derived from the mainstream ideology and connected science popularization practice with the needs of social production and construction, which resulted in a unified, centralized, mechanism of science popularization. The second model is science popularization or communication based on the authority of science and the public's ignorance of it. Science popularization in this model was to improve

Table 4.1 Models and standpoints of public communication of science

	Models	Standpoints
Traditional science popularization	Central broadcasting	Of nation (or party)
Public understanding of science	Deficit model	Of scientific community
Reflective science communication	Dialog (or participation)	Of citizen (or humanism)
Trends	*Feedback and participation*	*Multiple coexistence*

public scientific literacy and win public support for scientific work. The third model is characterized by diversity of audience and main bodies (those delivering the science), emphasizing public attitudes and the public's right to speak about social justice, fair distribution and so on. In Liu's opinion, science communication ideas in China are more like those between the first and second models, with a transition to the third model. None of the three stages or their respective models is better than the other two, and they can coexist to meet different demands. Experience tells us that science communication is a multidimensional concept in various shapes.

In concert with the three-stage argument, Liu Huajie analyzed science communication from the angle of 'first order' and 'second order' communications (Liu 2002). First order communication is the communication of scientific facts, developments and knowledge, while second order communication is the communication of scientific methods, processes, spirit, thought and S&T's influence on society. At the stage of traditional science popularization and public understanding of science, the parts of the main body and the audience cannot be changed: science popularization is a simple top-down delivery of knowledge. Moreover, science and technology are absolutely right and good, so there is no need to worry about what to develop and how to develop it. In contrast, science communication emphasizes second order communication other than first order communication. It emphasizes to some extent the interactions and discussions among participants, and the subjectival position of the public. Of course, first and second order communications overlap each other in many aspects.

Analyses of the three-stage division of science popularization revealed the humanistic perspective of science communication studies and answered the question about what to communicate (first or second order/first and second order) and why to communicate (people-oriented). At present, all of the three communication models are in use and overlap with one another in science popularization activities in China.

Influenced by theories of the public understanding of science and by the sociology of science, researchers began to reflect on the top-down science popularization ideas and model. They tried to abandon old ideas and carry out research based on foreign theoretical frameworks and ideas. To summarize a complicated development, the concerns of science popularization studies in this period were the content, purpose and mechanisms of communication. Research has shown the following:

- The communication contents (what to communicate) changed from mere 'positive' scientific knowledge, which was considered self-evident, to an

understanding of the importance of scientific method, thought and spirit, as well as doubt about the authority of science.
- The purpose of communication (for what) changed from meeting governmental needs to meeting the demands of the public.
- The communication mechanisms (how to communicate) took the form of discussions on communication models.

4.3.3 Studies of Public Scientific Literacy

During attempts to construct a theoretical framework for science popularization, and after reflection on traditional science popularization ideas, many researchers became interested in the study of public scientific literacy (PSL). Their interest was aroused in the 1990s after the translation into Chinese of research into the measurement of public attitudes to S&T by Jon D Miller.[7] Inspired by such investigations and the measuring framework of Professor Miller, China has since conducted national public scientific literacy surveys in 1992, 1994, 1996, 2003, 2005, 2007 and 2010.

Discussions on this issue grew from 1995 and blossomed especially after the turn of the twenty-first century (Tang 2003). Since 2001, more than 110 research papers per year with the keywords 'public scientific literacy' have been noted in the China Knowledge Resource Integrated Database. This coincided with what was happening in the research field overseas, where concerns about scientific literacy research also reached the climax (Huang 2004).[8]

Chinese researchers showed great interest in this issue, partly because of the release of two milestone government documents, the *Law of the People's Republic of China on Popularization of Science and Technology* (2002) and the *Outline of the National Scheme for Scientific Literacy* (2006). Some claim that science popularization shifted from traditional modes to the improvement of PSL in the 1994–2006 period, while the new stage of PSL promotion was launched in 2006 (Zheng and Gao 2008).

Academic research on PSL basically follows two orientations: one takes the empirical approach of sociology, making the assessment of PSL through large-scale

[7] In 1991, *Study on science popularization* (which was edited by CRISP and was not an officially published journal from 1987 to 2005) first published a translated part of the *Science and engineering indicators 1989* report that examined public attitudes to S&T in the United States (Shi 1991). Following that, the 1990 report on attitudes of the American public to S&T written by Jon D Miller was also translated into Chinese and published in the same journal (Li 1991).

[8] Huang analyzed research papers on science culture indices, focusing on PSL, and came to the conclusion that studies on this theme went through three periods: the beginning period, when the issue gradually came onto the research horizon (1950s–1970s); the mid-term period, when large numbers of papers came out (1980s); and the upsurge period, when PSL became one of the research focuses in this field (from the 1990s till now).

surveys or trying to find regularities behind PSL promotion practice (Ren et al. 2010; CRISP 2010). The other is more epistemically introspective, appearing in controversies over the feasibility and appropriateness of the use of Jon Miller's measuring framework in Chinese surveys. Researchers pointed out that PSL indicators designed for citizens in western countries are not appropriate for use in developing countries like China, which have different cultural and ideological institutions. Some called for the creation of our own set of indicators and variables to evaluate and monitor the status of and changes in Chinese PSL (Li 2006), taking into account the seriously unbalanced development of different regions in China (Liu, H. 2006). Because the borrowed PSL measuring tools had been based on the knowledge hierarchy of modern western science, someone ironically described the survey as the 'examination of western modern science knowledge for Chinese citizens' (Jiang and Liu 2004).

These two approaches to PSL research, constantly crashing and contending, facilitated the improvement of PSL surveys and studies in China and helped to disseminate and apply science communication ideas and theories in science popularization practices.

Researchers provided advice on how to modify the original measuring framework to fit the Chinese context, contributing much to the improvement of PSL surveys. Guo Chuanjie and Tang Shukun established a three-dimensional measuring structure for PSL, taking into account the knowledge, awareness and ability of the citizens (Guo et al. 2008; Tang et al. 2008). Li Honglin established an analytical framework to carry out a comprehensive inquiry into theories and practices of PSL measurement from the perspective of multidisciplinary research and the critical attitudes of science, technology and society, finally putting forward a hierarchical measuring model combining the examination of 'living science',[9] academic science and post-academic science for both domestic and international comparisons (Li 2009). The research group from CRISP (the China Research Institute for Science Popularization) explored the sampling, statistical method and measuring index of the surveys (Zhang et al. 2008).

In theoretical research, studies on this topic show a deep concern for the development of China; for example, the concept of 'living science' was proposed on the grounds of Chinese S&T development status and the real Chinese PSL level (Zeng and Li 2009), implying that science popularization should meet the living demands of ordinary people with general equality and benefit.

[9] 'Living science' is regarded as forming from the process of requiring, understanding, acquiring and using knowledge based on the demands of people's everyday lives. This knowledge is possibly derived in the systematic sense from academic science or post-academic science, and is probably the perceptual, intuitionistic and useful but unsystematic common sense formed in people's ordinary life (empirical cognition). Living science is connected tightly with basic living demands; giving importance to accessibility and perception; integrating with social knowledge; making instrumental and practical result the priority; and connecting inherently with cultural tradition deposition (Zeng and Li 2009).

Science communication theories are often used in PSL studies. Researchers have recognized that the 'deficit model' is still partly qualified for science popularization practice in China today. However, this does not prevent other communication models, such as 'participation' and 'dialog' models, being adopted. The possibility of interaction between the public and the government in policymaking has been discussed, including communication channels (mass media, networks etc.) and approaches (hearings, consensus meetings, community propaganda etc.)Some scholars even experimented with new approaches to science communication (Liu, B. 2006).

In the relatively new field of PSL research, multidisciplinary perspectives are being applied using a combination of sociology, statistics, the history of S&T, the philosophy of S&T, science journalism and science education. The actors being engaged in the discourse are not merely universities and research institutes: the government, scientists' communities, business enterprises and the public are gradually getting involved. This shows that PSL studies are not only a secret garden for planting scholars in, but also an open forum for all the voices of the public.

4.4 Conclusions

At the end of the 1970s, as S&T and science popularization grew in importance, science popularization studies in China entered a stage of conscious development. The idea of trying to construct science popularization as an independent discipline was put forward and discussed. With the introduction of theories of communication science and public understanding of science, reflections on traditional science popularization ideas and on the problems with the current science popularization model appeared. The relation between science and society was taken into consideration when researchers began to look for better working patterns and mechanisms for science popularization practices.

The development of science popularization studies in China has some particular features:

- Science popularization studies in China are undergoing the period of theoretical accumulation and integration. However, they are developing slowly, possibly because of the lack of a flexible and open framework and advanced research approaches.
- Science popularization studies is an open research field connected with many disciplines, such as communication, psychology, sociology, the history of S&T, anthropology and so on.
- Ideas about the public understanding of science and theories of communication science have exerted the greatest impact on science popularization studies in China.

Researchers pay attention to both theoretical research and applied research, but a macroscopic perspective is more often used, resulting in a lack of case studies.

Studies from the point of view of the main body (the science providers) greatly outnumber those from the angle of the target audience.

References

Chang Chih-tung (1900). *China's only hope: An appeal.* (Translated from the Chinese edition by Woodbridge, S. I.). New York/Chicago: Fleming H. Revell.
CRISP (China Research Institute for Science Popularization). (2010). *China science popularization report 2009.* Beijing: Popular Science Press.
Deng, X. (1978). Speeches at the Opening Ceremony of National Science Conference. In Administrative Office of CPC Central Committee (Ed.) (1992), *Speeches and lectures on science and technology development by government leaders* (pp.116–129). Beijing: Science Press.
Guo, C., et al. (2008). Civic scientific literacy: Concepts, measurements and problems. *Study on Science Popularization, 3*(2), 26–33.
Huang, Y. (2004). Statistical analyses of researches on scientific literacy in SCI and SSCI database. *Scientific Chinese, 11*, 14–15.
Jiang, X. & Liu, B. (2004, September 3). What indeed is scientific literacy? Retrieved from http://shc2000.sjtu.edu.cn/0408/kexues.htm.
Kwok, D. W. (1965). The doctrinal impact of modern science on Chinese thought. In D. W. Kwok (Ed.), *Scientism in Chinese thought, 1900–1950* (pp. 3–30). New Haven: Yale University Press.
Li, D. (trans.) (1991). Attitudes of the US public toward science and technology, 1990. *Study on Science Popularization, 5*, 1–83.
Li, D. (2006). Reviewing and thinking: Public understanding of science in China for 15 years. *Study on Science Popularization, 1*(1), 24–32.
Li, H. (2009). Research on theory and practice of measurement of public scientific literacy: Take Miller's system as a clue. Dissertation, Tsinghua University, Beijing.
Liu, B. (2006). The consensus meeting in the context of public understanding of science and its significance in China. 25 July. Retrieved from http://www.cast.org.cn/n35081/n35668/n35728/n36479/10190895.html
Liu, B., & Hou, Q. (2007). Research on domestic science communication: Theory and issue. *Studies in Dialectics of Nature, 20*(5), 80–85.
Liu, H. (2002). Integrating two kinds of traditions: Talking about science popularization in our vision. *Social Sciences in Nanjing, 10*, 15–20.
Liu, H. (2006). The measurement of citizen scientific literacy and its problems. *Journal of Beijing Institute of Technology, 8*(1), 12–18, (Social sciences edition).
Liu, H. (2009). Three models and three stages of science communication. *Study on Science Popularization, 4*(19), 10–18.
Rao, Z. (1981). Looking at science of science popularization from the angle of science history. In Chengdu Research Panel of Science of Science Popularization (Ed.), *Collection of science of science popularization* (pp. 87–97). Chengdu: Sichuan People's Publishing House.
Ren, F., et al. (2009). *Report on the development of science popularization studies in China (2002–2007).* Beijing: Popular Science Press.
Ren, F., et al. (2010). *Report on scientific literacy of Chinese citizens* (Vol. 1). Beijing: Popular Science Press.
Shi, S. (trans.) (1991). Public attitudes toward science and technology in America, 1989. *Study on Science Popularization, 4*, 1–49.
State Council (2006). *Outline of the National Scheme for Scientific Literacy (2006–2010–2020).* Retrieved from http://www.kxsz.org.cn/english.html
Sun, B. (Ed.). (1996). *Study on science and technology communication.* Beijing: Tsinghua University Press.
Sun, B. (Ed.). (1997). *A guidebook on science and technology communication.* Beijing: Tsinghua University Press.
Tang, S., et al. (2008). A study on the system of indicators for civic scientific literacy of China. *Studies in Science of Science, 26*(1), 78–84.

Tang, W. (2003). An analysis report about the literature on scientific literacy in the China National Knowledge [the China Knowledge Resource Integrated Database]. *Studies in Dialectics of Nature, 23*(3), 104–107.
Tian, S. (2004, March 5). How the standpoint of citizen could be possible. *Science Times*, B2.
Wu, G. (2000, September 22). From science popularization to science communication. *Science and Technology Daily*, 003.
Xinhua News Agency. (1978). National Science Conference being held in Beijing. *Renmin Ribao*, 19 March. Retrieved from http://scitech.people.com.cn/GB/25509/56813/57267/57268/4001562.html
Yang, W., & Wu, G. (Eds.). (2004a). *Introduction to modern science popularization.* Beijing: Popular Science Press.
Yang, W., & Wu, G. (Eds.). (2004b). *Course on modern science popularization.* Beijing: Popular Science Press.
Yuan, Q. (1989). *Introduction to science of science popularization.* Beijing: Academic Journal Press.
Yuan, Q. (2002). *Conspectus on science of science popularization.* Beijing: China Science and Technology Press.
Zeng, G. & Li, H. (2009). Living science and public scientific literacy. In Clark Glymour, Wang Wei & Dag Westerstahl (Eds.), *Logic, methodology and philosophy of science: Proceedings of Thirteenth International Congress* (pp. 6–17). London: Individual authors and King's College Publications, United Kingdom.
Zhai, J. (2007). Analyses of studies on communication of science and technology in China: Three perspectives and three problems. *Studies in Dialectics of Nature, 23*(8), 68–71.
Zhang, C., et al. (2008). The construction of 'citizen scientific literacy index' from monitoring and evaluation. *Study on Science Popularization, 3*(6), 51–58.
Zhang, D., et al. (1983). The concept and category of science popularization. In D. Zhang et al. (Eds.), *Conspectus on science writing* (pp. 1–6). Beijing: Peking University Press.
Zheng, D., & Gao, J. (2008). From science and technology popularization to citizens' scientific literacy construction: A historical review of measures for promoting citizens' scientific literacy. *Studies in Dialectics of Nature, 24*(11), 79–84.
Zhou, M., & Zeng, Q. (1981). Probing into science of science popularization. In Chengdu Research Panel of Science of Science Popularization (Ed.), *Collection of science of science popularization* (pp. 11–61). Chengdu: Sichuan People's Publishing House.
Zhou, M., & Song, Y. (2007). Three basic principles of science of science popularization. In M. Zhou & Y. Song (Eds.), *Science of science popularization* (pp. 123–132). Chengdu: Sichuan Science and Technology Press.

Chapter 5
Policy Perspective on Science Popularization in China

Shunke Shi and Huiliang Zhang

Abstract Science popularization is important in China. It is the Chinese term for the concepts of public understanding of science or public communication of science and technology now prevailing around the world. Science popularization in China has developed in an idiosyncratic way, as part of an organized and mobilized effort. This paper explores science popularization from a policy perspective. From about 1,000 policy documents on science popularization, the authors selected 100 of the most relevant. In this paper, the 100 policies are classified into four groups according to their operational effects. Three of the most important ones, which have in the past 15 years played decisive roles gaining funding and stimulating the advancement of the enterprise, are highlighted. The driving forces behind science popularization in China are many, but underlying ideology and imported advocacies have been especially significant. Because science popularization, by whatever name, has begun to appear on the agendas of the governments of many countries, policies reflecting government decisions in this area are becoming ever more influential, and need to be fully understood.

Keywords Popular science • Science popularization (SP) • Science communication • Science popularization literature • Science popularization policy

5.1 Introduction

On 1 October 1949, the People's Republic of China announced its birth to the world. Just a few days before the announcement, the first Plenary Session of the Chinese People's Political Consultative Conference (CPPCC) was convened. At the session,

S. Shi (✉) • H. Zhang
China Research Institute for Science Popularization (CRISP), 86 Xueyuan Nanlu, Haidian District, Beijing 100081, China
e-mail: shishunke@yahoo.com.cn; leanne2005@sina.com

B. Schiele et al. (eds.), *Science Communication in the World: Practices, Theories and Trends*, DOI 10.1007/978-94-007-4279-6_5,

a historic document laying the foundation of the new country was adopted—the *Common Program*, the precursor to the Constitution of the new country. As a legal document, the *Common Program* states in its 43rd article: 'To strive to advance natural sciences to serve the country in its industrial, agricultural and national defense construction, to reward science discoveries and inventions, and to popularize scientific knowledge' (CPPCC 1949). The statement marked the starting point of a national enterprise, science popularization (SP), that later evolved and expanded on all levels and to all corners of the country. It also implied that bringing science to the ordinary people is not only a recognized legal activity, but also a commitment of the government.

At about that time, the science community in China joined efforts to shape two national organizations based on earlier science societies: the All China Union of Natural Science Professional Societies and the All China Association for Popularization of Science and Technology. In 1958, the two organizations merged into the China Association for Science and Technology (CAST) (CAST 1994). In succession, the three organizations served as the main force disseminating the sciences to people in all walks of life. Government agencies and some other non-government organizations have also been engaged in SP as part of their responsibilities.

With a mandate made clear in the *Common Program* and a troop of manpower ready to go, SP would soon sweep the country in a new approach. Many documents reflect the story of SP in China, but some of the most interesting and important ones are policy dossiers. This chapter examines some of the SP policy records in SP to show what the thing we call SP in China looks like.

5.2 Science Popularization Policy Documents

In modern Chinese society, government policy plays a significant role. It is used as a primary tool to harness and guide people's activities, especially when those activities move and develop in an organized way.

Ever since the 1980s, more and more countries have come to realize the close links between public science communication and the sustainable development of the state. SP has been quickly put onto the agendas of many governments and has been supported by a growing number of policies. Although there have been many discussions elucidating public science communication in different countries, there has been little examination or comparison of the official SP policies of those countries.

To take stock of the literature in the field and to examine SP in China from the policy perspective, we worked with partners in a team to collect nearly 1,000 SP-related documents, selecting 100 of the most relevant and interesting of them for closer analysis. The 100 documents cover a span of 60 years, involve many government agencies and national organizations as actors, and cover all aspects of SP activities.

Before we get into the main discussion of the chapter, let us say a few words about why we use the term SP and what it is in Chinese.

5.2.1 Science Popularization

We need first to explain why we use the term 'science popularization'. SP is known as *kepu* in China. *Kepu* can be thought of as an acronym: *KE* stands for *kexue* ('science') and *PU* stands for *puji* ('popularization'). The term can also connote technology: we can say *kexue puji* or *kexue jishu puji* (*jishu* means 'technology').

The term *kepu* is extremely popular in China. Very few people have never heard of it, even though most might not know exactly what it means.[1] It is the pet phrase of the government and practitioners in the field. Scientists, researchers, public officials and science communication practitioners understand each other well when talking about *kepu.* It is a sound, a sign, a symbol indicating what might be known in many other countries as public science communication, public understanding of science, scientific culture, or scientific literacy. There was once a debate arguing for the replacement of the term with the fashionable phrase of 'science communication', but the old convention persists. *Kepu*, or SP, is the term that appears most often in the policy documents we collected.

5.2.2 Science Popularization Policy Sifting

How many Chinese SP policy papers have been issued since new China came into being in 1949 is hard to determine, and depends on how we define 'SP policy'. There are not many at the state level, especially if only those dealing exclusively with SP are counted. However, if we widen the frame, the number increases greatly.

With a time frame of the past 60 years and *kepu* as a keyword,[2] the team found around 1,000 policy-related documents, including laws, regulations, guidelines, plans, outlines, programs, reports, decisions, instructions, notifications and state leaders' speeches. The documents were mainly issued by the central government, government departments and national organizations. Local documents were not

[1] According to the findings of scientific literacy surveys of China, Chinese citizens have a high level of interest in science but a low level of scientific knowledge. In the public eye, *kepu* most often refers to popular science products such as books, lectures or TV programs, instead of the social phenomenon of interaction or communication between science and the public.

[2] The word for SP (*kepu*) appears in nearly all the titles of the documents that focus specifically on the public communication of science and technology. Fewer documents use the phrase 'science communication' in their titles, such as the *Long and Mid Term Program for the Development of Science Communication of the Chinese Academy of Sciences (2006–2020)*. That phrase, however, appears more and more often since 1990s in parallel with *kepu* in many official papers.

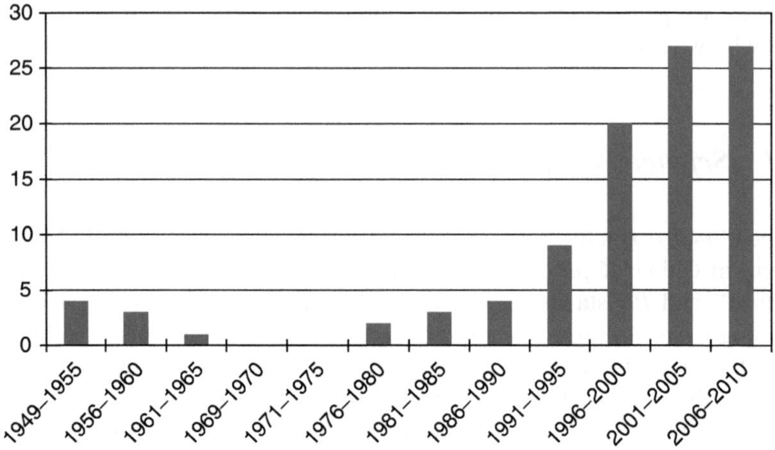

Fig. 5.1 Science popularization policy documents, distribution over 5-year intervals, 1949–1955 to 2006–2010

intentionally collected, but would number many thousands; nor were official documents from about 200 national professional science societies, associations or research institutes included, or documents from the industrial sector (although many big companies engage in SP, their voices are usually not loud).

The team read through the material to determine its relevance. To be deemed relevant, a document had to be one of the following:

- State policy (either special or comprehensive) with wide influence in China
- Departmental or national organization policy that directs SP activities, either on a nationwide scale or within the department or organization
- Departmental or national organization policy supporting large national projects in which SP is part of the objectives
- Policy jointly issued by many actors (government agencies and national organizations).

From the documents, we selected the 100 most relevant ones. Their distribution over the years illustrates efforts in SP for more than half a century (Fig. 5.1).

We found fewer policy documents from before 1966. A large decline coincided with the decade of the Cultural Revolution, which threw the country into turmoil. Government management of SP at that time was at its lowest point. A recovery began soon afterwards, reaching a peak at the turn of the century and remaining steady since then.

Figure 5.1 shows both the vicissitudes of SP policy generation and an increasing public investment in the enterprise, indicating the government's continuous and growing support for SP. The documents also suggest an ever more organized structure for public SP since the mid-1970s.

Table 5.1 Science popularization policies, grouped by actor

	National laws	Leaders' speeches	State policies	Departmental policies
Number	10	5	26	59

5.2.3 Grouping the 100 Policy Documents

There are many ways to look into the selected policy papers. Liu Li, a scholar at Tsinghua University, once classified SP policies into categories at three levels: state, central government agency, and local (Liu and Chang 2010). That classification could usually indicate how powerful a policy was: the higher the level, the more powerful the policy.

We divided the 100 selected policies into four groups to determine the type of authority they represented: national laws, state leaders' speeches, state policies and departmental policies (Table 5.1).

Of the 10 laws, four have special significance:

- The 2002 *Law of the People's Republic of China on Popularization of Science and Technology* (the *SP Law*), enacted especially for SP
- The *Common Program*
- The *Constitution of the People's Republic of China*, which superseded the *Common Program* in 1954 and was amended in 1982, legalizing SP with its 20th article (NPC 1982)
- The 1993 *Law of the People's Republic of China on Science and Technology Progress*, which states in its 6th article that 'The State shall disseminate scientific and technological knowledge to raise the scientific and cultural level of all the citizens' (NPC 1995).

The other laws are special laws regulating the energy saving, environmental protection, farming skills dissemination etc., which emphasize the dissemination of relevant science knowledge or skills.

Five speeches by state leaders were selected. They were all delivered at CAST conventions. The speeches focused especially on the country's SP as well as science and technology. The speeches of the highest state leaders gave prominence to SP and thus secured support from all social sectors.

The policies in the state group were mainly 5-year plans, science and technology advancement programs, and reports to the national congresses of the Communist Party of China (CPC). All these documents are national strategies in which SP has a substantial position. Two are of tremendous weight:

- The *Instructions on Strengthening Engagement in Science and Technology Popularization* (the *Instructions*)
- The *Outline of the National Scheme for Scientific Literacy (2006–2010–2020)* (the *Outline*—discussed further below).

Table 5.2 Programs supported by science popularization policies

Program	Target audience	Policy initiator	Date of issue
More ideal, more contribution	Enterprise employees	CAST	1987
Tutor science education	Normal school students	Ministry of education	1995
Bring health, culture and science to rural areas	Rural citizens	Publicity department, central committee of the CPC	1997
Enrich female farmers with science and technology	Female farmers	All-China women's federation	2000
SP benefiting farmers	Farmers	CAST	2006
Outreach of science museums onto campus	Children	Ministry of education	2006

Departmental policies are issued by an imposing array actors, such as the ministries of Science and Technology, Agriculture, Education, Culture, Health and Finance; government agencies in charge of environmental protection, seismology, meteorology, radio and TV broadcasting, the press and publications; the Chinese Academy of Sciences; and national organizations such as the Trade Union, the Women's Federation, the Youth League and CAST. Most policies in this group are special policies covering SP plans, projects, mass media, infrastructure, science museums, taxes, activities and workforces. For example, in 1996 a joint meeting system was established by the Ministry of Science and Technology to coordinate cross-border SP activities by 11 members, most of them among the above organizations (MST 1996). In 2003, the Ministry of Finance together with several other government agencies issued a notification regarding tax on SP articles. In 2008, the National Development and Reform Commission with three other partners brought out the *Plan for the Development of SP Infrastructure (2008–2010–2015)*. Many programs are supported by SP policies (see Table 5.2 for a selection).

About half of the 100 policies were specially created to govern SP activities. The rest are either comprehensive policies in which SP is included as one component of the whole, or specific policies targeting certain actions for which SP is mobilized as a supporting measure.

At least 10 SP policy papers are of extraordinary importance; three are examined further below.

5.3 Three Noteworthy SP Policies

The *Instructions*, the *SP Law* and the *Outline* are three very important policies. Chinese SP practitioners regard them as milestones in the history of the nation's SP enterprise. They are important not only because they were issued at the highest state level, but also because almost all later sub-level policies and numerous SP activities initiated by the state were framed by them. They are the benchmarks.

5.3.1 The Instructions on Strengthening Engagement in Science and Technology Popularization

The *Instructions* were the first-ever SP policy at the state level dealing comprehensively with all types of SP practice in China. They were jointly issued by the Central Committee of the CPC and the State Council at the end of 1994. At that time, China was at a critical moment in its ongoing reform in the economic and social domains. One year later, the *Law of the People's Republic of China on Science and Technology Progress* was enacted, restating the importance of SP.

According to the *Instructions*, SP had run into difficulty in two areas:

- SP work[3] had been losing priority among some local governments, resulting in a loss of SP strength and momentum since 1978.
- A superstitious craze had gathered strength since the 1980s, involving fortune-telling, 'extrasensory perception', 'magic' medical therapies and other false claims in the name of science.

The *Instructions*, in a short document only 5,000 Chinese characters long, were created to save the situation. The key points were as follows:

- Governments at each level were to place SP on their working agendas and play a part in its delivery. SP engagement was to be included in the state's forthcoming ninth 5-year plan and in local social, economic, and science and technology programs.
- Governments were to secure public input into SP and make sure that more money would be invested in it. The construction of basic SP infrastructure, such as science museums, science centers and public spaces for SP activities was to be supported.
- The mass media would be used to popularize science, and science-based institutions and scientists would be encouraged to contribute to the effort.
- Activities were to be targeted at youth, farmers and cadres in public posts.
- Sensationalist media reporting on superstitions and pseudoscience was to be opposed.

To help create a sound environment, the *Instructions* also put forward the following suggestions:

- Draw up special state laws or by-laws to govern SP practice.
- Set up a joint meeting system to integrate resources from different government sectors to run SP in a well-designed way.
- Formulate regulations or policies to encourage social or private organizations to do SP (CCPC–SC 1994).

[3]SP in China is a public business carried out for the people's wellbeing. A huge body of full-time public employees works in the area. Practice in this part of the social sector is normally considered to be SP work and a government-supported enterprise.

The *Instructions* gave rise to a series of nationwide activities in the following years, including the following programs and guidelines:

- Teacher's Action in Science and Technology Education (1995)
- Knowledge Project (1997)
- Bringing Science, Culture and Health to the Farmers (1997)
- *Guidelines for the Popularization of Science and Technology among Chinese Teenagers (2001–2005)*
- *Notification on Strengthening SP Propaganda Work* (1996), jointly issued by the Propaganda Department of the Central Committee of the CPC, the Ministry of Science and Technology, and CAST to guard against the reporting of superstitious beliefs and supernatural claims.

At this time, China began to import prevailing foreign notions, approaches and experiences in public science communication into domestic practices, greatly influencing many aspects of SP activities in China.

5.3.2 The Law of the People's Republic of China on Popularization of Science and Technology

The advent of the *SP Law* was an immediate outcome of the *Instructions*. A special group was formed to work on the drafting of the law. After years of investigation, consultation, discussion and repeated revisions, the draft was presented as a bill through the legislative channel and adopted by the state in 2002.

The *SP Law* is structured in six chapters with a total of 34 articles. The chapters cover general provisions; organization and administration; responsibility of the society; safeguards; legal responsibility; and supplementary provisions (NPC 2002).

The first two chapters of the law make it clear that SP (the law uses the phrase 'popularization of science and technology', shortened to PST) is a public welfare undertaking. Organizations and institutions, whether governmental or non-governmental, should engage in SP. Citizens have the right to participate in SP activities. On the state side, governments on each level should take up leadership in the administration of the work.

The third chapter legitimizes social sectors' accountabilities and involvement in SP activities. Schools, research institutions, media, national organizations (the Trade Union, the Women's Federation and the Youth League), enterprises, rural grassroots organizations, urban communities, and the managers of public places, parks, shopping centers, airports and the like are encouraged to perform SP.

The 'safeguards' chapter stipulates that public funding for SP activities should be secured at each level of government, favorable taxation should be applied to SP undertakings, and social donations from organizations and individuals at home or abroad are to be encouraged.

The 'legal responsibility' chapter specifies fines and punishments for those who violate the law in this regard or commit misdemeanors that damage SP.

The *SP Law* was not the first or only Chinese law protecting SP, but it was unique in its specificity, and has been the basis for by-laws covering particular matters.

5.3.3 The Outline of the National Scheme for Scientific Literacy (2006–2010–2020)

The *Outline* was promulgated by the State Council on 6 February 2006. It is the most ambitious SP scheme ever to be enacted in China, and greatly advances Chinese SP.

As the title shows, scientific literacy is the key concept. One of the arguments for the drafting of the policy was that Chinese citizens had been shown to have very low levels of scientific literacy.[4] The intervals 2006–2010–2020 mark the stages where milestones are set for improvement.

The major objective tasks of the *Outline* consist of four actions and four projects.[5]

The four actions are:

- Minors' Scientific Literacy Action
- Farmers' Scientific Literacy Action
- Urban Workforce Scientific Literacy Action
- Leading Cadres' and Public Servants' Scientific Literacy Action.

The four projects are:

- Science Education and Training Project
- SP Resources Development and Sharing Project
- SP Capacity Building Promotion Project for Mass Media
- SP Infrastructure Project (State Council 2006).

Each action includes a group of missions, and each project includes a group of measures. To enable the actions and projects to be accomplished smoothly, there are also 'supporting conditions', which emphasize subsequent policymaking, securing funding resources and cultivating human resources.

Immediately after the *Outline* was put into effect, a state leading group of 23 government departments and national organizations was formed. Tasks were divided among the 23 actors, and guidelines for each action and project were developed in detail. Local governments soon took up the *Outline* as a state guideline and developed a similar working pattern and a corresponding package of programs in each province.

[4] China began to conduct national civic scientific literacy surveys in 1992. According to the surveys, by 2005 less than 2% of Chinese could be considered to be scientifically literate, which was considered to threaten the sustainable social–economic advancement of the nation.

[5] Goals of the actions and projects were set to be fulfilled by 2010. New goals were established and the actions and projects were regrouped in 2011.

Table 5.3 Outputs from science popularization investments between 2004 and 2008

	S&T museums		Science	TV programs	Science		
Year	Number	Visitors (m)	books (titles)	(hours)	websites	Lectures	Exhibitions
2004	185	–[a]	2,523	74,959	995	381,345	70,583
2006	239	17	3,162	113,758	1,465	723,337	103,090
2008	380	23	3,888	219,168	1,899	955,142	115,339

[a]There is no individual number of visitors for a comparison with 2004. The total number of visitors who visited science centers, science and technology museums, youth centers or science stations in that year was less than 30 million

Source: *Science popularization statistics of China, 2009*, Department of Policy and Regulations of the Ministry of Science and Technology

Funding for SP soared. According to statistics from the Ministry of Science and Technology, the total investment in SP was about 2.4 billion yuan in 2004, 4.6 billion in 2006, 6.5 billion in 2008 and 8.4 billion in 2010 (DPR 2010). Table 5.3 sheds some light on the results.

Today, the *Outline* is a dominant policy for running SP across the country. As the document states, the principal strategy in carrying out the *Outline* is 'Government boost, mass participation, raising scientific literacy, and promoting harmony' (State Council 2006).

5.4 A Deeper Understanding of the Policies

SP policy output over the past 60 years roughly matches the evolution of SP activities in China. During that time, the SP enterprise has grown from poor beginnings to a much better condition today. With the support of the policies, China has invested heavily to build its capacity. It now has sound SP mechanisms, more practitioners, richer resources and better infrastructure, which enable SP activities to reach and benefit as many people as possible.

CAST is a typical organizational example. According to the *Statistical yearbook of CAST 2010*, in 2009 CAST:

- Organized 151,000 science lectures, which reached audiences of 77.3 million people
- Held 69,000 science exhibitions and showcases, which were visited by 138 million people
- Distributed 185 million leaflets and other printed items
- Trained 3.5 million working people in applicable techniques and skills (CAST 2010).

The package of 100 SP policies tells us many things (and there could be different ways to look into them), but some key points should not be overlooked: the main policy goals or priorities; the ideas and notions behind the policies; and the effects on local policies of inputs from abroad.

5.4.1 Attributes of the Policy Package

Two peculiarities of the selected policies are evident. The first is the number, which is huge (the major part produced after the 1980s). The second is the broad involvement of many actors.

The policies stress three main common areas of importance:

- First, SP engagement is consistently targeted at serving the economic development of the country. Many activities are organized with that aim in mind. For instance, each year a great number of workers and farmers are trained through SP channels to help raise their skills.
- Second, management and governance of the business are repeatedly emphasized in the *SP Law*, the state plans, and suggestions in the documents urging local governments to include SP in their working agendas. It is also noticeable that very large investments have been made in human resources, infrastructure (museums, centers, bases, galleries, caravans etc.) and information resources.
- Third, a matrix working pattern is being encouraged. Many government policies are jointly issued by different organizations, sometimes more than 10. This has especially been the case since the *Outline* was enacted.

All the policies were drafted with the aim of solving problems. While not all have been as effective as expected, they have nevertheless helped to shape SP in China today.

5.4.2 Ideology and Advocacy

SP is encouraged basically in two directions in China: to serve economic development, or to serve social development. The two domains are normally identified as 'material civilization' and 'mental civilization'.[6] In the latter domain, SP is valued for its ideological inspiration.

Science and technology offer strong support for a materialist understanding of the world, which is respected in China, and SP, as a useful tool in communicating science and technology, is considered beneficial in assisting people to learn things in a scientific way. In many of the SP policies, it is made clear that SP is important because it concerns people's views of the world. For pragmatic purposes, it is used to fight against superstition and pseudoscience, which is a persistent challenge for scientists and SP practitioners in China.

Some sayings recur in the policy documents and become government catchphrases, reflecting an attitude of science advocacy in SP. The most frequently used

[6]The two civilizations have now been expanded into five domains: material, mental, political, cultural and social. SP is related to all of them but in government reports is usually classified and addressed in the cultural domain.

terms are 'to learn science', 'to love science', 'to communicate science', 'to use science', 'to trust in science', 'to respect science', 'to rely on science' and 'to admire science'. These sloganistic jingles encourage people to stay far away from superstition and behave scientifically.

The most heavily advocated practice in approaching science nowadays concerns personal achievements in four aspects: scientific knowledge, methods, thoughts and ethos,[7] as stated in the *Outline*. Together with two abilities (to resolve practical problems and to participate in public affairs), the four aspects are used as a yardstick to assess whether a Chinese citizen is scientifically literate or not. From the late 1990s onwards, these themes have been repeated again and again in SP policy documents, and are now the highest goals in the pursuit of SP in China.

5.4.3 Imported Ideas and Experience Are Incorporated

For a long time, the Chinese SP community had little connection to the outside world and knew little about things happening beyond China's borders. There are few traces in pre-1990s policies of the influence of foreign ideas and notions on the practice of SP. Things began to change in mid-1990s. Soon after the Chinese scientific literacy survey of 1992, Chinese practitioners became aware of the American Association for the Advancement of Science's 2,061 Project and the Royal Society's *Public Understanding of science* report, along with many other such initiatives.

Researchers in the field and students in the universities turned towards the Western public science communication arena and began to import many fresh concepts, such as 'science communication', 'public understanding of science', 'science and society', 'scientific literacy', 'the deficit model', 'science consensus', 'dialog', 'bottom-up approach', 'hands-on activity', 'inquiry learning' and the like.

Chinese policies issued in recent years place more emphasis on the perception of science–society relations, public scientific literacy, and the engagement of the public. The policies now positively encourage interactive dialog, inquiry learning and hands-on practice. They reflect a strong effort in an organized pattern. The main underlying theme is to encourage people to love science and to use science, and the major effort is to bring science to the people on the streets following a top-down pattern.

It works. Policies are created to solve problems, and Chinese SP policies are created to solve Chinese SP problems. From this perspective, we say that Chinese SP is a business with a distinctively Chinese culture.

---

[7] Translated in the *Outline* as knowing some necessary knowledge of science and technology, mastering basic methods of science, building up science thoughts, and advocating science ethos (State Council 2006).

5.5 Conclusion

Why does China produce so many SP policies? The immediate answer is that the Chinese Government takes SP as one of its responsibilities from the first, because it is believed that science and technology belong to the people, are useful tools to build the country into a modern society, and bring great benefits to the citizens. For quite a long time, China suffered from poor education, low productivity, bad living habits, superstitious beliefs and ideas handed down from pre-1949 China. Science could help solve those problems, so SP was a job that had to be done.

How has China performed SP under the policy framework outlined here? SP is conducted through various networks in an organized way with state support. Science-related government agencies and national organizations are actively involved. Thousands upon thousands of full-time and part-time practitioners bring SP to all corners of the country in many forms. SP is practiced both within and outside the walls of schools, in newspapers and on TV, in factories and on farms, and in urban and rural communities.

What are the priorities of the policies? They are basically twofold:

- To engage workers and farmers to learn and use new techniques or skills in their daily productive activities
- To inform people of science knowledge, which today extends to encouraging them to be enlightened in science knowledge, methods, thoughts and ethos.

To achieve those goals, the main effort is devoted to building an SP mechanism that emphasizes the best allocation of human and material resources, and to constructing the required SP infrastructure.

The SP policy package presented here is already big enough, but there are still some areas for improvement. We should take fuller advantage of the media, encourage private sector involvement, evaluate input and outcome efficacy, understand how people approach science and technology of their own volition, and consider tensions between science development and public understanding.

Finally, because this chapter is a profile of Chinese SP from the perspective of policy, it tells the story only from a particular angle. To know Chinese SP, one needs to peep at it through many other windows.

Acknowledgement The authors wish to thank their colleague, Li Zhengwei, who did a lot of work in collecting and sorting the original documents referred to here. Their thanks also go to Professor Zhang Yizhong, who kindly provided legal references about SP in China.

References

CAST (China Association for Science and Technology). (1994). *China association for science and technology*. Beijing: Contemporary China Publishing House.

CAST (China Association for Science and Technology). (2010). *Statistical yearbook of CAST 2010*. Beijing: China Statistics Press.

CCPC–SC (Central Committee of the CPC and State Council) (1994). *Instructions on Strengthening Engagement in Science and Technology Popularization*. CCPC–SC.

Council, S. (2006). *Outline of the national scheme for scientific literacy (2006–2010–2020)*. Beijing: Popular Science Press.

DPR (Department of Policy and Regulations of the Ministry of Science and Technology of the People's Republic of China). (2010). *Science popularization statistics of China, 2010*. Beijing: Scientific and Technical Documents Publishing House.

Liu, L., & Chang, J., (2010). Category, system and evolution of Chinese science popularization policies. In China Research Institute for Science Popularization (Ed.), *Discussion on theories and practice of science popularization in China* (pp. 220–223). Beijing: Popular Science Press.

MST (Ministry of Science and Technology) (1996). *Notification on establishment of joint meeting system of SP activities*.

NPC (National People's Congress of the People's Republic of China) (1982). *Constitution of the People's Republic of China*. Retrieved 20 Feb 2011 from http://news.xinhuanet.com/english/2003-08/27/content_1047764.htm

NPC (National People's Congress of the People's Republic of China) (1995). *Law of the People's Republic of China on Science and Technology Progress*. Retrieved 20 Feb 2011 from http://www.most.gov.cn/eng/policies/regulations/200412/t20041228_18309.htm

NPC (National People's Congress of the People's Republic of China). (2002). *Law of the People's Republic of China on popularization of science and technology*. Beijing: Popular Science Press.

CPPCC (Chinese People's Political Consultative Conference) (1949). *Common Program of CPPCC*.

Chapter 6
Deliberation, Dialogue or Dissemination: Changing Objectives in the Communication of Science and Technology in Denmark

Maja Horst

Abstract This chapter provides an introduction to the Danish landscape of science communication, which is built on a deeply rooted culture of equality and anti-elitism. Within this cultural tradition, citizens have a moral right to question the testimony of authorities and to counter it with their own experiences of ordinary life. The tradition is described by a short introduction to one of its most influential proponents, the nineteenth century priest, poet and politician, N.F.S. Grundtvig, who promoted a particular educational philosophy in which citizens were expected to be able to reach consensus through deliberation about the life to lead in common. The teachings of Grundtvig were an important factor in the establishment of Danish deliberative institutions, such as the Danish Board of Technology and the Danish Council of Ethics, but the same anti-elitism has also been invoked in arguments to close them down. Describing how a change in government in 2001 had significant negative impacts on those institutions, the chapter demonstrates that the development of science communication in Denmark is less straightforwardly focused on dialogue and deliberation than many outside commentators believe. While the engagement agenda has grown in other countries since 2000, Denmark has moved in the opposite direction—towards a more traditional deficit model of public understanding of science. A legislative change in 2003 made it mandatory for universities to conduct outreach and science communication. Simultaneously, Danish universities increasingly find themselves in competition for resources, such as funding and well-qualified staff and students. In this situation, science communication is becoming an important ingredient of organizational branding. The chapter concludes with a discussion of how individual research managers in bio- and nanotechnology have adapted to this situation and how they describe their own communication practice.

M. Horst (✉)
Department of Media, Cognition and Communication, Faculty of Humanities,
University of Copenhagen, Njalsgade 80, DK-2300 Copenhagen, Kbh S, Denmark
e-mail: horst@hum.ku.dk

B. Schiele et al. (eds.), *Science Communication in the World: Practices, Theories and Trends*, DOI 10.1007/978-94-007-4279-6_6,

Keywords Anti-elitism • Branding • Consensus conferences • Danish Board of Technology • Danish Council of Ethics • Deliberation • Deficit model • Grundtvig • Institutionalization • Research managers

In 2002, the EU-funded STAGE network (Science, Technology and Governance in Europe) had its introductory meeting in Copenhagen. The network consisted of researchers from eight European countries interested in studying the governance of science in the light of recent developments towards deliberation. In line with a general trend in the study of public communication of science and technology, the normative underpinning for the network was the notion that science communication should be understood as a two-way process, in which the objective is not just to educate the public about science, but also to involve publics and citizens in the governance of science through dialogue about objectives and regulation (STAGE 2005).

Besides marking the beginning of the STAGE network, the 2002 workshop was supposed to provide insight into the particular Danish form of participatory governance of science and technology, which has been epitomized by the Danish-style consensus conferences (Andersen and Jæger 1999; Klüwer 1995; Seifert 2006). It was therefore no surprise that the agenda for the workshop included a presentation by Lars Klüwer, the director of the Danish Board of Technology, who was scheduled to talk about the work of the Board. However, the surprise came when Lars Klüwer rose to speak and announced that he had just had a message that the government was going to close down the Board as part of a restructuring of a number of advisory bodies. He therefore had to leave immediately.[1]

This chapter gives a short introduction to the particular Danish form of science governance and science communication, but it also looks at the background to the announcement of the closure of the Board of Technology. I argue that both of these features are deeply rooted in specific aspects of Danish culture and that the development of science communication in Denmark is less straightforwardly focused on dialogue and deliberation than many outside commentators believe. On this basis, the chapter describes the current framework for communication of science and technology and discusses the heterogeneous forces that influence working scientists when they engage in science communication. This Danish case study tells us something about the intersections between science communication and broader cultural traits, and also demonstrates how developments in science communication within a national framework can be very heterogeneous.

[1] The Board survived, although on 15 November 2011 it was informed that it would be closed down as a result of budget cuts. See http://www.tekno.dk/subpage.php3?page=forside.php3&language=uk (retrieved 30 November 2011).

6.1 A Dialogue Culture Based on Equality and Anti-elitism

The announcement of the intention to close the Board of Technology in 2002 came after an (in)famous New Year's speech by the prime minister, Anders Fogh Rasmussen. Following a decade of center-left governments led by the Social Democratic Party, Rasmussen's conservative and neoliberal alliance came into power in autumn 2001 based on support from the right-wing, anti-immigration party, the Danish People's Party (Dansk Folkeparti). For the first time in decades, a parliamentary majority did not require the support of the middle ground of Danish politics, and Rasmussen declared the 'battle of values' a political priority. That battle was a focus of his first New Year's speech in 2002, in which, among other things, he called for a 'confrontation [or showdown] with the arbiters of taste'. He announced that the new government intended to close a number of expert committees:

> Many of them have evolved into state authorized arbiters of taste, who decide what is good and right in different areas. There are tendencies towards a tyranny of experts, which threatens to oppress the free public debate. The public should not have to submit to raised fingers from so-called experts who think they know best. (Rasmussen 2002)

In his speech, the Prime Minister was referring to deeply held Danish cultural values that are best explained through a small historical detour to the teachings of one of the most influential cultural figures in Denmark, the priest, poet and politician N.F.S. Grundtvig (1783–1872) (see also Horst and Irwin 2010). Inspired by German idealism and British liberalism, Grundtvig was an active proponent of the creation of a nation state in which the Danish people would be united in a common history and a common mother tongue (Korsgaard 2004). For that purpose, he devised a special institution, the 'folk high schools', the task of which was education in knowledge about practical human life. The folk high schools were intended to be a school for life, in which 'the living word' would transform young people into citizens and members of a Danish people with a shared culture and a common destiny. He saw the schools as much more important for society than the universities, which he perceived as teaching 'dead' knowledge to individual scholars (Knudsen 2001:99–105). Grundtvig was fiercely opposed to one-way teaching and envisioned folk high schools as open and anti-authoritarian institutions dedicated to the achievement of educational dialogue. His ideal of dialogue was founded in a belief that 'the living word' would transform both teacher and student and unite them in a sense of shared culture (Korsgaard 2004:225–7).

Grundtvig had an explicitly anti-elitist perception and regarded the ordinary people as far more knowledgeable about the common life of man than any of the authorities in society (Knudsen 2001:104). Knowledge, in his perception, came from experience of an ordinary life, a shared culture and a common destiny as members of the national community. In one of his songs, an often cited line reads: 'And the sun rises with the farmer, not at all with the learned' (Grundtvig 1839).[2] Ordinary

[2] All quotes from Danish sources have been translated by the author.

folk (particularly farmers) were seen as better connected with the knowledge of practical life than learned people in universities. They should therefore not listen to authorities and think that elites know better than them. Rather, they should find their own standpoints through common deliberation among themselves.

The folk high schools became an integrated, although informal, part of the Danish educational system, as it became common for young people (particularly those from the countryside) to spend a year at a folk high school before they settled into more adult life. The educational ideals of the schools were based on dialogue and community building, and the schedule included practical topics as well as issues of general enlightenment. Following industrialization and the development of new urbanized lifestyles, the educational content in the folk high schools progressed, but the core objective has continued to be the development of the democratic skills and identities of the students (Korsgaard 2004). Mejlgaard summarizes the influence of the schools:

> As such, the people's high schools have been influential beyond providing training in S&T skills by promoting a wider discourse of 'active humanism' …, by institutionalizing a principle of 'life-long learning', which has become very important in Denmark, and by stimulating an environment of active appropriation of science and technology in a Danish context. (Mejlgaard 2009:488)

The teachings of Grundtvig and the backdrop of the folk high schools were important factors in the anti-authoritarian, left-wing critique of science and technology that developed in the aftermath of the student revolts in the late 1960s in Denmark: 'A large part of these oppositional arguments drew upon a challenge to modernity, industrialization, capitalist exploitation and—not least—hierarchical antagonism' (Horst and Irwin 2010). Grundtvig was invoked as a founding father of a culture in which experts were envisioned to be no more competent than so-called laypeople in making decisions about the life to be led in common. Technical experts were often described as having a particular interest in the development of a given technology, and they were therefore less able to speak for the common good than citizens with no specific or vested interest in the issue. In this way, Grundtvig's anti-elitism and his distinction between the dead knowledge of universities and the shared knowledge about common life developed through dialogue was explicitly invoked as a foundation for the discussion of public engagement with science and technology.

6.2 The Institutionalization of an Engagement Culture

Prior to the early 1970s, direct public participation in science and technology policy had been modest if not completely absent (Mejlgaard 2009). However, the oil crisis in 1973 led the Danish Government to publish a nuclear program, which suggested that nuclear power plants would be constructed in Denmark. This sparked an intense public debate and generated widespread NGO activity on the energy issue. Grassroots organizations were conducting practical experiments with alternative energy, such as wind power and biogas, and were also very actively engaged in the production of knowledge about alternative energy. In this way, they were 'laying a foundation for what would later become a profitable Danish alternative energy industry' (Mejlgaard 2009:489). The widespread public debate on nuclear energy came to a conclusion in 1982, when a referendum led to the abandonment of nuclear power in Denmark, but public discussions about energy systems and environmental issues continued.

On the basis of those developments, Danish science governance took a more deliberative turn in the 1980s. Inspired by National Institute of Health and the Office of Technology Assessment in the United States, the first Danish consensus conference was arranged in November 1983. It focused on the early diagnosis of breast cancer, and was arranged in collaboration between the Danish Medical Research Council and a public research organization called DSI (Danish Health Institute). Although it was inspired by the American experience, the Danish version was intended to focus less on the relationship between research and clinical practice and more on the connection between science and the public. An evaluation report of the first consensus conference states that it was intended to help promote 'democratic decisions' and 'competent judgments in the common interest ... on the basis of constructive interaction between experts and non-experts' (Agersnap et al. 1984:7).

Torben Agersnap, who led the evaluation, had a background in organization studies, and he and his colleagues had worked extensively on democracy and participatory processes in workplaces. In his Department of Organization and Work Sociology at Copenhagen Business School, a number of researchers were engaged in the study of democratic approaches to technology (Agersnap 1992). Besides the Grundtvigian influence, those approaches were also supported by the Danish trade unions, which promoted workers' participation in technological decisions and the introduction of technology in the workplace (Cronberg 1995). What was later to be known as the Scandinavian 'participatory design' tradition in studies of information technology was based specifically on two assumptions:

- That the involvement of users will make better systems and reduce the risk of resistance to change
- That democracy is a goal in itself (Vikkelsø 2003:31).

In this way, the Grundtvigian ideals were merged with the objectives of the trade unions. The merger was documented and encouraged by organizational scholars who made a strong case for the development of participatory engagement with technology (Joss and Durant 1995).

This background, as well as general expectations of major future change brought about by emerging information technology and biotechnology, led to demands for a more institutionalized way of dealing with new technology and its effects on society, organizations and individuals (Lassen 2004). The result was suggestions to parliament that Denmark should follow the example of other countries and create a public body that could foster and develop public debate and assessments of emerging technologies. But politicians did not simply want to copy a solution from other countries. Instead, they wanted a specifically Danish model derived from the Danish democratic tradition of equality and dialogue (Klüwer 1995:41). Consequently, the first version of the Danish Board of Technology was established by parliament in 1985.

Its objectives were to 'follow and initiate comprehensive assessments of the possibilities and consequences of technological development for society and citizens [and to] support and encourage a public debate on technology' (Klüwer 1995:41).

The Board was intended to be an inclusive force and to encourage interactions between a number of different stakeholders in society. It was organized with a secretariat that took care of daily operations, while a board of governors was responsible for directions and strategic decisions. Finally, a board of representatives from a large number of different public, private and third sector organizations served as a group of important mediators to stakeholders. During its years of operation, the Board developed a number of specific formats for technology assessment, including expert reports, but the specific format of the participatory consensus conference is most widely known (Einsiedel et al. 2001; Horst 2008; Seifert 2006).

The 1980s, however, also saw the birth of a different body, the Council of Ethics, designed to assist political decision-making and encourage public debate about technology. The Council did not grow out of discussions about participatory technology assessment, but out of controversies about reproductive medical technology that gained momentum in the beginning of the 1980s. The controversies were sparked by the fact that fetal diagnostics had been a standard offer in the Danish health care system after 1977, as well as by the birth of the first Danish IVF baby in 1983. The combination of the technical possibilities with images of cloning and genetic engineering led to much public speculation about designer babies and a general fear of the erosion of moral norms.

In 1984, a report on the social and political responses to these new technical options suggested the formation of the Council of Ethics. The assumption in the report was that it was possible to achieve consensus about the regulation of biomedical technologies through deliberation among people who, due to their professional experience and personal integrity, could speak for the 'common human condition' (Koch and Horst 2007). Although that suggestion received immediate widespread support from policymakers, it took the Danish Parliament 2 years to agree on the composition and mandate of such a body. The resulting law stated that the Council should make recommendations to policymakers and health authorities, and also follow and initiate public debate about ethical issues.

During the parliamentary discussion about the Council's formation, the forms of knowledge and competencies to be represented in the Council were crucial issues. Several attempts had been made to specify the types of expertise that members of the Council should possess (Lund and Horst 1999). However, the resulting law merely specified that the minister would appoint half the members on the basis of their general knowledge about relevant ethical, cultural and social aspects; a particular parliamentary committee would appoint the other half. In practice, members have been medical experts as well as people with ethical, social, judicial and religious expertise. It also seems that members are appointed to represent a broad spectrum of opinions on ethical issues.

Another point of discussion in the parliamentary negotiations about the Council was the value foundation for the work of the Council (Kappel and Lykkeskov 2007).

When the legislation was finally approved in 1986, the small Christian Democratic Party had managed to insert an introductory clause stating that the Council should base its work upon a belief that life begins at procreation, although that line was clearly at odds with Danish legislation covering abortion. That small addition can be seen to point to a major fault line in the debates in the Council. As it turned out, most of the policy advice from the Council has been marked by disagreement and majority–minority recommendations (Koch and Horst 2007), and disagreement on the status of the fetus has often been the basis for those conflicts.

Compared to the Board of Technology, the Council of Ethics must be regarded as an expert committee, although the unspecific definition of the kinds of expertise the members are expected to have has been criticized (Kappel and Lykkeskov 2007).

In addition to the Council's 'expert role', however, the legislation states explicitly that the Council has an obligation to generate public debate. Throughout its history, the Council has therefore not only produced reports of advice to politicians, but also arranged numerous open meetings, conferences, web forums and other engagement activities to elicit public debate and form opinions about new biotechnologies. Due to its history of split recommendations, the Council has probably had most influence on the governance of science and technology in Denmark through its role as an initiator of public debate.

Interestingly, the parliamentary processes involved in the creation of the Board of Technology and the Council of Ethics took place in the same years, but without any apparent interconnection. One reason is that the two organizations were answers to questions about technological development raised from two different bodies of knowledge. The Board of Technology grew partly out of the focus on the participatory design of new technology and deliberative democracy in workplaces and was primarily focused on information technology and questions of environment and energy. The Council of Ethics, on the other hand, grew out of controversies about medical ethics and the wider impact of medical technology on societal norms. Another difference is that the Board of Technology engaged researchers in organization and democracy, as well as workplace and environmental activists, in shared knowledge production about participatory technology assessment. In contrast, the Council of Ethics very much epitomized a controversy between medical doctors and scientists who were in favor of the use of the new technologies, and social scientists, ethicists and religious scholars who were opposed to the use of those technologies.

Despite those differences, however, both organizations embodied a strong ideal about deliberation and engagement, which has often been explicitly based on the cultural tradition developed on the basis of Grundtvig's writings. In participatory consensus conferences, for instance, the citizens take center stage. It is their task to listen to the testimony of the experts and then decide which aspects of that testimony are relevant for a shared understanding of the technology and a consensus agreement on its future development. The Board, therefore, makes a specific effort to identify the 'right' type of citizen: on a general level, they should be engaged and interested in the issue at hand, but they should not have specific, vested interests. Strong previous engagement or specialist knowledge in the field is not an advantage for an ideal member of the citizens' panel in a consensus conference. In this way, participatory consensus conferences build directly on the Grundtvigian ideal of the common sense of the ordinary person (Horst and Irwin 2010).

Similarly, the Council of Ethics was founded on an expectation that people with different opinions would be able to reach consensus through deliberation based on experience of the common human condition. And, although the Council must be described as an expert body, the expertise of its members was not bound to a specific academic set of skills but rather to more general competencies developed through professional experience and personal integrity.

In this way, it can be argued that Grundtvigian ideals have had a strong influence on the governance of science and technology in Denmark. Furthermore, since dialogue and deliberation feature so strongly in this form of governance, the communication of science and technology has been an implicit part of these governance structures. The important characteristic of the Danish model, however, is that public understanding of science was not seen to be solely a question of diffusion of knowledge, but part of a larger culture of debate and enlightenment. Science communication was not just a question of dissemination, but part of a larger process of sharing knowledge about the life to be led in common.

6.3 Institutions in Decline: The Reinvention of PUS

The Danish model has been an inspiration in other countries as they have sought to develop more participatory forms of public engagement with science (SCST 2000). However, seen from the inside, the foundational institutions in the Danish model seem to have lost their momentum. Ironically, the 2001 prime ministerial speech about the 'tyranny of experts' used the anti-elitist elements of Danish culture to argue against bodies such as the Board of Technology, which epitomized the free debate between citizens, but that incongruity was lost in the general 'battle about values'. In general, however, the announcement of the decision to close the Board of Technology was not the first blow to the institutions of public engagement. Funding had withered since the 1990s (Lassen 2004), and media attention and support for the activities of the institutions also seemed to diminish (Lund and Horst 1999). Consequently, as the engagement agenda seemed to grow in other countries, Denmark moved in the opposite direction towards a more traditional public understanding of science agenda. As one observer has put it:

> In the 2000s, however, public participation seems to have lost its status as the dominant tool for holding science and technology accountable to society. A new 'fiscal' regime of public accountability is rapidly gaining momentum, and meanwhile, policies on science and society increasingly support strategies of science dissemination and public education rather than public participation. (Mejlgaard 2009:484)

In 2003, the Danish law on university governance was revised. The amendment introduced an obligation to disseminate knowledge as a third mission of the universities (Videnskabsministeren 2011). Commentaries on this change made it clear that the aim was to increase the application of knowledge in order to foster increased innovation in companies and improve the performance of the Danish knowledge economy. The minister also appointed a think tank on public understanding of science with a mandate to suggest policies on science communication in order to 'give the Danish population an understanding of the importance of this area for our future welfare, environment, health and growth' (Ministry of Science, Technology and Innovation, 2004, cited in Mejlgaard 2009). The mandate was connected to the general objective of increasing innovation through the dissemination of applicable knowledge, but there was also a specific worry about the declining numbers of youngsters choosing to be educated in science and engineering.

The mandate quite straightforwardly built on a deficit model of public understanding of science (Irwin and Wynne 1996), but it is unclear how much the decision to establish the think tank was fuelled by a sense that public skepticism towards, for instance, biotechnology might be a roadblock to the innovation agenda. In this context, it should be noted that a number of consensus conferences on biotechnology and numerous other engagement activities had not led to Danes being less skeptical about biotechnology than their European neighbors around the turn of the millennium (Suine and Mejlgaard 2001). However, developments in public opinion before that time are impossible to discern, as knowledge about public attitudes was not collected in any systematic fashion before the establishment of the Center for Studies in Research and Research Policy at Aarhus University in the late 1990s. The center mainly employs political scientists and, among other things, has been responsible for the Danish Eurobarometer surveys.

The 2003 think tank on public understanding of science had 23 members, of whom 8 were from the mass media, 2 were from universities (one of whom was a communication director) and the rest were from public and private organizations with an interest in knowledge dissemination. Interestingly, nobody from the Board of Technology or the Council of Ethics was part of the think tank, just as none of the researchers previously involved in research on science communication or public understanding of science was included. Although the final report by the think tank did mention the Board of Technology, it appeared more or less totally oblivious to the Danish tradition of engagement with science. The report noted the value of dialogue and two-way communication of science, but subsequently focused a great deal on mass media, without explaining how they can increase dialogue (Videnskabsministeriet 2004).

Several of the think tank's suggestions were implemented in the first decade of the new millennium, including a research communication prize; funding for research and experiments in science communication; an annual festival of research; an internet portal for science communication; and a special task force for communication of science to children. But neither the Board of Technology nor the Council of Ethics regained its former strength. The Board managed to fight off the threat of closure by mobilizing substantial support among supporters nationally and internationally (Mejlgaard 2009), but its funding was sparse. The last consensus conference was held in 2005, and only happened because the Board received funding from international sources.

Meanwhile, the introduction of the obligation to disseminate knowledge, as well as an increasing sense of competition between them, has led Danish universities to strengthen their professional capabilities in communication. Researchers are offered courses in science communication, and most universities also have communication units to help disseminate stories about research results and new research projects.

The result of the developments in the beginning of the new millennium has therefore been a change in the general framework for science communication. Systematic science communication is now less connected with the institutions of democratic participation and more connected to the branding of organizations and research groups. This framework has also influenced the context for individual scientists and their efforts to communicate publicly about their knowledge production.

6.4 The Role of the Individual Scientist

Before 2003, science communication by publicly employed scientists was based completely on individual initiative. Scientists were under no formal obligation to participate in communication, just as they would not expect the university to have an opinion about how they chose to organize their communication activities as long as they stayed within the general professional norms of science. Science communication would therefore be undertaken by individuals who found it worthwhile, and to the extent that they could catch the attention of an audience. Sometimes, Danish scientists' communication activities would be part of the particular institutions mentioned above, such as the folk high schools and the consensus conferences, but they have also used books, news media and public meetings (Kragh et al. 2008). The Grundtvigian tradition has not been the only factor shaping the communication of science and technology by Danish scientists: a large part of their communication activities has been guided by a traditional ideal of dissemination similar to that found in many other countries (Gregory and Miller 1998).

Following the change in the university law, the organization of science communication is slowly changing. It is no longer left exclusively to individual initiative, as universities now have strategic interests in the nature and extent of communication activities. Danish universities are increasingly finding themselves in competition with each other, and science communication has become an ingredient in organizational branding. Also, individual researchers and research groups find that visibility might increase their ability to attract resources, such as funding and well-qualified staff.

In order to discuss how such changing circumstances influence researchers' behavior and sense-making in relation to science communication, I now draw on a specific analysis of the communication practices of research managers in biotechnology and nanotechnology (Horst under review). The analysis is based on 20 semi-structured interviews with research managers in those fields. The interviews were conducted as part of a research project on research management, communication and risk, and were focused on respondents' views on their own communication activities and those of the people in their research groups. The overall impression from the interviews is that there is great variation in the perception of science communication and the extent to which interviewees see it as a strategic activity.

Overall, the research leaders perceive the objectives of communication activities in three different ways, corresponding to three different modes of science communication.

First, there is communication on behalf of a discipline. When researchers communicate in this mode, they represent a certain body of knowledge and they speak as experts in a particular field. Their goal is to share their expertise with an audience, which is often perceived as a target group with a certain set of characteristics. When communicating in this mode, some informants find it useful to utilize the services of the communication professionals at their university, while others do not. The practice of communicating in this mode is very much perceived as something that 'comes with the job'. As part of a community of expertise, one is expected to share one's knowledge. This is also part of what younger researchers are expected to learn: just as they have to acquire a number of other skills to grow into independent researchers, they have to learn how to talk to different target groups. Those skills are acquired as part of the collective work that takes place in the group.

Second, there is communication, which is undertaken more as a representative of the entire institution of science. Scientists who reflect on their own communication activities in this mode often describe those activities as part of their personality. Their objective is focused on enlightenment and is often described as a personal choice motivated by a wish to educate citizens in scientific rationality. Rather than representing a specific area of expertise, these scientists represent science as that institution in society which produces truth, and they seem to regard their own role as that of a kind of guardian for the institution of science. When science communication is done in this mode, communication professionals are not seen to be helpful at all; rather, they are perceived as an unnecessary barrier between the scientist and the citizens. Journalists are often also seen in the same way, as problematic intermediaries who simply distort the communication process. Because this mode of communication is closely linked with personality, it is not perceived as a set of skills that can be acquired as part of professional training, but rather as an individual ethos that is gained partly by experience and partly by choice. In this context, the research leaders seem to perceive themselves as examples that their younger staff can choose to follow, if they are so inclined.

The third mode is communication on behalf of the research organization, such as the research center or the entire university. When communicating in this mode, the focus is on branding in relation to the organization's stakeholders. Scientists communicating in this mode are very aware of the necessity of creating external legitimacy in order to attract resources and funding, and their aim is to represent their university as a professional research organization. To that end, they often find it useful to have professional communication expertise available and they will readily use it, although they do sometimes criticize the professional quality of the assistance. Scientists who primarily reflect on communication in this mode tend to consider communication skills as part of a set of managerial competencies that young researchers need to learn in order to develop into independent research leaders. Several of these informants also argue that the acquisition of these skills cannot be left to chance, but has to be systematically managed through professional training.

Reviews of the interviews show that the specific Danish tradition of engagement and dialogue does not seem particularly pertinent. The culture of anti-elitism poses problems for several of these research leaders. They are indignant about a culture that does not seem to value their competencies and/or share their view of the importance of using scientifically produced knowledge as the basis for decisions. However, other informants seem to enjoy the Danish debate culture and believe that experts should be challenged to give reasons for their judgments. In general, research leaders from the biotech area are most at ease with the idea of public engagement with science. It is tempting to conclude that the engagement of many of them in public controversies for several decades has influenced their attitude. In comparison, the nanotechnologists in the sample, whose background is primarily in physics and chemistry, generally seem less willing to entertain the idea that the public should take part in the governance of science.

6.5 Conclusion

The Danish model of participatory governance in science and technology is founded upon a cultural tradition of dialogue. Based on their experiences of a common life and a sense of shared destiny, citizens are expected to engage in deliberation with the aim of identifying consensual solutions for the common good. The communication of science and technology has been part of this general shaping of competent citizens through deliberation, but it has always had a distinctive anti-elitist flavor. Within this tradition, citizens have always had a moral right to question the testimony of authorities and to counter it with their own experiences of ordinary life. The deliberation tradition has been institutionalized in bodies such the Board of Technology, and has gained attention outside Denmark as an inspiration for moves towards dialogueical and participatory forms of science communication.

During the past decade, however, the Danish model has been in decline in Denmark. A new government turned the focus away from participatory technology assessment towards innovation as a driver for the knowledge economy and rebooted the tradition of science communication. In the new millennium, science communication is mainly perceived as a process of dissemination that is intended to enhance the public understanding of science and technology.

References

Agersnap, T. (1992). *Konstruktiv teknologi*. København: Nyt fra Samfundsvidenskaberne.

Agersnap, T., Jakobsen, G., & Kempinski, J. (1984). *Konsensuskonferencer i danmark*. København: DSI.

Andersen, I., & Jæger, B. (1999). Danish participatory models. Scenario workshops and consensus conferences: Towards more democratic decision-making. *Science and Public Policy, 26*(5), 331–340.

Cronberg, T. (1995). Do marginal voices shape technology? In J. Simon & J. Durant (Eds.), *Public participation in science: The role of consensus conferences in Europe* (pp. 125–134). London: Science Museum.

Einsiedel, E. F., Jelsøe, E., & Breck, T. (2001). Publics at the technology table: The consensus conference in Denmark, Canada, and Australia. *Public Understanding of Science, 10*(1), 83–98.

Gregory, J., & Miller, S. (1998). *Science in public: Communication, culture, and credibility*. New York: Plenum Trade.

Grundtvig, N. F. S. (1839). 'Er Lyset for de Lærde Blot' [Is the light just for the learned?]. *The Danish Folk High School Song Book* (17th ed.), song number 462, Odense: Foreningens Forlag.

Horst, M. (2008). In search of dialogue: Staging science communication in consensus conferences. In D. Cheng, M. Claessens, T. Gascoigne, J. Metcalfe, B. Schiele, & S. Shi (Eds.), *Communicating science in social contexts* (pp. 259–274). New York: Springer.

Horst, M. (under review) *The discipline, the organization, or science itself?* Working paper, unpublished.

Horst, M., & Irwin, A. (2010). Nations at ease with radical knowledge. *Social Studies of Science, 40*(1), 105–126.

Irwin, A., & Wynne, B. (Eds.). (1996). *Misunderstanding science? The public reconstruction of science and technology*. Cambridge: Cambridge University Press.

Joss, S., & Durant, J. (1995). *Public participation in science*. London: Science Museum.

Kappel, K., & Lykkeskov, A. (2007). *Etik i tiden: 20 år med det etiske råd.* Det Etiske Råd.

Klüwer, L. (1995). Consensus conferences at the Danish Board of Technology. In S. Joss & J. Durant (Eds.), *Public participation in science: The role of consensus conferences in Europe* (pp. 41–49). London: Science Museum.

Knudsen, T. (2001). Da demokrati blev til folkestyre: Dansk demokratihistorie I. Akademisk Forlag A/S.

Koch, L., & Horst, M. (2007). Fra almenhed til pluralitet: Forestillinger om konsensusskabelse i det etiske råds historie. In K. Kappel & A. Lykkeskov (Eds.), *Etik i tiden: 20 år med det etiske råd* (pp. 143–166). København: Det Etiske Råd.

Korsgaard, O. (2004). *Kampen om folket—et dannelsesperspektiv på dansk historie gennem 500 år*. København: Gyldendal.

Kragh, H., Kjærgaard, P. C., Nielsen, H., & Nielsen, K. H. (2008). *Science in Denmark: A thousand-year history*. Århus: Aarhus University Press.

Lassen, J. (2004). *Changing modes of biotechnology governance in denmark.* STAGE discussion paper 3. Retrieved 13 Nov 2011 from http://www.stage-research.net/STAGE/documents/3_BiotechnologyGovernanceDK_final.pdf

Lund, A. B., & Horst, M. (1999). *Den offentlige debat—mål, middel eller mantra*. København: Fremad.

Mejlgaard, N. (2009). The trajectory of scientific citizenship in Denmark: Changing balances between public competence and public participation. *Science and Public Policy, 36*(6), 483–496.

Rasmussen, A. F. (2002). *Prime minister's New Year's speech.* Retrieved 13 Nov 2011 from http://www.stm.dk/Index/dokumenter.asp/_p_7354.html

SCST (House of Lords Select Committee on Science and Technology). (2000). *Science and society*. London: House of Lords.

Seifert, F. (2006). Local steps in an international career: A Danish-style consensus conference in Austria. *Public Understanding of Science, 15*(1), 73–88.

STAGE (Science, Technology and Governance in Europe). (2005). Website and final report. STAGE network. Retrieved 13 Nov 2011 from http://www.stage-research.net/STAGE/index.html

Suine, K., & Mejlgaard, N. (2001). *Public understanding of science—Perceptions of and attitudes toward biotechnology in Denmark*. Aarhus: Analyseinstitut for forskning.

Videnskabsministeren. (2011). *Universitetsloven.* Retsinformation. Retrieved 13 Nov 2011 from https://www.retsinformation.dk/forms/r0710.aspx?id=137835

Videnskabsministeriet. (2004). *Forsk og fortæl.* Ministeriet for Videnskab, Teknologi og Udvikling.

Vikkelsø, S. (2003). *Electronic patient records and medical practice reorganization of roles, responsibilities, and risks*. København: Samfundslitteratur.

Chapter 7
Social Sciences and the Communication of Science and Technology in France: Implications, Experimentation and Critique

Michèle Gellereau, Yves Jeanneret, and Joëlle Le Marec

Abstract This chapter details how attention to issues of communication has influenced science studies research in France, in an interdisciplinary convergence of different theoretical trends in the social sciences. Three of those trends are the long-standing attention paid to discourse analysis and to media products; a sociological tradition concerned with relations of legitimacy; and new currents in ethnography focusing on familiar and contemporary cultural practices. The contributions of communication and information sciences to the conceptualization of science in society are connected with the question of symbolic and material mediations that organize the production and life of knowledge in society, especially in the public sphere. A link has been established in France between research on science communication and research on cultural mediation. While the profession has developed mediation as a way of connecting separate spheres, another trend of academic and professional researches theorizes social communication as a continuous, collective creation of references, practices and objects. This chapter also emphasizes the tension between academic research, involvement in cultural production and involvement in the market for communication devices and expertise. It opens on to issues concerning theoretical and critical understandings of the world, the managerial production of devices for functional communication and, finally, the cultural and social questioning of the links between academic and lay knowledge. Collaboration between researchers and

M. Gellereau (✉)
University of Lille, Lille, France
e-mail: michele.gellereau@univ-lille3.fr

Y. Jeanneret
Paris Sorbonne University, Paris, France
e-mail: yves.jeanneret-celsa@paris4.sorbonne.fr

J. Le Marec
University Paris-Diderot, Paris, France
e-mail: jlemarec@neuf.fr

B. Schiele et al. (eds.), *Science Communication in the World: Practices, Theories and Trends*, DOI 10.1007/978-94-007-4279-6_7,

actors in scientific and technical culture is strengthened by a common desire to defend a democratic ideal of public service and sharing of knowledge, which is quite different from the marketization of access to services and knowledge. The major results of science communication research might be a displacement of the symbolic border stretching between academic knowledge associated with functional models of expertise, and practical knowledge. A new border might separate, in a more fundamental way, technical models of expertise and complex models of the life of knowledge.

Keywords Communication and information sciences • Science and technology • Knowledge mediation • Public space • Model discourses • Expertise • Functional communication

In this chapter, we detail how attention to issues of communication has influenced science studies research in France.

We first describe the convergence of different theoretical trends in the social sciences and, in particular, the growing interest in the complex and dynamic nature of social phenomena. The sciences of communication play an important role in this interdisciplinary convergence.

Such a convergence, and more particularly the role that the communication sciences have played in it, are not, of course, exclusive to France, although they have taken shape in France in a particular way corresponding to a specific academic tradition and political context.

In terms of academic traditions, we notice the coming together of three trends: first, the long-standing attention that literary scholars have paid to discourse analysis and to media products; second, a certain sociological tradition concerned particularly with relations of legitimacy and their evolution; third, the development of a new current in ethnography focusing on familiar and contemporary cultural practices, or the 'ethnology of proximity'.

Regarding the political context, a certain conception of public service partly unifies the different contexts in which activities in research and culture take place. We can add to this a confrontation between very affirmative government policies for the democratization of access to cultural masterpieces and to scientific knowledge and the emergence of more situated visions of culture and knowledge promoted by academic research and appropriated both by civil society and be market forces in the name of social differentiation.

Second, we describe the contributions of communication and information sciences to the conceptualization of discourses and practices of science by connecting those contributions to a more general question: what are the symbolic and material mediations that organize the production of knowledge as objects circulating in society?

We mention in particular the peculiar way that a link has been established in France between two fields that are rarely associated: research on science communication and research on cultural mediation. It must not be forgotten that issues of the social sharing of scientific knowledge were previously implicitly restricted to the fields of the physical and biological sciences.

Finally, we emphasize one of the problems facing contemporary research on sciences in society. That problem is linked to the tension between academic research, involvement in cultural production and involvement in the market for communication devices and expertise.

This leads us to open up issues concerning theoretical and critical understandings of the world, the managerial production of devices for functional communication and, finally, the cultural and social questioning of the links between academic and lay knowledge.

7.1 Theoretical Convergences

A French intellectual tradition concerns the relations between communication and science. It is marked by certain theoretical milestones, by changes (which involve research) in the symbolic economy of public space, and by the implication of the economic and political worlds in issues of science and communication.

The communication approach to science has developed within a highly interdisciplinary framework, one that is also very often linked to political and social demands.

This framework is chiefly characterized by the following four factors:

- **The development of linguistic sciences** and their increased interest in social customs and in specific kinds of discourses issuing from different social worlds, including the world of science. Scholars have observed an evolution in what socio-discursive analyses have aimed to bring to light (that is, what is inscribed into textual and semiotic material so as to become an object of study). Such investigations have ranged from studies of scientific vocabularies to inquiries into the diversity of discourses in different disciplines or fields of speciality, as in the work of Jean Peytard, Daniel Jacobi and Sophie Moirand (Moirand 2004). They have spread to the analysis of interdiscursive dynamics in science, media and culture. Scholars in communication sciences have, for instance, followed the development of controversies in the heterogeneous spaces of media and research around the Sokal affair (Jeanneret 1998; Jurdant 1998).
- **The rise of sociology and anthropology of scientific activity**, which aim to account for the social and political dimensions of research. We shall not elaborate here on the abundant work in sociology and in the history of sciences that has helped bring to light many dimensions of research which had long been underestimated, such as the fact that research is shaped by power relations, professionally managed and run by political authorities (Latour and Callon 1991; Licoppe 1996; Vinck 2007; Pestre 2006, 2010; Berthelot 2005).
- **The theorization of popularization** as a fundamental activity for the emergence of the very notion of science itself and for the analysis of communication strategies in the scientific world. Science popularization has been shown to be a means to

generate reflexivity, which is then revealed to be a necessary component of the sciences themselves (Jurdant 2009). This point of view on popularization practices is closely connected to other studies that purport to reactivate epistemological issues, such as the critique of the ideology of unity in the sciences (Lecourt 1981) and the critique of the idealization of method (Feyerabend 1979; Hacking 1983), both foregrounding the irreducible heterogeneity of the production of knowledge.
- **The critique of the political role played by science in social life** (Lévy-Leblond 1981), which can be linked to contemporary efforts to question literary, cultural and academic practices (Barthes 1971; Bourdieu 1984; Fayolle 2009). Such a critique questions relations of legitimacy, effects induced by authority and certain categories such as 'works' and 'authors'. This critique can also be related to historical inquiries concerning the links between legitimate forms of knowledge production, which include science, and ideological frameworks such as those produced by literature and the media (Bensaude-Vincent and Rasmussen 1997; Mattelart 1994).

7.2 The Communication Approach

Within this general interdisciplinary framework, the communication sciences have developed in France in several broad areas of inquiry: social representations, or the social knowledge being actualized in situations of communication (Moscovici 1961); how the media functions, and how mass-media discourses are produced (Veron 1986; Davallon 1999; Schiele 2001); the symbolic economy of public space (Habermas 1991/1962; Miège 1989); and intellectual technologies and the evolution of reading and writing practices (Souchier et al. 2003).

All of these studies have confronted questions of the popularization of the sciences and of the relations between science and society.

Moscovici, for example, developed the notion of social representation by linking the study of the social dissemination of specialized knowledge (psychoanalysis) with an examination of the way in which non-specialized media function. Following this, social representations have increasingly been the object of attention in the study of the mediation of science in museums and the media (de Cheveigné 2000; Schiele 2001; Le Marec 2007).

As for the media, it constitutes a heterogeneous social space, where multiple actors intervene in the production of polyphonous discourses by mobilizing communication technologies which are themselves not simple tools, but complex apparatuses incorporating technical and intellectual genealogies (Perriault 1989). The usage of technologies also activates norms and knowledge that are external to the spaces in which they are used, but which structure how all those implicated in the conception of such technologies intervene there. Journalists are very bound by rhythms of production (stemming, for example, from the division of labor or from logical frameworks implicit in certain projects) or by production formats (often linked to technologies of writing) that they do not completely grasp and that are imposed upon them by factors and decision-makers that never appear in the text or in the image.

The media thus provides an ideal point of view from which to problematize the 'scholar'/'layman' distinction without posing it as either factual or illusory, but rather seeing it as a social representation that structures both the production of discourse and the relation to a public. That is perhaps why popularization studies and research on how media function have so strongly nourished one another.

The media are also an important object in the elaboration of theories and in the reconfigurations of public spaces since the eighteenth century: how, for example, can we observe and describe public debate without referring to models of 'face to face' interaction, or using the representation of confrontation between antagonistic positions?

Conceptions of the circulation of discourses or of the dynamics of debate based on metaphors or analogies that are either too easy or too reductive, such as the linear model of communication, or the metaphor of the arena of the media, have been overthrown by a poetics specific to texts and to instruments of science communication (Jacobi 1986, 1999; Jeanneret 1994).

Strong attention to the complexity and heterogeneity of relations with science has raised many daunting methodological problems: how should we observe and what should we look at, once everything has been declared complex and dynamic? In terms of methodology, communications can be called on, this time within empirical studies for the analysis of practices (social, professional) and as a means to roam through spaces, to explore the dynamic way quite differentiated social spaces operate: things do not move from one space to another, they are transformed. Above all, communication does not imply that different social spaces will merge; on the contrary, communication can become autonomous in its own spaces where specific norms are observed. In France, for example, the National Ethics Advisory Committee found that opinions communicated through press conferences did not 'transit' from the confines of the committee towards the media via the press conference with the journalists. The social spaces that are the committee, the press conference and the newsroom are too differentiated. The members of the ethics committee, acknowledging this problem, have called upon a public relations agency in order to promote the circulation of opinions! (Kapitz 2007)

Monitoring the discourses, spaces and participation of actors has also led to observing how models and practices of professional communication in research institutions (project-based work, modelization of effects, rationalization of exchanges) have even engendered a near-complete autonomization of communication services, which prescribe the norms for research laboratory production (Babou 2008; Le Marec and Babou 2003, 2005; Hert and Paul-Cavallier 2007).

Within a more institutional framework, rather than a theoretical or methodological one, France has also witnessed the convergence between part of the communication sciences that have become established as an academic discipline, and part of the so-called 'STS' (science, technology and society) community that has never truly been institutionalized.

7.3 Dynamics of Public Space

The emergence of ways of analyzing the relations between science communication and society is inseparable from developments within the symbolic economy of public space that are continuously displacing research issues. Research on communication, its issues, its financing, its objects, cannot be outside this public space, which nevertheless remains one of its favored objects of study.

We thus find that early theories of popularization are difficult to divorce from the desire to rationalize and optimize social communication. This desire is expressed in the ideal of the 'third man' (Moles and Oulif 1967), a conciliator between different worlds. The image of the translator, elaborated at the dawn of the Enlightenment by Fontenelle, might be considered the ideal type of this operation of mediation. The industrial development of the mass media has been organized around this figure of the third man, whose action is modelized according to Lasswell's program (1948): Who says what to whom in what channel with what effect?

The professional sphere of communication associated with the professions of journalism, mediation, institutional communication and communication technology is also a domain for the application and development of the sciences of communication. But is it also a movement linked to the ideology of functional communication, which the scientific approach must deconstruct (Fayard 1988).

The originality of European thought on this issue lies in its resolute opposition to this functional model and its foregrounding of the legitimizing functions of popularization and of the fictional character of the opposition between the learned and the ignorant, as well as of a poetics of popularization.

Such a renewed theoretical framework has led a generation of scholars to analyze the relations between science and the media in particular, looking for a balance between the specificity of scientific questions on the one hand and the specificity of the media on the other, with a focus on discourses, mechanisms, practices and publics (Babou 2004; Chervin 2000; Pailliart 2005). The emergence of so-called 'socio-scientific' questions (Simonneaux and Legardez 2010) in the biological and climate sciences has contributed to questioning the partition between 'purely scientific' questions and modes of reasoning and what are readily called the 'social dimensions' of problems, as in the case of the climate sciences (Dahan-Dalmedico 2007; Urgelli 2008) or genetics. Conversely, a real temptation exists to consider all scientific controversy to be political or media-related and thus of no academic import.

In addition to this abundant research on media, scholarly work has also clarified other types of social communication on the sciences. We might mention, for example, research that concerns literary and theatrical forms (Jeanneret 2003; Raichvarg 2000), museums and exhibits (Davallon 1999; Le Marec 2007), forms of sociability and communication between scientists, or the social circulation of forms of knowledge issuing from the social sciences (Cordonnier 2007; Flon 2005; Tavernier 2004).

The continuous development of this field of research has relied both on maintaining critical independence from the managerial logics of communication and on having a concrete familiarity with the practical logics of institutional actors, professionals, creators/artists and activists. The strong critical impetus, which was necessary in order to identify the ideological framework around popularization or the media, has today given way to an attention to the complex dynamics of research on science and communication. This field is informed by its own questioning of the borders between academic and practical or lay knowledge.

The critical perspective on ideological discursive notions such as 'the information society', 'the knowledge economy' and 'sustainable development' relates closely with sustained attention to the debate between scholars, institutional actors of scientific and cultural mediation, and all actors interested in the links between science and society (amateurs, community organizations, associations). This movement connects with an experimental exploration of collaborations between scholars and actors on grounds other than those of applied research for industrial production.

One crucial result of developments in this field is a certain lack of differentiation between research involved in practice and practice inspired by research. Many actors aim for more than the completion of a program developed according to a functional managerial model. They often seek to integrate attention to complexity similar to that which inspires scholars and researchers. The professionals involved, who often have extensive academic training, are the first to recognize and integrate the complexity that they experience in practice and social scientists observe and formalize in research. Forms of reflexivity are being generalized not only in the academic domain, but in all intellectual professions.

This complicates matters somewhat, as people may be inclined to feel they are protected from the pitfalls of thinking that is too functionalist or that they are vaccinated against the power of ideologies and thus exempt from needing to develop a critical stance that is incorporated into practices. The widespread acknowledgement of the attention to be given to complexity differs according to the meaning attributed to research or practice.

We may say that a consensus has been reached on the need to take distance from the model of popularization as emblematic of the relations between science and society. But such agreement masks profound divisions on the meaning to be given to this critique of the linear model of popularization. Divisions are also apparent in the type of work that follows such critical acknowledgements.

Having critiqued functional conceptions of popularization, research in communication has steadily abandoned the critical commentary of the didactic, transitive model to produce new forms of knowledge through identification, through discussion with social actors about communication forms that were previously less visible: the quarrels and conflicts expressing themselves within different media that constitute together an 'intermedia space' (Jeanneret 1998; Jurdant 1998), the definition of roles such as those of the expert or the public (Chevalier 1999), the professional relations between the scientific, political and communication professions, the translation of knowledge in documentary practices (Couzinet 2000; Després-Lonnet 2000), the forms of communication linked to social, managerial, cognitive logics of daily work, and the modalities of implementing research as an activity.

There are a number of competing issues relating to the understanding of the different mediations of knowledge and science. One is a new theoretical framework incorporating both the necessary critique and the implication in the material production of knowledges and culture, and thus in the political dimension of the life of such knowledges (Jeanneret 2008).

Another quite different issue resides in the formalization of these new mediations and new modalities of association between scholars and managerial actors. There are few defenders of a model of popularization considered to be pure transmission of information, and managed as such. But many researchers in communication participate in engineering social debate in the name of a critique of the normative character of that model of popularization. Such engineering of debate also calls for the development of critical research.

There are also some scholars and researchers who participate in cultural or scientific activity, not in order to contribute to a market of expertise in the management of social sciences, but in order to understand and describe the plurality of objects and logics inherent in all social activity.

Finally, the collaboration between researchers and actors in scientific and technical culture is strengthened by a common desire to defend a democratic ideal of public service and sharing of knowledges, in opposition to the marketization of institutional spaces and of access to services and knowledge.

Science studies lead to a plurality of legitimate points of view on the sciences (Quet 2009). Similarly, communication sciences elicit a plurality of implications in forms of social communication that are at times quite contradictory.

The field of scientific or cultural mediation represents in a particular way this permanent tension between the critical impetus, the development of precise knowledge on the complexity of social communication, experimentation through involvement in practical programs, collaboration with managerial engineering of social communication, and a strong implication in cultural or pedagogical projects within a public service framework.

7.4 Mediation Between Theoretical Construction and the Professional Sphere

Research calls up the project of theorizing social communication as a continuous, collective creation of references, practices and objects that are activated during every interaction. This collection of references or shared culture constitutes the 'symbolizing third' of all communication (Quéré 1982). This theoretical conception of mediation is radically opposed to the linear model of communication as the establishment of a relation and the transport of information between two points.

However, the notion of mediation in everyday life very often designates an intermediary whose role is to remedy problems that affect the relation or transport of information between two points. A prosperous professional sphere has developed that proposes to mediate between the public and institutions, products and knowledges, according to a very linear conception of communication.

Yet existing practices and many artifacts in this professional field of mediation demonstrate that this linear conception is inoperative, and they generate theoretical conceptions of communication as a complex process of creation. Museology has been a space of convergence for research on popularization and on cultural mediation, and has also provided a place for confrontation between the critical impetus, experimentation and the development of a field of communication engineering (Davallon 2004; Gellereau 2006).

The notion of cultural mediation is framed by three lines of reflection: the democratization of culture and of scientific knowledge (that is, issues of legitimacy and power), the involvement of publics and citizens in the interpretation of arts and culture, and attention to technological developments.

In France, a desire to democratize art and culture as well as the recognition of the social construction of relations to culture and judgments of taste (Bourdieu et al. 1966) led to the development of this concept. The fundamental questions are the same as those that are asked about the dissemination of science. Experiences in the new Centres of Scientific and Technical Culture also inspired certain lines of inquiry, notably on issues related to the negotiation of meaning. The centers were constructed in opposition to 'formal education' to promote the idea of mediation as a way of legitimating a plurality of interpretations (Caillet and Lehalle 1995) and, later, a plurality of knowledges.

A link exists between those in social science research who criticize linear models of popularization or of equal access to cultural works, and the implementation of forms of cultural action that are stimulated or justified by this critique of dominant models and by the desire to counter the relations of legitimacy or authority implied in such linear models.

Researchers and scholars encourage the implementation of those 'alternative' practices, all of which rely upon a legitimation of what was supposed previously to have been ignored or underestimated. Cultural mediation since the 1980s covers a wide range of cultures, practices and knowledges that can be expressed in the public sphere.

The role of the mediator is seen as that of converting those different cultures, practices and knowledges into cultural products. The public to which the mediators turn becomes the reservoir of a number of possible interpretations and meanings.

An example of this process is the way we can consider exhibition scenarios or narratives not simply as a way of translating scientific discourse and bringing it within reach of the public, but as mediation that is circumscribed within a process involving both the producers and recipients of the narration (Gellereau 2005). This means that we must take into consideration not only the manner in which the storytelling brings to life scientific objects or characters, provides metaphors as to how to understand scientific texts, or reorganizes them, but also how narratives involve the public's experience, enact ways of interpreting the world, and serve as the basis of discursive thinking and collective action and as ways of structuring knowledge and the social representation of science.

The mediator proposes procedures and devices that allow for exchanges between different publics—amateurs and specialists. His or her purpose is also to involve visitors in a shared space, providing biographical substance to scientific experiments and contributing to building imagination (Raichvarg 2000). The activity of mediation is one of creating knowledge and disseminating it through representations (Jeanneret 2008).

There is a more or less implicit alliance between research and action in the cultural sphere. Research contests the transparency of discourses and emphasizes the multiple layers that intervene in the elaboration and reception of works and knowledges. Action in the cultural sphere creates procedures and devices that objectify those different mediations, which are implicit or unseen up to that point. The alliance between research and cultural action is political and cultural. It is political inasmuch as the pluralizing of discourse and the countering of relations of legitimacy are meant to serve a shared conception of democratic life. It is cultural inasmuch as the social sciences find in the field of mediation the possibility of empirically and culturally inscribing theoretical and critical notions. At the university level, professionally or non-professionally oriented bachelor's and master's level programs in 'cultural mediation' began to appear in the 1990s. Interdisciplinary programs were organized around the concept of mediation, rather than around a discipline, such as 'communication', for example, and were created as a response to new careers in 'mediation' appearing in institutions, cities and regions.

These developments were supported in the 2000s by the strong challenge that public policies posed to the legitimacy of academic knowledges in the name of the necessary plurality of the forms of production of knowledge. We must add to this the rapid explosion of the professional field of communication, which grew to encompass part of the sphere of cultural action and popularization. What previously belonged in minority and alternative spaces came to be strongly backed by institutional power.

Research communities in the communication sciences then made an ideological inversion, bringing a critical impetus to bear on objects and processes that communication research contributed to producing.

As well as studying active involvement in the production and inscription of discourses, the communication sciences also describe processes of withdrawal or displacement or, inversely, processes of top-down direction of inscriptive practices.

At times, the analysis of what it means to belong to the public in different institutional, political or media spaces implies that the public gives up the possibility of inscribing something, thus putting a relation of trust to the test (Le Marec 2007). The results of research on museum publics or on the publics of cultural mediation, related for example to the relations between doctors and their patients, show weakening sensibility in the social sciences to implicit pacts, in favor of the overvaluation of anything related to production and activity. The injunction to participate, to become active, is more and more present, damaging relations based on tact and on a certain wisdom of contact that is steadily disappearing.

The push towards using technologies, and particularly social networking technologies, in cultural mediation practices is often accompanied by a discourse on the possibilities such technologies provide for the development of mediation. But it involves new actors in this dynamic whose power is not acquired by possibly inscribing something from their own point of view, but rather by pushing others to act without themselves appearing in the discourse field.

7.5 Critique of the Ideological Models Underlying the Development of a Professional Field of Communications

The demand for critical vigilance can apply to at least three objects. First, there is the abovementioned engineering of procedures and devices for deliberation and the associated applied research. Second, there are the fundamental modifications that the development of information and communication technologies have introduced into the relations of knowledges with technologies.

A third object is particularly worthy of attention: this is the increasing formalization of environmental issues into a collection of procedures for 'sustainable development'.

This type of global, political and industrialized project is twofold. It makes concrete the heterogeneity of knowledges and poses the question of their respective legitimacy, but at the same time it also requires research to conform to standards of efficiency and to measure its communication performance.

We see the growth of research development and expertise with strong ties to political and industrial managerial teams. The field includes specialists in procedures of deliberation, in the use of information and communication technologies, in digital democracy, in sustainable development and so on.

The fact that they seem to relay the critique of a supposed dominant order fuels their development. But the positive image of project-based work masks their very normative character (Piponnier 2010). Indeed, the model of the project appears as a counterpoint to the exercise of authority: it promotes the involvement of many organizations and actors in a production process presented as multiple and dynamic, operating for change that is taken as representing progress, openness and development.

Sustainable development thus appears as a general, holistic project, capable of dissolving all contradictions between different or antagonistic logics (market, nature, society). The project cancels itself out as a model arising from communication engineering.

This situation in research has caused unexpected transformations. The result of previous research in science studies was to recognize the heterogeneity and pluralism of different kinds of knowledge in society.

That recognition led to the conception of instrumental devices and models to promote dialog and participatory communication. This is most explicit in the field of environmental issues, with its plurality of discourses, actors, disciplines and representations. These have now been translated into managerial models of 'sustainable development'.

This creates new conditions for research on the natural sciences and society, as well as on the uses of knowledge produced by the social sciences. There is now polarization among social scientists between two major trends. The first fosters and supports new conceptions of communication as if they still constituted a critical approach. In this case, the normative model is no longer transitive but participative (Schiele 2008). The strength of the norm, however, disappears because it adds legitimacy to what appeared to be alternative. The second trend tries to think of the critical dimension of research and the attention paid to the practices and cultural products of research itself as intimately linked. Our view is that this does not mean trying to reconcile opposites but rather trying to find a way to describe practices as they emerge, or trying to take into account the practical logics of actors without sacrificing the critical analysis of ideological frames. In our opinion, the participatory frame, even if it is appears to be a generous, alternative and critical position, must be analyzed as a potentially hegemonic norm.

Our evaluation of the situation in is no way consensual within the French scientific community of science studies and information and communication sciences.

To close, we would like to insist on some distinctive features of the French research tradition. First, there is a continuous commitment to providing precise descriptions of the relations that obtain between discourses, objects and situations constituting communication phenomena.

Second, there is the idea that it is impossible to analyze communication without questioning the social sciences and their broader involvement in the circulation and normalization of social knowledge.

This perspective might seem quite barren at a time when decades of critical theoretical work on models of communication in the sciences seem to have led to a general consensus: to recognize the plurality of forms of knowledge; to question the hierarchical relations structuring the relations between 'scholars' and 'laymen'; to pay attention to controversies in the continuing elaboration of socio-scientific questioning. Such a consensus seems to mark a general interest in more dialog, openness and democracy in social relations with science. It stimulates and legitimates the promotion of models, such as participatory debate and sustainable development, and communication procedures and devices that put such models into effect. The enthusiasm and critical justification that these models receive, due to the fact that they still seem 'alternative', can weaken critical vigilance, which is thought to be no longer necessary. Indeed, it is presumed that such vigilance has won and has been integrated once and for all into practice. However, critical vigilance remains necessary precisely in the name of the plurality of points of view and dynamics. It has to be practiced, however, in a context very different from that which structured the polarity between theoretical elaboration and applied research in the 1970s.

The communication and science research community produces texts, representations and forms of debate. And, as is the case for the practical knowledge of the actors engaged in mediation, it shows a strong sensibility to the complexity and reflexivity of the knowledges at stake. From this point of view, one of the major results of this research might be a displacement of the symbolic border stretching between academic knowledges associated with functional models of expertise, and practical knowledge. A new border might separate, in a more fundamental way, functional technical thinking as applied to the production of expertise and managerial models from the knowledges of complexity.

References

Babou, I. (2004). *Le cerveau vu par la télévision*. Paris: PUF.

Babou, I. (Ed.). (2008). *Images et sciences* (Communication & langages, Vol. 157). Paris: A. Colin.

Barthes, R. (1971). Réflexions sur un manuel. In *Œuvres complètes*, 3, 945–951. Paris: Éditions du Seuil.

Bensaude-Vincent, B., & Rasmussen, A. (Eds.). (1997). *La Science populaire dans la presse et l'édition (XIXème et XXème siècles)*. Paris: Éditions du CNRS.

Berthelot, J.-M. (Ed.). (2005). *Savoirs et savants: Les études de science en France*. Paris: PUF.

Bourdieu, P. (1984). *Homo academicus*. Paris: Éditions de Minuit.

Bourdieu, P., Darbel, A., et al. (1966). *L'amour de l'art*. Paris: Éditions de Minuit.

Caillet, E., & Lehalle, E. (1995). *A l'approche du musée, la médiation culturelle*. Lyon: PUL.

Chervin, J. (2000). *Les modes de présentation de la science dans les journaux télévisés français, de 1949 à nos jours*. Ph.D. dissertation, University of Paris 7.

Chevalier, Y. (1999). *L' expert à la télévision: Traditions électives et légitimité démocratique*. Paris: CNRS éditions.

Cordonnier, S. (2007). *L'exposition des savoirs dans l'art contemporain: Les formes de mobilisation des sciences humaines dans l'institution de l'art contemporain comme champ et comme formation discursive*. Ph.D. dissertation, ENS-LSH

Couzinet, V. (2000). *Médiations hybrides: Le documentaliste et le chercheur en sciences de l'information*. Paris: ADBS.

Dahan-Dalmedico, A. (Ed.). (2007). *Les modèles du futur. Changement climatique et scénarios économiques: Enjeux politiques et économiques*. Paris: La Découverte.

Davallon, J. (1999). *L'exposition à l'œuvre: Stratégies de communication et médiation symbolique*. Paris: L'Harmattan.

Davallon, J. (2004). La médiation: La communication en procès. *Médiation & information (MEI), 19*, 37–59.

de Cheveigné, S. (2000). *L'environnement dans les journaux télévisés: Médiateurs et visions du monde*. Paris: CNRS éditions.

Després-Lonnet, M. (2000). Contribution à la conception d'interfaces de consultation de bases de données iconographiques. Ph.D. dissertation, University of Lille 3.

Fayard, P. (1988). *La communication scientifique publique: De la vulgarisation à la médiatisation*. Lyon: Chronique sociale.

Fayolle, R. (2009). *Comment la littérature nous arrive*. Paris: Presses de la Sorbonne Nouvelle.

Feyerabend, P. (1979). *Contre la méthode: Esquisse d'une théorie anarchiste de la connaissance*. Paris: Éditions du Seuil.

Flon, É. (2005). La patrimonialisation de l'archéologie: La mise en scène des vestiges dans l'exposition. Ph.D. dissertation, University of Avignon.

Gellereau, M. (2005). *Les mises en scène de la visite guidée: Communication et médiation*. Paris: L'Harmattan.

Gellereau, M. (2006). Pratiques culturelles et médiation. In S. Olivesi (Ed.), *Sciences de l'information et de la communication: Objets, savoirs, discipline* (pp. 27–42). Grenoble: Presses Universitaires de Grenoble.

Habermas, J. (1991/1962). *The structural transformation of the public sphere: An inquiry into a category of bourgeois society*. Cambridge: MIT Press.

Hacking, I. (1983). *Representing and intervening, introductory topics in the philosophy of natural science*. Cambridge: Cambridge University Press.

Hert, P., & Paul-Cavalier, M. (Eds.). (2007). *Sciences et frontières: Délimitations du savoirs, objets et passages*. Bruxelles: EME.

Jacobi, D. (1986). *Diffusion et vulgarisation: Figures du texte scientifique*. Besançon: Annales littéraires de l'Université de Besançon.

Jacobi, D. (1999). *La communication scientifique: Discours, figures, modèles*. Grenoble: Presses universitaires de Grenoble.

Jeanneret, Y. (1994). *Écrire la science: Formes et enjeux de la vulgarisation*. Paris: PUF.
Jeanneret, Y. (1998). *L'affaire Sokal ou la querelle des impostures*. Paris: PUF.
Jeanneret, Y. (2003). L'objet technique en procès d'écriture: La scène impossible de la science appliquée. *Alliage, 50–51*, 21–34.
Jeanneret, Y. (2008). *Penser la trivialité: La vie triviale des êtres culturels*. Paris: Hermès-Lavoisier.
Jurdant, B. (Ed.). (1998). *Impostures scientifiques: Les malentendus de l'affaire Sokal*. Paris: La Découverte/Alliage.
Jurdant, B. (2009). *Les problèmes théoriques de la vulgarisation scientifique*. Paris: les Archives contemporaines. Ph.D. dissertation, Strasbourg, 1973.
Kapitz, C. (2007). *L'éthique, une nouvelle figure du lien entre science et société? Approche socio-discursive d'un phénomène de communication*. Ph.D. dissertation, ENS-LSH.
Lasswell, H. D. (1948). The structure and function of communication in society. In L. Bryson (Ed.), *The communication of ideas* (pp. 37–51). New York: Harper.
Latour, B., & Callon, M. (1991). *La science telle qu'elle se fait. Anthologie de la sociologie des sciences de langue anglaise*. Paris: La Découverte.
Le Marec, J. (2007). *Public et musées: La confiance éprouvée*. Paris: L'Harmattan.
Le Marec, J., & Babou, I. (2003). De l'étude des usages à une théorie des 'composites': Objets, relations et normes en bibliothèque. In E. Souchier, Y. Jeanneret, & J. Le Marec (Eds.), *Lire, écrire, récrire: Objets, signes et pratiques des médias informatisés* (pp. 233–299). Paris: Editions de la BPI.
Le Marec, J. & Babou, I. (eds) (2005). *Sciences, médias et société*, Lyon: ENS LSh/Laboratoire 'Communication, Culture et Société'—Proceedings of the conference, ENS LSH, 15–17 June 2004' (digital publication).
Lecourt, D. (1981). *L'Ordre et les jeux: Le positivisme logique en question*. Paris: Grasset.
Lévy-Leblond, J.-M. (1981). *L'Esprit de sel* (Science, politique, culture). Paris: Fayard.
Licoppe, C. (1996). *La formation de la pratique scientifique: Le discours de l'expérience en France et en Angleterre (1630–1820)*. Paris: La Découverte.
Mattelart, A. (1994). *L'Invention de la communication*. Paris: La Découverte.
Miège, B. (1989). *La société conquise par la communication*. Grenoble: Presses universitaires de Grenoble.
Moirand S. (2004). De la médiation à la médiatisation des faits scientifiques et techniques: Où en est l'analyse du discours? *Proceedings of the Sciences, Médias et Sociétés Conference* (pp. 71–99), École Normale Supérieure Lettres et Sciences Humaines.
Moles, A., & Oulif, J. (1967). Le troisième homme, vulgarisation scientifique et radio. *Diogène, 58*, 29–40.
Moscovici, S. (1961). *La psychanalyse, son image et son public*. Paris: PUF.
Pailliart, I. (Ed.). (2005). *La publicisation de la science: Exposer, communiquer, débattre, publier, vulgariser*. Grenoble: PUG.
Perriault, J. (1989). *La logique de l'usage: Essai sur les machines à communiquer*. Paris: Flammarion.
Pestre, D. (2006). *Introduction aux science studies*. Paris: La Découverte.
Pestre, D. (2010). Penser le régime des techno-sciences en société: Production, appropriation, régulations des savoirs et des produits techno-scientifiques aujourd'hui. In J. Le Marec (Ed.), *Les études de sciences: Pour une réfléxivité institutionnelle*. Paris: Éditions des archives contemporaines.
Piponnier, A. (2010). Publier la recherche en mode projet: Objets et pratiques entre autonomie et normalisation. In *Édition et publication scientifiques en sciences humaines et sociales: Formes et enjeux* (pp. 193–203), Proceedings of the Conference, March 2010. Avignon: Éditions de l'UAPV.
Quéré, L. (1982). *Des miroirs équivoques; aux origines de la communication moderne*. Paris: Aubier-Montaigne.
Quet, M. (2009). *Politiques du savoir: Une approche communicationnelle des rapports entre sciences, technologies et participation en France (1968–1983)*. Ph.D. dissertation, EHESS.

Raichvarg, D. (2000). *Science et spectacle*. Paris: Delagrave.
Schiele, B. (2001). *Le musée de sciences. Montée du modèle communicationnel et recomposition du champ muséal*. Paris: L'Harmattan.
Schiele, B. (2008). On and about the deficit model in an age of free flow. In D. Cheng et al. (Eds.), *Communicating science in social contexts* (pp. 92–117). Dortrecht: Springer.
Simonneaux, L., & Legardez, A. (2010). The epistemological and didactical challenges involved in teaching socially acute questions. *Journal of Social Science Education, 9*(4), 24–35.
Souchier, E., Jeanneret, Y., & Le Marec, J. (Eds.). (2003). *Lire, écrire, récrire: Objets, signes et pratiques des médias informatisés*. Paris: Èditions de la BPI.
Tavernier, A. (2004). *Paroles d'experts: Rhétoriques journalistiques de recours aux paroles extérieures. Journaliste et sociologue: La construction d'un référentiel*. Ph.D. dissertation, University of Lille 3.
Urgelli, B. (2008). Éducation aux risques climatiques. Premières analyses d'un dispositif pédagogique interdisciplinaire. *Aster*, 46, L'éducation à l'environnement ou au développement durable, July, pp. 97–121
Veron, E. (1986). *La sémiosis sociale: Fragments d'une théorie de la discursivité*. Saint-Denis: Presses universitaires de Vincennes.
Vinck, D. (2007). *Sciences et société: Sociologie du travail scientifique*. Paris: Colin.

Chapter 8
The Recent Public Understanding of Science Movement in Germany

Markus Lehmkuhl

Abstract This paper deals with the most recent history of five current or past initiatives to popularize science in Germany. The initiatives are roughly divided into those that primarily target science journalism and those that see science and the humanities as major agents for the improvement of public understanding of science. The initiatives are described and analyzed in terms of their potential societal impacts. It is preferable to focus on initiatives rather than individual activities because, at least to a certain extent, the design and choice of individual activities manifests the practitioner's views about the public understanding of science.

Keywords Public science in Germany • Science journalism • Science in Dialogue • Recent history of science popularization

8.1 The Meaning of the Public Understanding of Science in Germany

This paper deals with the most recent history of current or past initiatives to popularize science in Germany. The term 'popularization of science' is preferable to 'public understanding of science' because there is no direct German translation of the latter, even though German science, by naming a recent initiative 'Public Understanding of Science and Humanities' (PUSH), has followed the British practice. In Germany, 'popularization of science' stands in the tradition of the political notion of populism,

M. Lehmkuhl (✉)
Research Center Jülich, Freie Universität Berlin, Berlin, Germany
e-mail: m.lehmkuhl@fz-juelich.de

B. Schiele et al. (eds.), *Science Communication in the World: Practices, Theories and Trends*, DOI 10.1007/978-94-007-4279-6_8,

which in the nineteenth century meant 'condescending to the populace' or the attempt to 'curry favor with the common people' (Daum 2002:39). Today, in an analogy with the term 'public understanding of science' that is particularly used in Britain, the German term 'popularization' serves to group single activities or initiatives (groups of several activities) that share the same problem of perception as well as the same objectives.

Pivotal to all individual activities and initiatives is their critical perception of public understanding of science and technology as insufficient in the light of the great social significance of science and technology. In this context, the little word 'public' may, depending on the focus, refer to a rather undefined lay audience or to specific target groups, such as politicians, whose understanding of science is considered deficient. It may also refer to mass media, which are regarded as agents of information.

The word 'understanding' has a twofold meaning in Germany. It can denote understanding *of* as well as *for* science and technology. The latter meaning leads us to the most important objective, pursued through various activities. At the core, it is about securing or achieving social acceptance for science and technology, which are considered major agents of social change. The focus here is often on economic change; that is, the innovative powers of the economy. Following this logic, popularization is thought to be vital to create acceptance, which in turn is crucial to innovative power and economic growth. The popularization of science is thus thought to be at the service of public welfare.

In addition to securing the social acceptance of science, another motive is to educate through science, to explain the world to the common man or woman with the help of scientific knowledge. These efforts are ultimately guided by the conviction that scientific knowledge must be considered superior to other forms of knowledge. The objective is to combat pre- or unscientific views or traditions, or to establish 'scientific' world views ahead of competing world views. Another goal is to reveal the magic of the world to mass audiences or to target specific groups (which today are often teenagers and children) to show how that magic translates into scientific findings, or to create a fascination for science. In a historical perspective, that motive was found to be the ultimate force driving the popularization of science in nineteenth-century Germany (Daum 2002:14).

There is a striking continuity in the motivational patterns that are at the root of activities to popularize science. Individual analyses (Daum 2002, 2008; Kohring 2005) suggest that the motives for securing acceptance and education are, in a way, constant factors that have driven efforts to popularize science in Germany since the beginning of such efforts at the turn of the eighteenth century. The concrete design of initiatives and their general direction, on the other hand, vary over the course of history. I trace those variations below, concentrating on the most recent developments in the popularization of science in Germany, with a focus on larger scale initiatives that each bundle a sizeable number of individual activities.

It is preferable to focus on initiatives rather than individual activities because, at least to a certain extent, the design and choice of individual activities manifest the practitioner's mind-set about public understanding. Also, broader initiatives have always offered opportunities and impetus for reflection about the relationship between science, the public sphere and society at large. That is why initiatives are well suited to describing both change and continuity in the popularization of science in Germany.

8.2 How Can German Initiatives Be Differentiated?

A good way to differentiate individual initiatives is to ask which agents are considered vital to improving public understanding of science. While the first initiative analyzed below—the program for science journalists run by the Robert Bosch Foundation—started out in the 1980s primarily targeting journalists, that focus broadened as the number of initiatives increased over time. More recent ventures, such as the PUSH initiative by the Stifterverband für die deutsche Wissenschaft (Donors' Association for the Promotion of Sciences and Humanities in Germany) in the late 1990s, gear their activities primarily towards science itself. Others, like the 'Early Education' initiative of the Telekom Stiftung (Foundation), target certain segments of the German educational system, such as primary schools and, in particular, nursery schools.

It is safe to say that over the past 30 years, emanating from the promotion of science journalism, there has been a great diversification of initiatives pushing towards greater social acceptance of science, better education through science, or both. As a result, we see a multitude of activities in today's Germany to improve the public understanding of science, ranging from scholarship programs for journalists, to generously endowed communicators' awards for scientists, to fitting out nurseries with science kits. In the following sections, I describe initiatives targeting science journalism, science, and the public at large as agents to improve the public understanding of science. Table 8.1 lists five initiatives and their main characteristics.

8.3 Initiatives Targeting Science Journalism

The promotion of science journalism plays quite a significant role in the most recent history of science popularization in Germany. Between 1979 and December 2011, three large-scale initiatives endowed with 2–3 million euros each primarily targeted science journalism. Private foundations and BASF Inc. launched and implemented the initiatives. It is important to differentiate between the basic theoretical assumptions that motivated the establishment of the initiatives and the practical activities that were conceived and implemented in the context of the initiatives to generate specific outcomes.

All three initiatives share the basic assumption that acceptance of and education through science can be improved by way of more intensive and more competent journalistic coverage of science and technology. Problems with the acceptance of

Table 8.1 Initiatives targeting science popularization in Germany

Founder	Title	Focus	Duration
Robert Bosch Foundation	Aid Programme: Science Journalism	Science journalism	1979–1993
Bertelsmann Foundation Volkswagen Foundation BASF Inc.	Qualification Programme: Science Journalism	Science journalism	2002–2007
Robert Bosch Foundation Donors' Association for the Promotion of Sciences and Humanities in Germany BASF Inc.	Science Journalism Initiative	Science journalism	2008–2011
Donors' Association for the Promotion of Sciences and Humanities in Germany	Public Understanding of Science and Humanities (PUSH)	Science	1999–2003
Donors' Association for the Promotion of Sciences and Humanities in Germany Several science organizations (e.g. DFG, Helmholtz, Max Planck) Federal Ministry of Education and Research	Science in Dialogue	Science and the public	Ongoing since 1999

science, or rationality 'deficits', that were detected among the public were fundamentally interpreted as a crisis in the mediation of science—a crisis that journalism was able to remedy or at least alleviate. These initiatives therefore expected that journalism could be harnessed to bring about acceptance and education.

That attitude, at least in the 1980s, was intellectually fuelled by empirical data and theoretical reflections by social scientists as well as science journalists working for large national newspapers, who implicitly or explicitly conceived of journalism as a sort of 'partner' of science, as an 'interpreter' with the ability and the duty to mediate between science and the public. In his analysis of the scientific literature of those years, in particular the output of the social sciences, Matthias Kohring (2005) came to the conclusion that the central theoretical message boiled down to a 'normative purpose-bound programming' of journalism.

These theoretical notions about the relationship between science, journalism and the public are important in order to plausibly explain why the foundations directed their attention specifically to journalism. From today's perspective, it is not self-evident that journalism should be deemed central to achieving a primary objective of securing acceptance for and understanding of science. The theoretical background is also important in order to understand the donors' expectations in connection with their promotion of science journalism. However, that background only partly explains the practical design of the initiatives (that is, the choice of particular promotional tools).

When it came to choosing concrete promotional tools, the first initiative by the Robert Bosch Stiftung, in particular, showed itself to be directly influenced by a debate, arising in the early 1970s, on establishing journalism as a degree program at some West German universities. The newly created programs were (and still are) designed to render the profession of journalism more academic and, in addition, to inspire practical journalism with scientific, reflectional knowledge. The programs were thus to be seen as an academic service that basically addressed practical journalism, and which was meant to help improve journalism in general.

On the one hand, this 'service' consisted of training journalists, and on the other, of generating practically relevant social scientific knowledge about journalism, which was supposed to make its mark in practice via newly trained young journalists (Ruß-Mohl 1987; Brosda 2010). Accordingly, the basic idea of the degree programs was a close interconnection of practical journalism and social scientific penetration and reflection. It found its prototypical expression in the 10-semester journalism degree program of the Technische Universität Dortmund, which integrated practical journalist training with a 12-month internship with a publishing house or a broadcasting station. This concept continues to be fundamentally the same until today, even though the transition of the German university system to the Anglo-American system of bachelor's and master's degrees required some structural modifications.

The impact of this mind-set can be observed in the first initiative by the Robert Bosch Stiftung, masterminded mainly by Stephan Ruß-Mohl, who was later to become a professor of journalism. But it also influenced the third initiative, which is being implemented by Technische Universität Dortmund's Institute of Journalism. The most important tool of the first initiative, launched in 1979, was a scholarship program for young scientists. They were given the opportunity to complete internships with scientific editorial offices, which was supposed to qualify them to work as specialized science journalists within a year. Work at the editorial offices was complemented by workshops, which taught the scholarship recipients reflectional knowledge about journalism that was deemed necessary. By the end of the 1980s, approximately 200 scholarship recipients had been sponsored in the context of this program.

The initiative also promoted social–scientific reflection on the interplay of science, journalism and the public. It organized a total of four scientific conferences to compile the current state of research on the subject. The results were published in conference volumes. In this way, the initiative not only provided practical training in science journalism, but also stimulated social–scientific reflection on science journalism. These efforts were inspired by the hope that science journalists, in particular, would be more receptive of social–scientific findings about their own professional field than journalists with other areas of specialization (Ruß-Mohl 1985).

Practice-oriented training and social–scientific reflection were institutionalized when an endowed professorship at the Freie Universität Berlin was established at the close of the program in 1990. The promotion of science journalism now had a firm base for long-term support in Germany through training at the individual level as well as general social–scientific reflection. Winfried Göpfert served as professor for science journalism at the Freie Universität Berlin until 2006. He was succeeded by Alexander Görke in 2008.

The two other initiatives, which followed that of the Robert Bosch Stiftung, were also characterized by:

- Training or professional development opportunities at the individual level
- Activities that served to mediate or acquire scientific, practically relevant knowledge, which stemmed from social sciences, in particular, and which was fed back into practice through appropriate media
- Efforts to institutionalize and thus stabilize the promotion of science journalism in Germany by supporting chairs at universities and universities of applied sciences (The second initiative thus supported the establishment of two more departments for science journalism at the Technische Universität Darmstadt and the Technische Universität Dortmund.)

Despite marked differences in the choice of tools, these two initiatives also applied the double approach of concrete training and professional development efforts on the ground in combination with chiefly social–scientific analyses and reflections on the interplay of science, journalism and the public. These were then channeled back into practice by integrating them into journalist training and via book publications.

The first initiative by the Robert Bosch Stiftung generated the publication of the first books on the field, of which one in particular was broadly used in practice. It is a manual, revised and updated five times since 1987 (Göpfert 2006), containing contributions by social scientists as well as by practicing journalists. It caters for newcomers to the profession of science journalism and combines concrete advice with various theoretical reflections that have emerged from recent academic work on science journalism (for example, Lehmkuhl 2006; Peters and Jung 2006; and Meier 2006).

The ‘Qualification Program: Science Journalism’, significantly promoted by the Bertelsmann Stiftung from 2002 to 2007, produced a manual with a similar thrust. Compared to the first initiative’s manual, however, it is a much broader and more systematic treatise on science journalism (Hettwer et al. 2008).

In 2004, as a new platform for exchanges between science journalism, public relations and (social) scientific research, this initiative established the ‘Wissenswerte’ conference, which is now an annual event in Bremen. That the conference has been fully booked, with over 400 participants, each year since its inception indicates science journalists’ interest, particularly in reflecting on their own work and receiving new stimuli.

The third initiative, which started in 2008 and ended in December 2011 and was another undertaking of the Robert Bosch Foundation, among other sponsors, was a seamless continuation of this approach. Like the first initiative, it was more focused on the training of science journalists than was the case with the second initiative. A new edition of the scholarship program channels about 10 young natural scientists a year towards a career in science journalism.

However, the third initiative also placed some new emphases, such as a sporadic departure from the focus on specialized science departments, which had until then been considered to be the major agents for the enhancement of science journalism. In the debate on possible means to promote science journalism, a new awareness had already begun to emerge during the second initiative: that efforts ought not to be restricted to rather small but well-established departments at press and broadcasting institutions. Instead, the acquisition of relevant journalistic skills and the integration of scientific knowledge into reporting were increasingly considered a necessity that affected the entire editorial staff, particularly the politics and economics sections. This way of thinking led to experiments with wholly new training tools that no longer primarily targeted the individual journalist, but entire editorial teams.

8.3.1 The Impact of the Science Journalism Initiatives

All three of these initiatives, while still running, distanced themselves from the theoretical notions that originally motivated their donors to establish them. The idea that the relationship between science and the public was a crisis of mediation, caused partly by poor journalistic coverage and remediable by manipulating journalism, proved to be inadequate, as was the attempt to harness journalism as a partner of science.

It turned out as early as the 1980s that those notions only partly reflected the professional self-image of science journalists, who did not think of themselves as partners of science as much as independent observers and critics. Already at the third conference on science journalism, held in 1988, the idea that science journalists are mere interpreters (a notion voiced particularly by German communication scientists) was considered antiquated and removed from practical reality (Waldner-Stiefelmeier 1990). It quickly became clear that the theoretical notions, which had, after all, been the primary objective of the initiatives, could only be partly aligned with practical reality.

An increasing number of social scientific analyses and reflections produced insights that also invalidated certain ideas about media coverage, including, for example, the assumption that media coverage of technological risks would have an 'alarming' effect and thus damage the social acceptance of technology. This notion, fed in particular by a large-scale study by Kepplinger (1989), was contradicted by Michael Schanne at the fourth and last Robert Bosch Stiftung conference on science journalism in 1992, which was dedicated to the topic of risk communication. The study featured a meta-analysis of 52 empirical works, which showed that media coverage of risks usually had a 'reassuring' effect. 'Whenever news is more alarming than reassuring it is related to alarming official announcements and press releases' (Schanne 1998:60).

Thus, the first initiative enhanced awareness that there was a dearth of adequate empirical and, above all, theoretical descriptions of the relationship between science, journalism and the public—a deficiency that these initiatives were little able to remedy due to their focus on journalist training. Only the 600-page manual by Hettwer et al. (2008), which was made possible by the second initiative, testifies to the editors' ambition to describe the theory of science journalism more adequately. Right at the beginning, in its introduction, it rejects as theoretically and empirically inadequate any notions that posit journalism as a mere service provider for science (Lehmkuhl et al. 2008). The manual, oriented towards practical application, thus starts out by emphasizing the fundamental distinction between science and science journalism, which contradicts any subjugation of journalism to science. In doing so, the manual retraces a paradigm shift that had been brought about in particular by Matthias Kohring (1997, revised in 2005).

This work influenced reflections on science journalism in Germany, especially among academics. It was no coincidence that German social scientists, in particular, abandoned science-centered descriptions and analyses of science journalism and turned to media-centered analyses instead (for example, Lehmkuhl et al. 2010; Lublinski 2004). However, they were rather sporadic research activities with no systematic references to each other. The fact that there are now three departments for science journalism and one for technical journalism has not changed that fact. One reason for this may be that all of those positions were filled with former practicing journalists who focus their work on training journalists, not on an analytical penetration of the subject area.

8.3.2 Conclusion

In the overall evaluation of all three of the initiatives discussed in this paper, it seems that their central assumption was that science journalism was the agent of the public understanding of science. Their efforts not only helped to establish programs for training science journalists, but also spilled over into and stimulated an analytical penetration of the subject area. This helped to raise awareness that there must be a distinction between science and the journalism that covers it, even though that certainly was not one of the central objectives of the donors, in particular. On the one hand, the social sciences provided significant impetus for an in-depth analytical penetration of the subject, and the initiatives then helped to validate individual findings and theoretical concepts in practical journalism. On the other hand, the systematic treatment of the practical problems faced by science journalism also proved stimulating.

To condense it all into one formula, the initiatives certainly served to promote science journalism far beyond securing acceptance and a scientific rationalization of social issues. However, they also raised awareness that appropriating Western journalism for science's purposes could not and should not be the way to achieve either of those objectives.

8.4 Initiatives Primarily Targeting Science and the Humanities

Germany did not give rise to a 'public understanding of science movement' comparable to that in the United Kingdom until 1999. This does not mean that prior to that time Germany was not also aware of the issues pertaining to the British movement, as shown by the 1979 Robert Bosch initiative described above. In its motivation and objectives, it was essentially comparable to those in the United Kingdom. The relatively late arrival of German science must be seen as a result of its fragmentation—the fact that it had no institutional body that might organize German science from the top down (Hornbostel and Olbrecht 2008). It took an initiative by the pro-business Donors' Association for the Promotion of Sciences and Humanities in Germany to prompt German science organizations to sign a memorandum with the title 'Dialogue Science and Society', which declared that the popularization of science was a task incumbent on science itself. Modeled on 'public understanding of science' in the United Kingdom, it is described as a 'movement' driven by science, business and politics (Stifterverband 2000:59).

In large part, the problem of perception expressed in the memorandum reflects the way of thinking briefly outlined in the introduction. The rationale, for example, is stated as the public's unwillingness to appreciate natural sciences and technology as a cultural achievement, despite their eminent importance. Its diagnosis is that this is a communication problem, ascribable to the high degree of specialization in science, but also related to the fact that science is no longer only problem-solving, but also problem-causing. It thus also contains clear references to more recent developments; that is, the diagnoses by Ulrich Beck, which gained international currency under the term 'reflexive modernization' (Beck 1986, 1992; Beck et al. 1994).

These publicity-geared efforts by the science organizations had two explicit goals. First, they were to secure acceptance for science, which ought to be in the interests of both business and politics for the sake of Germany's innovative power. The second objective was education, deemed necessary to enable democratic participation in debates of scientific relevance on the one hand, and to kindle enthusiasm for science on the other. In order to achieve those goals, the aspiration was to 'establish a permanent dialogue between science and society' (Stifterverband 2000:59).

The memorandum addressed mainly scientists. They were asked to present their work in a manner that a layperson could understand. To strengthen their commitment, a suitable incentive system that would make dialog with the public an additional hallmark of scientific excellence was sought. Scientists were supposed to learn these skills through appropriate forms of training. The tone of the memorandum was more binding with regard to the public relations work of each individual science organization. Public relations departments were supposed to synchronize and coordinate their efforts in the future (Stifterverband 2000:60).

Two interrelated initiatives directly followed the memorandum: The Donors' Association for the Promotion of Sciences and Humanities in Germany launched an action program entitled 'Public Understanding of Science and Humanities' (PUSH). It was a competition for ideas, calling upon scientists to submit their suggestions for popularizing scientific knowledge in the spirit of the memorandum. Up to 2003, 67 projects were selected from among 514 applications and won 1.2 million euros in grants. Due to the relative vagueness of the call for submissions, the grant-winning projects are difficult to group. They form a conglomeration of extremely diverse activities (exhibitions, field trips, lecture series etc.), for the most part related to natural sciences, generally designed to spark a fascination for individual disciplines or research areas, and making direct contact with their target group, with a primary focus on children and teenagers (Conein 2004). Some activities picked up on the dialog idea mentioned in the memorandum. The action program featured a series of citizens' conferences that were supposed to initiate a dialog between scientists and the lay population.

Also immediately following the publication of the memorandum, Science in Dialogue, a non-profit organization, was founded; it is still in existence today. Science in Dialogue can be seen as an institution in which the memorandum goal of coordinated PR measures takes a tangible form. It is financed in large part by the Federal Ministry of Education and Research and directed by a formal steering group that includes the highest representatives of the German science organizations, the pro-business Donors' Association, and the Federal Ministry of Education and Research. In practice, the association is managed by the heads of the respective PR departments.

In the past 10 years, this body has acted as the organizer of nationwide activities, which are mainly attempts to ignite a fascination for science. The so-called 'Years of Science' served as a common denominator binding the diverse activities. They initially focused on individual scientific disciplines or scientific fields (physics, chemistry, life sciences and computer science). Large events, such as the 'Summer of Science' or a floating science center, as well as many smaller activities, were assembled under the respective slogan of each Year of Science. From 2009, the orientation towards scientific disciplines gave way to more topic-based Years of Science, such as energy in 2010 and health research in 2011.

The shift was a departure from a science-centered thematic approach to the Years of Science towards a more problem-centered one, showing an orientation towards the presumed interests of the public. The problem orientation then created opportunities to experiment with more dialog-based activities, such as 'consensus conferences', at which a small number of citizens formulated recommendations to decision-makers on future energy solutions.

The annual Forum on Science Communication, established in 2008, primarily aims at reflecting on these publicity-related efforts. The conference is a platform created by Science in Dialogue to share experiences among professional science communicators. In addition, the conference might have the potential to become a medium for integrating reflectional knowledge—usually generated by the social sciences—about the popularization of science. However, the list of lectures and discussions shows that social scientists make little use of this forum to present relevant results.

It would be a misconception to ascribe science's activities to popularize itself exclusively to Science in Dialogue, which does more than conceive and organize events, exhibitions and conferences. Its significance also lies in its bundling of all kinds of relevant activities by individual science organizations. As stipulated in the memorandum, the science organizations have created Science in Dialogue as a platform that provides transparency about everyone's activities. To that extent, the initiative is not the instigator of intensive measures to popularize science. Rather, it is an attempt to inspire and steer the many existing activities to popularize science in order to increase their potential efficacy.

8.4.1 The Impact of Science in Dialogue and PUSH

This leads us to one of the general trends that characterize the more recent history of science popularization in Germany. 'Science' today, unlike even just 15 years ago, presents itself to the public as the organizer of large-scale events and 'science experiences' as a sort of unit, which is mainly due to Science in Dialogue and was one of the major political objectives that motivated that institution's founding and promotion.

This initiative must therefore be seen in the light of other recent political efforts to prompt science organizations to join together in collaborative research centers in order to give 'science' as a socially relevant force more public and political prestige. One such effort was the failed political endeavor to amalgamate several of the academies of science into one 'National Academy of Science', which was then supposed to gain influence as the 'voice of science' on socially relevant issues, such as regulating pre-implantation diagnostics in reproductive medicine. Currently, the concrete reality of science popularization is that influencing socially relevant discourse is almost completely beyond the practical reach of what we understand by 'science popularization'.

This is a result of the fact that the role of journalism has become a great deal less significant in organized science popularization over the past 10 years. That development was both made clear and catalyzed by the PUSH program and the Science in Dialogue initiative. Without the mass media, wielding any influence in social debates with the aim of instilling scientific rationality is inconceivable. Among the many activities bundled in Science in Dialogue, there is not one with a significant focus on mass media. The situation is similar for the PUSH program. More recent trends in public understanding of science in Germany characteristically strive to address the lay population directly. This results in a remarkable diversification of science popularization, yielding a large number of different formats that mainly aim to tickle one's fascination, offer fun experiences and provide information.

Behind this is an attitude that defines the public more as an assembly of people who receive advertising messages with diffuse intended effects, rather than as a social system in the sociological sense that fulfills certain social functions and that (at least, this was the assumption in the past) to a large extent can only be created by mass media. For lack of pertinent analyses, it is hard to gauge whether science ever even considered this concept of 'the public' in the strategic orientation of science popularization. However, one can say with certainty that, despite its name, Science in Dialogue has failed to help make the memorandum goal of scientifically rationalizing social debates any more achievable. What started out as a genuinely serious venture in science popularization has degenerated into a sort of image campaign that does more to mask the true nature of science than to reveal it.

However, it would be too narrow to blame this failure exclusively on the trend towards direct communication of messages to a targeted audience. Science, or more precisely, methodically acquired knowledge, can be seen as a source capable of helping to rationalize social debates through public discourse, as conceived by Habermas, and as envisioned in the memorandum. There will, however, be major doubts about whether this vision can be approached through clustered PR efforts on the part of science *organizations*, which ultimately are bound only by their own organizational logic.

One must also doubt whether the vision can be approached by mobilizing scientists, especially as science has not even come close to the memorandum's objective of making dialog with the public an integral part of scientific credentialing. Partly thanks to the memorandum, it can be assumed that scientists today recognize the significance of science popularization more than they might have 20 years ago, but it is not part of their scientific credentials. And, incidentally, the incentives, such as the generous 50,000 € Communicators' Award, are designed to reward not scientific participation in social discourse, but easy-to-understand presentations of exciting science.

As this description of the most recent trends in science has shown, social scientific findings and reflections on the interplay of science and the public have a rudimentary influence at best and are limited to the evaluation of concrete projects. Theoretically and analytically weighty works, such as those by Peter Weingart (2001) and Friedhelm Neidhardt (1993), not to mention critical analyses on public understanding of science in the United Kingdom (Wynne 1992, 1995), have not yet had any noticeable impact either on activities or—more importantly—on reflections on those activities. One can merely note a strengthening of dialog elements, as envisioned in the concept of 'scientific citizenship' (Felt 2003), even though in most cases the word 'dialog' is used where the term 'self-serving manipulation' would be appropriate.

The reason for this is not primarily ignorance on the part of individual actors, but can be traced back to the way that science communication was professionalized. The past three decades have seen the partial formation of the profession of 'science communicator', even though important characteristics necessary to typify a profession in the sociological sense are certainly not present (Haller 2008). For example, there is no clearly defined set of training objectives, nor any distinct path to reach them. There are, however, associations and networks that resemble professional associations in their characteristics. A consistent job description is also beginning to emerge, resulting from the relatively homogeneous demands that science organizations make on their PR staff. This homogeneity is due to the very similar tools that universities and research institutions use for science popularization. It can be assumed that the ability to use all the currently developed tools in practice will be the point of departure for formulating a consistent job description (Meier and Feldmeier 2005).

A social-scientifically sound reflection on professionalized science popularization is definitely not part of the core job specifications. This is mainly due to the lack of pertinent training opportunities compared to, say, German degree programs in journalism, which logically combine social–scientific reflection and practical training.

8.4.2 Conclusion

Looking at these current efforts on the part of German science to foster the 'public understanding of science', one may have doubts about their effectiveness. This is generally due to the concepts of science and the public, which have been aptly described under the term 'deficit model' (Gregory and Miller 1998). It is, however, also due to German science's focus on creating responses through publicity effects, revealing a structural inability to integrate any unintended side effects of those activities into its own concept. Individual analyses by observers nourish the suspicion that the manner in which German science currently drives forward its own popularization is creating the very monster it set out to combat: problems with acceptance and the legitimacy of its science communication (Lehmkuhl 2009a, b; Schnabel 2009).

References

Beck, U. (1986). *Risikogesellschaft. Auf dem Weg in eine andere Moderne*. Suhrkamp Frankfurt am Main.

Beck, U. (1992). *Risk society: Towards a new modernity*. London: Sage.

Beck, U., Giddens, A., & Lash, S. (1994). *Reflexive modernization: Politics, tradition and aesthetics in the modern social order*. Stanford Polity Press.

Brosda, C. (2010). *Diskursiver Journalismus: Journalistisches Handeln zwischen kommunikativer Vernunft und mediensystemischem Zwang*. Wiesbaden VS Verlag.

Conein, S. (2004). Public understanding of science: Entwicklung und aktuelle Tendenzen. In S. Conein, J. Schrader, & M. Stadler (Eds.), *Erwachsenenbildung und die Popularisierung von Wissenschaft* (pp. 20–31). Bielefeld Bertelsmann.

Daum, A. W. (2002). *Wissenschaftspopularisierung im 19 Jahrhundert. Bürgerliche Kultur, naturwissenschaftliche Bildung und die deutsche Öffentlichkeit, 1848–1914*. München Oldenbourg Verlag.

Daum, A. W. (2008). Geschichte des Wissenschaftsjournalismus. In: H. Hettwer, M. Lehmkuhl, H. Wormer, & F. Zotta (Eds.), *WissensWelten: Wissenschaftsjournalismus in Theorie und Praxis* (pp. 155–175). Gütersloh Bertelsmann.

Felt, U. (2003). Scientific citizenship: Schlaglichter einer Diskussion. *Gegenworte. Hefte für den Disput über Wissen, 11*, 16–20.

Göpfert, W. (Ed.). (2006). *Wissenschaftsjournalismus: Ein Handbuch für Ausbildung und Praxis*. Berlin Econ.

Gregory, J., & Miller, S. (1998). *Science in public: Communication, culture and credibility*. Cambridge Plenum Press.

Haller, M. (2008). Wie viel Wissenschaft braucht der Wissenschaftsjournalismus. In H. Hettwer, M. Lehmkuhl, H. Wormer, & F. Zotta (Eds.), *WissensWelten: Wissenschaftsjournalismus in Theorie und Praxis* (pp. 531–549). Gütersloh Bertelsmann.

Hettwer, H., Lehmkuhl, M., Wormer, H., & Zotta, F. (Eds.). (2008). *WissensWelten: Wissenschaftsjournalismus in Theorie und Praxis*. Gütersloh Bertelsmann.

Hornbostel, S., & Olbrecht, M. (2008). Wer forscht hier eigentlich? Die Organisation der Wissenschaft in Deutschland. In H. Hettwer, M. Lehmkuhl, H. Wormer, & F. Zotta (Eds.), *WissensWelten: Wissenschaftsjournalismus in Theorie und Praxis* (pp. 45–63). Gütersloh Bertelsmann.

Kepplinger, M. (1989). *Künstliche Horizonte: Folgen, Darstellung und Akzeptanz von Technik in der Bundesrepublik Deutschland.* Frankfurt, New York Campus.

Kohring, M. (1997). *Die Funktion des Wissenschaftsjournalismus.* Wiesbaden VS Verlag.

Kohring, M. (2005). *Wissenschaftsjournalismus: Forschungsüberblick und Theorieentwurf.* Konstanz uvk.

Lehmkuhl, M. (2006). Defizite im Wissenschaftsjournalismus. In W. Göpfert (Ed.), *Wissenschaftsjournalismus: Ein Handbuch für Ausbildung und Praxis* (pp. 14–24). Berlin Econ.

Lehmkuhl, M. (2009a). Die Öffentlichkeitsarbeit der Wissenschaft und ihre Rationalität. *Geographische Revue, 11*(2), 22–26.

Lehmkuhl, M. (2009b). Zehn Jahre PUSH und die Irrationalitäten öffentlicher Wissenschaft. *wpk-Quarterly. Quartalszeitschrift der Wissenschaftspressekonferenz, 7*(3), 25–28.

Lehmkuhl, M., et al. (2008). Was ist Wissenschaftsjournalismus? Eine Einleitung. In H. Hettwer, M. Lehmkuhl, H. Wormer, & F. Zotta (Eds.), *WissensWelten: Wissenschaftsjournalismus in Theorie und Praxis* (pp. 13–24). Gütersloh Bertelsmann.

Lehmkuhl, M., Karamanidou, C., Mörä, T., Trench, B., Petkova, K., & AVSA-Team. (2010). *Science in audio visual media: Production and perception in Europe, Final report for the European Commission.* Retrieved from http://www.polsoz.fu-berlin.de/en/kommwiss/v/avsa/Downloads/finalreport_avsa_2010.pdf

Lublinski, J. (2004). *Wissenschaftsjournalismus im Hörfunk.* Konstanz uvk.

Meier, K. (2006). Medien und Märkte des Wissenschaftsjournalismus. In W. Göpfert (Ed.), *Wissenschaftsjournalismus: Ein Handbuch für Ausbildung und Praxis* (pp. 37–53). Berlin Econ.

Meier, K., & Feldmeier, F. (2005). Wissenschaftsjournalismus und Wissenschafts-PR im Wandel: Eine Studie zu Berufsfeldern, Marktentwicklung und Ausbildung. *Publizistik, 50*, 201–224.

Neidhardt, F. (1993). The public as a communication system. *Public Understanding of Science, 2*, 339–350.

Peters, H. P., & Jung, A. (2006). Wissenschaftler und Journalisten: Ein Beispiel unwahrscheinlicher Co-Orientierung. In W. Göpfert (Ed.), *Wissenschaftsjournalismus: Ein Handbuch für Ausbildung und Praxis* (pp. 25–36). Berlin Econ.

Ruß-Mohl, S. (1985). Journalistik-Wissenschaft und Wissenschafts-Journalistik: Anmerkungen zur Theorie und Praxis der Wissenschaftsvermittlung. *Publizistik, 30*(2–3), 265–279.

Ruß-Mohl, S. (1987). Hochschulgebundene Journalistenausbildung: Von der Problemverstaatlichung zur Problemlösung? *Publizistik, 32*(1), 5–22.

Schanne, M. (1998). Media coverage of risk: A meta analysis of 52 content analyses. In: W. Göpfert, & Bader, R. (Eds.), *Risikoberichterstattung und Wissenschaftsjournalismus* (pp. 53–69). Stuttgart, Robert-Bosch.

Schnabel, U. (2009). Der Schatten der Popularisierung. *wpk-Quarterly. Quartalszeitschrift der Wissenschaftspressekonferenz, 7*(3), 28–29.

Stifterverand für die deutsche Wissenschaft. (ed.). (2000). *Dialog Wissenschaft und Gesellschaft.* Essen. Retrieved from http://www.stifterverband.org/publikationen_und_podcasts/positionen_dokumentationen/push_2000.pdf

Waldner-Stiefelmeier, R. (1990). Über die Praxisdefizite der Kommunikationsforscher. In S. Ruß-Mohl (Ed.), *Wissenschaftsjournalismus und Öffentlichkeitsarbeit. Tagungsbericht zum 3.* Gerlinge Colloquium, Stuttgart, Robert-Bosch.

Weingart, P. (2001). *Die Stunde der Wahrheit? Zum Verhältnis der Wissenschaft zu Politik, Wirtschaft und Medien in der Wissensgesellschaft.* Weilerswist Velbrück Wissenschaft.

Wynne, B. (1992). Public understanding of science research: New horizons or hall of mirrors? *Public Understanding of Science, 1*, 37–43.

Wynne, B. (1995). Public understanding of science. In S. Jasanoff, G. E. Markle, J. C. Petersen, et al. (Eds.), *Handbook of science and technology studies* (pp. 361–388). London, New Delhi.

Chapter 9
Public Understanding of Science: Glimpses of the Past and Roads Ahead

Gauhar Raza, Surjit Singh, and P.V.S. Kumar

Abstract This chapter gives a historical account of turning points that mark the ever-changing relationship between modern science and the public at large. Scholars recognize the importance of that, but assert that there is a growing gap between the scientist's way of configuring nature and the people's world view. This led to an intense debate about the science–technology–society relationship. The chapter then examines the development of science communication in India and similar countries, the science–technology–society relationship followed a trajectory that was not rooted in the Enlightenment and the Industrial Revolution. Modern science was alien and had to be learned and mastered by natives. The realization that science and technology are essential to improve conditions for Indians marked the first phase of science popularization. In the second phase, propagating science among the general public and building scientific institutions were seen as essential parts of the national struggle for freedom. In the third phase, science and technology and their acceptance among the masses were considered necessary for building a modern and self-reliant nation. The chapter then gives an account of the past 30 years of research experience in the Indian context and discusses the cultural distance model for analyzing public understanding of science. It also discusses the efficacy and limitations of empirical methods of measuring cultural distance.

G. Raza (✉)
Head, SCM Division, National Institute of Science Communication and Information Resources, Dr KS Krishnan Marg, New Delhi, India
e-mail: gauhar_raza@yahoo.com

S. Singh
Scientist, SCM Division, National Institute of Science Communication and Information Resources, Dr KS Krishnan Marg, New Delhi, India
e-mail: ssdabas@yahoo.com

P.V.S. Kumar
Scientist, SCM Division, National Institute of Science Communication and Information Resources, New Delhi, India
e-mail: kumarpatnam@gmail.com

B. Schiele et al. (eds.), *Science Communication in the World: Practices, Theories and Trends*, DOI 10.1007/978-94-007-4279-6_9,

Keywords Public understanding • Science communication • Literacy • Cultural distance • Movement • NGOs • Freedom movement • Scientific temper

9.1 Introduction

Jacob Bronowski—a scientist and a mathematician with a deep interest in literature—pointed out more than 50 years ago that 'We are aware now that somewhere within the jungle of valves and formulae and shining glassware lies a content; lies, let us admit it, a new culture. How are we to reach that culture, across its jargons, and translate it into a language that we know?'(Bronowski 1953:2). Contrary to the apparently popular stereotypical image of science and scientists prevalent at that time, Bronowski placed 'newness' at the center of the content and language of the 'new culture of science'. He was not alone in recognizing this newly emerging culture: Bernal (1939), Snow ([1959] 1993) and many other scholars were grappling with the relationship between science and larger society (Levis 1962; Swann and Aprahamian 1999). Each of them helped to shape the newly emerging area of 'science communication' (Munns and Rajan 1995).

Historians tell us that the eighteenth century saw the emergence of newspapers (Vilanilam and Varghese 2004:1, 3). The trajectory of development had many twists and turns in both imperialist and colonized countries (Kundra 2005:2). It is interesting to note that the history of newspapers is enmeshed with the history of advertising in the Western countries. Initially, the ruling classes reacted sharply to the idea of paid advertisements[1]; however, despite resistance, advertising continued to grow (Cook 2001:5). By the first quarter of twentieth century, advertising had emerged as a new industry. Interestingly, almost at the same time significant developments were taking place in mathematics. Statistics had emerged as a separate branch of investigation. Scientists and social scientists dealing with large datasets benefited from the cumulative work of A. Quetelet, Sir Francis Galton, K. Pearson, W.S. Gosset and W.E. Deming.[2] Subsequent work by J. Neyman provided a 'logical foundation and mathematical rigor to statistical methodology encouraging scholars to apply statistical and survey methodologies in analyzing social phenomena'.[3] These attitudinal studies prepared the ground for future survey studies to probe the level of scientific literacy among people all over the world.

The stages through which science communication has grown over the past two centuries have also affected research on science communication (Bensaude-Vincent

[1] *Early eighteenth-century newspaper reports: A sourcebook compiled by Rictor Norton* (http://rictornorton.co.uk/grubstreet/learning.htm) shows that in first two decades of that century a number of advertisements in British newspapers covered science subjects such as innovations, education, popular lectures, and geographical and natural phenomena.

[2] http://www5555.morris.umn.edu/~sungurea/introstat/history/indexhistory.shtml

[3] http://www.morris.umn.edu/~sungurea/introstat/history/w98/Jerzy_Neyman.html

2001:99). Initially, factors such as the spread of communication channels, the appearance of scientific information as part of advertisements and the emergence of survey methodologies indirectly affected science communication activities (Crump 2001). However, World War II, the development of the atomic bomb and an environment of increasing technological competitiveness were stronger and more direct influences. Whereas scientists had previously played a passive role in science communication, after the war those researchers seeking funds from military, government or industry sources had to justify the expenditure and in the process had to communicate science to non-experts. Governments had to communicate science to seek public support for increased funding for science education and research (Dove 2002). Public debates became increasingly scientific and technical in their content. First the 'peace' movement and later environmental organizations had to communicate scientific information, data and principles to the public to garner support and enlarge their mass base.

In short, the objectives of government, industry, business and those critical of science and technology converged, although for different and sometimes even divergent reasons, to take science to the public. Science communication had emerged as a distinct and well-recognized activity.

9.2 Scientific Literacy Versus Public Understanding of Science Debate

The term 'scientific literacy', by analogy with language literacy and numeracy, is most commonly used in surveys on the public understanding of science (NSF 1993, 1996, 1998a, b, 2000). However, scientific literacy presupposes that the human mind has no comprehension of natural phenomena outside the information obtainable from formal structures that transmit scientific information. The science and engineering indicators used by the United States National Science Foundation and the work of Miller et al. (1996) use that approach, which is categorized as the *deficit model* of science communication and which was the subject of much criticism.

Scholars' suspicions about the purpose and intent of the deficit model eventually led to the 'end' of the model, at least in the United Kingdom and among many academics researching the public understanding of science (SCST 2000). However, Steve Miller (2001:118) pointed out in a critical article that 'the demise of the deficit model does not mean there is no "knowledge deficit" among laypeople.'

Debates on the deficit model gave rise to yet another approach: the *contextual model*. It was argued that the generation of scientific knowledge and its appropriation by the public were essentially a form of social organization of culture. The contextual model expresses the science–technology–society relationship in terms of two concentric circles (Godin and Gingras 2000:53). Further work on the model recognized that the large universal set of 'culture' in any given society consists of smaller subsets, each one protected by a boundary that demarcates it from other subcultures.

The boundary membrane is permeable, but not entirely so: ideas generated in one cultural setting face resistance in diffusing into diverse cultural subgroups (Raza et al. 2002). Scientific knowledge generation and its appropriation should therefore be viewed as a multidimensional socialization processes in any subculture (Lee et al. 1995). The diffusion of new scientific discoveries across cultural boundaries should enrich the social consciousness of citizens, so the model of public understanding should be sensitive to the multiplicity of those boundaries.

The above discussion puts into a global perspective the path that science communication and research on science communication has followed in India. Since the epicenter of modern science and therefore science communication was located in the European countries during the nineteenth and twentieth centuries, ideas percolating to the colonized cultures, which were still subjugated, took time to assimilate. However, the trajectories of science communication in the West and the developing world have distinct points of intersection and departure. In colonized India, the ideas introduced from the West evoked three diverse reactions among both the newly emerging Indian scientific community and the political leadership: some totally rejected the ideas; some argued that all the 'new' knowledge was rooted in Indian culture and that the colonizers had simply developed it further; and some saw an opportunity to acquire and develop the knowledge and use it to sharpen the political struggle.

9.3 Science Communication in India

The first efforts to communicate modern Western scientific ideas to the Indian public were made during the second half of the nineteenth century. Small 'science societies' developed in various parts of the country. Parallels could be drawn between what was happening in Europe in the seventeenth and eighteenth centuries and the efforts to disseminate science in India in the latter part of nineteenth and first half of twentieth century (Venkataswaran 2011:39). The number of people involved in those activities was quite limited, and no serious effort was made to bring them together into a large-scale 'science movement'.

The freedom movement in India, although primarily political, also operated as a carrier of modern scientific ideas.[4] India achieved its independence from British rule in 1947, and the plans of the emerging ruling classes and political elite to build a

[4] Almost all political leaders of Indian freedom movement and social reformers repeatedly emphasized the need to integrate modern science into Indian culture. For example M.K. Gandhi as early as in 1937, emphasizing the need to introduce *Nai Taleem* (New Education) wrote 'Only every handicraft has to be taught not merely mechanically as is done today, but scientifically …' (*Harijan*, 31 July 1937). J. Nehru, who became the first Prime Minister of India in 1947, in his book *Discovery of India* introduced and defined the phrase 'scientific temper' and popularized it by repeatedly using it in his speeches. Nehru became member of the Science and Social Science Relations Committee of the Indian Science Congress in 1940 and in 1947 was elected as General President of the Indian Science Congress. It is remarkable that scientific and political leadership were quite close to each other in pre- and post-independence India.

modern, industrialized India recognized the need for a wider acceptance of scientific ideas in society. Phrases such as 'scientific temper', 'broad scientific outlook', 'scientific belief system' and 'scientific method' echoed in various public forums.

In order to train the younger generations within the country, the education system was revamped. A large number of new books for teaching science were written in regional languages. Translating English texts offered no easy solutions. Besides structural problems, the vernacular vocabulary was limited. There was a shortage of standard technical terms in Indian languages, so they had to be developed. At times that was done mechanically, leading to obscurity or, in some cases, the incorporation of anglicized terms into the local languages.

9.4 Mass Science Campaigns and the Political Left

A few members of the Communist Party of India and some social reformers who were politically positioned on the Left had also realized the importance of communicating science to the people in their own tongue (Joshi 1993), not only for the important objective of raising the consciousness of the people but also to help the Left reach larger sections of society, especially the younger generations. They saw possibilities for enlarging the mass base of the political movement itself. Efforts in Kerala in 1957 to organize science writers were undertaken, based on those twin objectives. Both the Indian ruling classes and those who constituted the resistance movement were conscious of the importance of science communication, quite obviously for very different reasons.

9.4.1 A Nationwide Campaign

In 1962, the *Kerala Sastra Sahitya Parishad* (KSSP) was formed. Its regular activities included publishing books and journals on science, running science clubs throughout the year and organizing public lectures by scientists over a period of about a month all over Kerala. In 1970, after its annual conference, the KSSP organized the first '*science jatha*' (science procession) in the city of Ernakulam.[5] *Jathas* subsequently became an integral part of the KSSP's yearly activities.

[5] Traditionally, in India, people move together in processions (originally on foot but now using vehicles) on various social and religious occasions. During the freedom movement, this tradition and many other religio-cultural forms of gathering were used quite effectively for anti-British political mobilization. In 1970, the KSSP mobilized people for lectures on scientific issues for the first time, and the mobilization naturally took form of processions and was called '*jatha*'. In the following years, realizing the potential of this form, the KSSP decided to mobilize scientists and artists, once a year, to spread science awareness on specific issues through mass contact. Street plays, skits, and songs written and performed on science issues gradually became an integral part of the *jathas*.

Groups of scientists traveled in cars from one city to another. They were received by local reception committees responsible for organizing the lectures in their area. Often, *jathas* reached the venue of the public meeting late, so the local committees slowly started producing short plays and songs on various science issues to keep the gatherings engaged. Gradually, a shift in the conceptual framework came about, and *science jathas* were transformed into *kala jathas* (in English, 'cultural processions').

The name change reveals a shift in the framework of science communication. Science was now communicated in a planned manner through traditional as well as modern art forms, such as songs, street theater and poster exhibitions. For that purpose, artists, teachers, writers, students, social workers and unemployed youth were mobilized in addition to scientists. The first *kala jatha* performed at 900 places, contacted more than half a million people and sold pamphlets worth more than 25,000 rupees. Thereafter, *kala jathas* became a regular and significant activity of the KSSP (Isaac & Ekbal 1988:15).

It was only at the beginning of the 1980s that the need to create some sort of national network of these organizations was felt strongly (KSSP 1984). This was later to become the basis for long-term cooperation among NGOs involved in science communication. In 1985, a workshop in Delhi discussed the possibility of implementing the KSSP's *jatha* blueprint at the national level. Representatives of various science NGOs, which were supported by the Communist Party of India (Marxist), participated in the workshop, which identified issues of national concern: 'self-reliant technology policy of the country', 'peace and nuclear disarmament', 'Bhopal gas tragedy and popularization of science'.

Ideas developed during the freedom/independence movement still influenced the ideological basis of the emerging people's movement. The national coordination committee began calling itself the People's Science Movement (later renamed as the All India Peoples Science Network). The People's Science Movement initially comprised only a handful of organizations: the KSSP, the Delhi Science Forum (Delhi), Eklavya (Madhya Pradesh), *Vigyan Sabha* (Madhya Pradesh), the Pondicherry Science Forum (Pondicherry), *Lok Vidnyan Sangathana* (Maharashtra), the Tamil Nadu Science Forum (Tamil Nadu) and *Karnataka Rajya Vijnana Parishath* (Karnataka). Its base expanded rapidly.

9.4.2 Bharat Jan Vigyan Jatha: A National Science Movement

In 1986, a meeting of the Co-ordination Committee of the Campaign for Peace and Against Nuclear War proposed staging an all-India *jatha*. During preparations for the *jatha*, many new groups were formed all over the country and older inactive organizations were energized. Twenty-six new and old science NGOs carefully worked out a detailed plan of action in a 2-day meeting held in April 1987. A detailed perspective plan was also prepared at the meeting, based on three slogans: Science for the People, Science for the Nation and Science for Discovery.

In the following months, workshops were held on particular themes; lectures, play scripts and songs were written; slide shows and poster exhibitions were prepared; films were produced; and science toys, games, experiments and exhibits were collected or created. Volunteers were drawn from all over the country. In order to cover all the 500 districts of India, 5 teams of scientists and 5 teams of cultural activists were constituted and trained to perform 3–5 hours programs. Five such processions began on 2 October 1987 (Mahatma Gandhi's birthday) in different parts of the country, to culminate at Bhopal City on the first anniversary of the Bhopal gas disaster. Each *jatha* covered about 5,000 km in 37 days. On average, each day they performed in three different districts, where reception committees had already been set up. Millions of people were exposed to scientific information and issues of social relevance. Thousands of science activists joined the Peoples' Science Movement. As a result of this intensive mass campaign, new local and regional science organizations came into being.

The *Bharat Jan Vigyan Jatha* ('campaign for taking science to the public') was the largest science communication mass mobilization program in the world, and also resulted in many spin-offs. The All India Peoples Science Network, currently a network of about 40 NGOs, holds the National Peoples Science Congress each year. Other mass movements, such as 'Anti Arrak' (anti-alcoholism), the National Literacy Mission (Raza 1994), the *All India Gyan Vigyan Samiti*, Joy of Learning, the Solar Eclipse Campaign and the Anti-Superstition Campaign are but a few offshoots of the *Bharat Jan Vigyan Jatha*.

9.5 Indian Research on the Public Understanding of Science

A small group of scientists at the Council of Scientific and Industrial Research working on science and technology policy and its interface with society were associated with the conceptualization of *Bharat Jan Vigyan Jatha* from its inception. During the *jatha*, it was realized that some of the software packages used for science communication were very popular throughout the campaign, whereas others did not evoke the same enthusiasm among the public. This raised a serious question: what kind of scientific information do the people receive the best, and why?'

A review of the available literature on the subject helped a little. Investigations on the subject, carried out mostly in the West, aimed to measure the scientific literacy of the target population. Those methods were clearly inappropriate in India, where a large, illiterate population was embedded in non- or semi-industrialized social and cultural contexts. The group decided to develop a methodology of surveys that would be more appropriate in India.

9.5.1 Public Understanding of Science Survey Studies

In most developed countries, specialized agencies carry out surveys. The research scholars who later analyze the data are completely delinked from the execution of the survey. In developing countries such as India, South Africa and China, the research team handles the entire survey operation, from designing the questionnaire and selecting a representative sample to analyzing the data and producing reports (Raza et al. 1995). This gives the team a deeper understanding of social reality and helps them arrive at their conclusions.

The first survey of public understanding of science in India was conducted in 1989, and administering surveys subsequently became a regular activity. In Western countries, because of their relative cultural, economic and social homogeneity, smaller samples could be used to statistically represent the entire national population. Working out a statistically representative sample in a large, complex and culturally rich and diverse country such as India is a convoluted exercise (Raza et al. 1996). The challenge lies in selecting a nationally representative sample that takes into account all the social, economic, linguistic, religious and cultural variations (Shukla 2005). Therefore, to capture differences arising from the social and cultural heterogeneity of the 'Indian' public, the first two surveys were conducted at sites that represented culturally, economically and socially different milieus (Raza et al. 1991a, 1996).

9.5.2 The Cultural Distance Model of Public Understanding of Science

Retrospective analysis of developments in the public understanding of science and 'scientific temper' research shows that the objective pursued in the West was quite different from the agenda that Indian researchers followed. In India, since research into the public understanding of science was rooted in the People's Science Movement, the objective was to help to improve the efficacy of science communication and identify gaps in understanding and knowledge. Developing internationally comparable indicators was not a significant part of the work.

The search for a conceptual model that could explain Indian reality and answer pressing questions led to the identification of four areas of scientific knowledge that constituted a scale on which 'astronomy and cosmology' could be placed at the highest degree of complexity of abstract knowledge and the lowest degree of impact on human life. 'Health and hygiene' could be placed at the lowest level of complexity but at a very high level of direct impact on daily life. 'Geography and climate' and 'agriculture' occupied spaces between those two extremes.

In short, the scale defines the distance of an area of scientific investigation from the immediate experience of ordinary people. It was realized that, when demographic factors are kept constant, the most important determinants of cultural distance are factors inherent in the scientific knowledge system, such as the complexity involved in explaining the phenomenon, the duration of its lifecycle, the control that an individual or a collective can exercise, and the intensity with which the phenomenon could influence the life of common citizens. It was also observed that the pace of scientific knowledge propagation slows as the inherent mathematical obscurity of the phenomenon and the conceptual gymnastics needed to explain it increase (Raza et al. 1996).

9.5.3 *Survey Methodology*

Surveys using written, mailed questionnaires could not be conducted in India, and in any event would have excluded the illiterate population from the sample. Nor were telephone interviews an option, since few had phones and because the context of the population is difficult to ascertain from conversation at a distance. Scheduled person-to-person onsite interviews were the most suitable method available (Raza et al. 1991a).

Social reality on the one hand and the emerging conceptual model on the other were instrumental in the construction and selection of indicators of public understanding of science. The focus was on recording responses and thereby developing categories that could help in mapping out the cognitive structure of the target populace. Eventually, a four-point scale was constructed, with the following categories of response (Raza et al. 1991b):

1. *Scientifically correct or closest to scientifically correct responses*: Responses that revealed an understanding of the issue, not necessarily conveyed in scientific terminology, but coming close to the scientifically correct explanation.
2. *Naturalist or secular but scientifically incorrect responses*: Intuitive explanations that were not correct but did not resort to metaphysical constructs.
3. *Extra-scientific responses*: Explanations that called upon divine powers, mythology, etc.
4. *Don't know*.

These are not discrete categories, but form a band. Statistical techniques were used to place individual answers at appropriate places in that band.

9.6 Cultural Distance, Science Understanding and Education

The objective of the project was not to compare various sections of the population but to find out where different segments of society stood vis-a-vis various scientific concepts that the activists were trying to propagate. It was more important for the research team to find out which notions were prevalent among the groups that did not subscribe to scientific explanations of natural phenomena. 'Which scientific tenets will require more effort to become part of the peoples' cultural thought complex, and why?' was an important question. It was important to test whether there was a natural distance between the culture of science and the peoples' culture. If there was, could it be measured empirically (Raza et al. 2002).

Mapping the responses using the four categories led to the construction of an empirical model that could be used for determining the 'cultural distance' of a scientific notion or area from the everyday life of ordinary citizens. Scholars have not been able to arrive at a universally acceptable definition of culture (Geertz 1999), and most advise not to 'try to circumscribe it within a strict definition' (Godin and

Gingras 2000:43). The notion of culture notoriously evades clarity and forces each scholar working on 'cultural studies' to elaborate on what they mean by 'culture' (Sardar and Loon 1997:4–7).

For the purpose of this investigation, 'culture' was defined in terms of culturally determined explanations offered by common citizens, and 'cultural distance' was defined as the time it would take for a world view, attitude, perception or idea generated in one cultural context to be democratized within the thought structures of another cultural group. Various ideas generated within the scientific realm of thought could then be placed at varying cultural distances from the quotidian life of a set of respondents (ordinary citizens).

9.7 Indicators of Cultural Distance

A simple method was proposed for determining the cultural distance of a given scientific explanation from the everyday life of a common person. The method designated years of socialization in modern education as a proxy for cultural distance. Since data were not normalized for any other demographic variable, it was assumed that 'years of schooling' would also reflect the influence of all other demographic variables, to varying degrees.

Although the suggested empirical method can be applied to multiple response categories, to explain the point we consider a dichotomous response variable in which valid scientific answers constitute the first category and all invalid responses are grouped together to form the second category (for a detailed description, see Raza et al. 2002).

For example, if, in a population of 100, Y_v give scientifically valid answers, then because responses are dichotomous variables ($\Phi_{iv} = 100 - \Phi_v$) is the number of respondents who give scientifically invalid answers. Both Φ_v and Φ_{iv} are plotted on the y-axis, and the years of formal schooling that the respondent has received is plotted on the x-axis. Since the response variable is dichotomous, the two curves would always intersect each other at a point where 50% of the total of those interviewed offered valid scientific explanations, or vice versa. Let us call this point the 'index of democratization' (i_d) of a concept. If we take the responses to a question (say, 'How did the humans come into being?'), the i_d point corresponds to about 9 years. Beyond that point, it is obvious that more than 50% of the population under consideration subscribe to the valid scientific explanation. Conversely, between the origin and i_d, less than 50% of the populace has incorporated the scientific explanation within their cultural world view. We could argue that a piece of scientific information or concept i has to travel on the cultural distance scale for χ_i years to achieve the threshold level i_d at a given point of economic and sociocultural development (Raza et al. 2002).

9.8 Conclusions

We have outlined the beginnings and development of science communication in India. Science and the public were construed as having two different cultures, which are characterized by different styles of argumentation and vocations. After World War II, Indian scientists and the Indian Government, which funded the science, needed public support for continued research and development activities—which led to greater emphasis on science communication. Developments in communication channels and their reach—television and later the internet—necessitated survey research to measure the reach of science among the public. Initial efforts to spread science to the public and to measure the effects of those activities through 'scientific literacy' surveys implied a 'deficit' model of public understanding of science. Later developments in the theoretical domains of public understanding of science refined those efforts into 'contextual' models and still later into 'dialogical' models.

Public understanding of science in India largely followed a deficit model through the efforts of NGOs—such as *Kerala Sastra Sahitya Parishad* and others—as well as government agencies such as the National Council for Science and Technology Communication and *Vigyan Prasar*. Our earlier work showed that the percolation of scientific ideas differs among sections of the population depending on their cultural socialization and their vocations. Abstract ideas, such as those in astronomy and mathematics, take a much longer period of socialization than scientific ideas on agriculture or health. These insights are codified under the cultural distance model developed and validated through a series of surveys in India.

References

Bensaude-Vincent, B. (2001). A genealogy of the increasing gap between science and the public. *Public Understanding of Science, 10*, 99–113.

Bernal, J. D. (1939). *The social function of science*. London: George Routledge.

Bronowski, J. (1953). *The common sense of science*. Cambridge: Harvard University Press.

Cook, G. (2001). *The discourse of advertising*. London: Routledge.

Crump, T. (2001). *A brief history of science*. London: Constable & Robertson.

Dove, A. (2002). It takes a village to raise an idiot: Fixing US science education. *Journal of Clinical Investigation, 110*, 1057–1060. doi:10.1172/JCI200216959, Retrieved 25 Apr 2006 from http://www.jci.org/cgi/reprint/110/8/1057.

Geertz, C. (1999). *A life of learning*. 1999 Charles Homer Haskins Lecture, American Council of Learned Societies, occasional paper no. 45.

Godin, B., & Gingras, Y. (2000). What is scientific and technological culture and how is it measured? A multidimensional model. *Public Understanding of Science, 9*, 43–58.

Isaac Thomas, T. M., & Ekbal, B. (1988). *Science for social revolution: The experience of Kerala Sastra Sahitya Parishat*. Trichur: KSSP.

Joshi, P. C. (1993). Development perspectives in rural India: The techno-economic transformation. In R. Prasad (Ed.), *Science, technology and rural development* (pp. 2–3). Agra: Y.K. Publishers.

KSSP (Kerala Sastra Sahitya Parishad). (1984). *Science for social activism: Reports and papers on people's science movements in India.* Trivandram: Kerala Sastra Sahitya Parishad.

Kundra, S. (2005). *Introduction to advertising and public relations*. New Delhi: Anmol.

Lee, O., Fradd, S. H., & Sutman, F. X. (1995). Science knowledge and cognitive strategy use among culturally and linguistically diverse students. *Journal of Research in Teaching, 32*(8), 797–816.

Levis, F. R. (1962). *Two cultures? The significance of C.P. Snow*. London: Chatto and Windus.

Miller, S. (2001). Public understanding of science at the crossroads. *Public Understanding of Science, 10*, 115–120.

Miller, J. D., Pardo, R., & Niwa, F. (1996). *Public understanding and attitudes toward science and technology: A comparative analysis of Canada, the European Union, Japan, and the United States*. Paper presented at a World Association for Public Research (WAPOR) Regional Conference, 9 Nov 1996, Tokyo.

Munns, J., & Rajan, G. (1995). *A cultural studies reader: History, theory, practise*. London/New York: Longman.

NSF (National Science Foundation). (1993). *Science and engineering indicators*. Washington, DC: National Science Foundation.

NSF (National Science Foundation). (1996). *Science and engineering indicators*. Washington, DC: National Science Foundation.

NSF (National Science Foundation). (1998a). *Science and engineering indicators*. Washington, DC: National Science Foundation.

NSF (National Science Foundation). (1998b). Science and technology: Public attitudes and public understanding, Chapter 7. In *Science and engineering indicators*. Arlington/Virginai: National Science Board.

NSF (National Science Foundation). (2000). *Science and engineering indicators*. Washington, DC: National Science Foundation.

Raza, G. (1994). Street theater and literacy mission, *Saksharta*, *XVII*(4), 21–24. National Literacy Mission, New Delhi.

Raza, G., Dutt, B., & Singh, S. (1991a). *Scientific attitude among masses: A report of the survey conducted at Allahabad during Kumbh Mela in 1989* (Vol. I, pp. 5–9). NISTADS-REP-107(AV)91.

Raza, G., Dutt, B., Singh, S., & Wahid, A. (1991b). *Prototypes of the forms of scientific cognition: A survey of cultural attitudes to natural phenomenon* (pp. 5–6). NISTADS REP-108(AV) 91.

Raza, G., Singh, S., & Dutt, B. (1995). Peoples' attitude to scientific knowledge: The context of culture. *Journal of Scientific and Industrial Research, 54*, 108–121.

Raza, G., Singh, S., Dutt, B., & Chander, J. (1996). *Confluence of science and people's knowledge at the Sangam*. New Delhi: Ishtihaar.

Raza, G., Singh, S., & Dutt, B. (2002). Public, science, and cultural distance. *Science Communication, 23*(3), 293–309.

Sardar, Z., & Loon, B. V. (1997). *Introducing cultural studies*. New York: Totem Books.

SCST (House of Lords Select Committee on Science and Technology). (2000). *Science and society*. London: House of Lords.

Shukla, R. K. (2005). *India science report: Science education, human resource and public attitude towards science and technology*. New Delhi: NCAER.

Snow, C. P. [1959] (1993). *The two cultures* (Reprint). Cambridge: Cambridge University Press.

Swann, B., & Aprahamian, F. (Eds.). (1999). *J.D. Bernal: A life in science and politics*. London: Verso.

Venkataswaran, T. V. (2011). Infrastructure for public communication of science. In G. Raza et al. (Eds.), *Constructing culture of science: Communication of science in India and China* (p. 39). New Delhi: NISCAIR.

Vilanilam, J. V., & Varghese, A. K. (2004). *Advertising basics! A resource guide for beginners* (pp. 1–3). New Delhi: Response Books.

Chapter 10
Whose Science? What Knowledge? Science, Rationality and Literacy in Africa

Hester du Plessis

Abstract Drawing on the intersection between philosophy and science, this chapter demonstrates the diversity of influences on the development of public communication of science studies in South Africa. In broader terms, the problematic notions of African rationality and (western) ideas of science literacy that are currently considered to be the result of the colonial subjugation of Africa are contextualized against that colonial background. The historical process of colonial-style political subjugation of indigenous science in South Africa under the National Party's apartheid regime is briefly discussed as the main reason for the lack of development of representative science communication. Both the philosophical argument for a re-evaluation of the notion of rationality and the sociopolitical efforts to manipulate the knowledge of a complex society into a western mold serve to answer the initial question: whose science and what knowledge are communicated to the public(s)? The chapter is in five sections. Section 1 contextualizes the lack of development of science communication in the South African past against the progressive notion of modernity that excluded the so-called 'primitive' African knowledge systems. Section 2 looks at the colonial efforts of the British, who established racial segregation between white and black populations, with far-reaching impacts on the development of scientific research in South Africa. Section 3 shows how segregation was implemented by the

H. du Plessis (✉)
Senior Research Specialist, Research Use and Impact Assessment,
Human Sciences Research Council (HSRC), Pretoria, South Africa

Sustainable Energy Technology and Research (SeTAR) Centre, Faculty of Science,
University of Johannesburg, Johannesburg, South Africa

Mapungubwe Institute for Strategic Reflection, Johannesburg, South Africa
e-mail: hesterdup@gmail.com

B. Schiele et al. (eds.), *Science Communication in the World: Practices, Theories and Trends*, DOI 10.1007/978-94-007-4279-6_10,

Afrikaners during the apartheid regime, thereby marginalizing the knowledge systems of the African population. Section 4 examines post-apartheid reconstruction, and Section 5 details current efforts to popularize and communicate science.

Keywords Colonialism • Apartheid • Knowledge systems • Public understanding of science

10.1 The Notions of Scientific Rationality and Science Literacy in Africa

The popularization and communication of science in South Africa are in a nascent stage. Its underdevelopment must be seen against a problematic political background that spanned a period of several decades. To fully understand the slow development of science communication and the lack of a comprehensive assessment of the relationship between science and a highly stratified public(s), it is necessary to revisit the lasting ideals of the European Enlightenment of the seventeenth and eighteenth century in its consideration of modernity—seen by Habermas (1987) as the 'project' of the Enlightenment—as a benchmark of scientific progress and social transformation. Disciplines such as science communication, as a result, developed globally in line with the ideals and ideas promoted by modernity. However, in the reciprocity of scientific information between western and non-western science, we encounter an asymmetry in power relations. The balance, still half-acknowledged by current social movements such as post-colonialism, and half-denied by current academia, is undeniably in favor of the west. The consequential 'unbalance' is pronounced in previously colonized countries in Africa.

Despite efforts to effectively communicate science, South Africa lacks a comprehensive structure of science communication and public understanding of science (PUS) programs. The nation has highly stratified public(s), and large-scale communication of science is slow in development. The reason for this is complex and can be found in the legacy of the European Enlightenment's 'idea of progress', in the legacy of British colonialization and in the sociopolitical development of the education systems and manner of governance up until the apartheid era.

To contextualize the sociopolitical colonial legacy of South Africa it is necessary to consider, within the broader context, the promotion of (western) science as part of the European Enlightenment ideal. Globally, humankind has used the scientific interpretation and understanding of our world for centuries. However, this practice of sharing scientific knowledge intensified during the European Enlightenment enterprise (Descartes 1637, 1644). In this regard, Massimiano Bucchi and Brian Trench (2008:1) inform us that 'communicating ideas or insights drawn from scientific research to a wider public was part of the enlightenment enterprise of the eighteenth century'. Organized and structured science research became associated with the specific 'idea of progress'—which was measured against the growth of new scientific inventions originating from the west.

As a result, the European Enlightenment vision of being 'scientifically literate' presupposes a universal civilization—with science as the principal force that shapes knowledge in all human societies. However, an anthropological notion of 'race and culture' competed with and challenged the general validity of that claim to universalism as cultures started to interact and share scientific knowledge. From the perspective of the modern western philosophical condition, it was the perceived absence of (western) science that was used to justify the difference and inferiority of the African mind (intellect). The perceived absence of science led to the general conclusion that the African mind is incapable of producing anything of epistemic significance.

Following the lead of cultural anthropologist Lévy-Bruhl (1910), researchers of the 'African mind' accepted a conceptual distinction between pre-logical (primitive) culture (as representing the pre-modern age) and the modern western culture as representative of the superiority of western rationality in and through science. Lévy-Bruhl argued that the mentality of the so-called 'primitive' people is radically different from that of the scientific mind of western rationality, in so far as the former shows little or no evidence of an understanding of the principle of logical consistency or the process of discursive reasoning. From this perspective, the African race was designated as the 'other of reason' and the 'marker' of cultural difference.

Lévi-Strauss (1966) followed these ideas of Lévy-Bruhl by attempting to reveal the underlying structure—the deep grammar of mythical thought—that he believed could explain the endless multiplicity of culture-specific meanings and forms. So-called 'primitive societies' identified scientific meaning through a method of *bricolage* (Torgovnick 1990).[1] Commenting on the significance of the role of rationality—the so-called 'cultural turn'—in contemporary philosophical discourse, Emmanual Eze (2008:151) correctly points out that since culture shapes our practices of freedom and agency and a culture bridges the gap between implicit and explicit social knowledge, we are provided with the complex means to 'become that which makes us who we are by making ourselves into that which we believe makes us who we are'.

In Eze's (1997, 2008) efforts to analyze the characteristics of human rationality, he refers back to the best way to philosophically define rationality—through demonstration. Such demonstration will require 'amassing empirical or scientific evidence for the rational, and reflecting on the concept of evidentiality. It is only from such demonstrative acts that we can explore what is at stake in the activity itself' (Eze 2008:xii). What is already implicitly comprehended in rational action is required to be rendered explicitly as the abstractness of philosophy-as-epistemology. The ultimate quest is to literally 'gain the world' by acknowledging the value and contribution of the everyday experience on its most intuitive and reconstructive levels (Eze 2008). Since reason is

[1] Lévi-Strauss (1966) defines a *bricoleur* as someone who works with his hands using different means, like those of an artisan. Mythical thought is therefore a kind of intellectual *bricolage* used by 'raw' or 'naive' art. The (western) engineer will always try to make his way out of and beyond the constraints imposed by a particular state of civilization, while the (non-western) *bricoleur*, by inclination or necessity, always remains within his own domain.

a series of mental acts that involves perception, understanding and explanation within the framework of philosophy-as-epistemology, Eze (2008:xvii) developed a framework for categorical discussion that embraces, for example, the formal or logical, the hermeneutic or interpretative, some phenomenological models, the empirical–probabilistic, the skeptical and the political.

This effort to make use of empirical (scientific) evidence in support of rationality is not new. Different researchers have proved that is difficult to link theory with praxis (Macamo 2005; Masolo 1994). The spectacular efforts and intellectual endeavors of Max Horkheimer (1939), Herbert Marcuse (1964) and Theodore Adorno (1970) of the Frankfurt Institute for Social Research (later referred to as the Frankfurt School[2]), failed in exactly this regard due to their inability to take their 'critical theory of society'—in its ambition to establish 'knowledge and understanding of social life in its totality'—into praxis (Slater 1977).

However, what is clear is that the interpretation and definitions of 'reason' and 'scientific rationality' are dependent on a number of factors and, most conspicuously, require diversity. This includes cultural diversity. Richard Rorty (1980:269) drew attention to the fact that the relation of the concept of rationality to the so-called 'philosophical dogma of the day' reflects the Kantian endeavor to present a permanent neutral framework for culture whereby the 'framework is built around a distinction between inquiry into the real—the disciplines which are on the "secure path of a science"—and the rest of culture'. Thus Rorty's (1980) apt challenge to philosophy: in its distinction between science and non-science, philosophy is endangering all current formulations, endangering philosophy itself and with it, rationality.

In the case of Africa, the African philosophers had to persistently justify the existence of a 'vaguely collective unconscious' and are blamed for not having anything remotely akin to science or philosophy. This accusation of having a 'collective unconsciousness' disregarded individuality and prompted a racially informed idea that African conceptual systems are the product of collective work rather than the elaborations of individuals (Karp and Masolo 2000; Mudimbe 1988). The colonial system, in furthering this notion, placed undue value on the thoughts of political leaders by replacing tradition with authoritarianism. The traditional scientific knowledge of the public, as well as individual contributions to science, was thereby easily marginalized.

Turning to contemporary African philosophers, it becomes clear that the notion of modernity in its adoption of the idea of progress is currently receiving considerable attention. African philosopher Kwame Gyekye (1997) argues that the western-inspired understanding of modernity has led to a problematic distinction between culturally rich traditional societies, which have been identified as practicing 'irrational and backward thinking', on the one hand, and modernity, which has been identified with 'progressive thinking', on the other. From this perspective, modernity

[2] The Frankfurt School was opened in 1924 by the first Director, Carl Grünberg (1861–1940) to oppose the prevailing socio-economic order in Germany through the application of Marxist theory (Slater 1977).

functions on the assumption that progress can only be established within the paradigm of western scientific thinking and measured as *science literacy*, with the west claiming the exclusive ability to be scientifically rational.[3] This problematic claim to scientific and philosophical universalism has blinded its followers to its androcentric[4] assumptions on the one hand and its ethnocentric arrogance on the other. Modernity is currently still defensive of its original assumption regarding a priori knowledge, with man, as rational being, as the privileged center of philosophical universalism—beyond the constraints of culture and tradition.

The social movements of feminism and post-colonialism contributed significantly to the questioning of such western assumptions of philosophical, epistemological and scientific universalism. The Eurocentric model of epistemology has, from this perspective, correctly been challenged in its claims as the one and only valid epistemology, thus opening up the debate to the possibility of alternative (and different) forms of knowledge. Michael Cloete (2008:65) aptly states that, with the denial of other non-western forms of knowledge and rationality:

> the possibility of other non-western 'subjects of reason', committed to different forms of inquiry, and equally committed to an explication of the rational grounds of legitimation and validation of the knowledge claims, raised within their own philosophical systems of thought, is therefore ruled out.[5]

Therefore, the questions: Whose science? What knowledge? If we argue that practitioners and advocates of the epistemological contribution of traditional indigenous knowledge systems (IKSs) are rational and that their activities are based on scientific reason, we enter into a contested debate. However, there is sufficient confirmation that sound scientific knowledge is embedded in traditional knowledge systems (Raza and du Plessis 2002, 2003, 2004; Riana and Habib 2004; Sardar 1998, 2002). Local communities' understanding of science influences indigenous design

---

[3] The notion of 'scientific rationalism' was introduced by the western 'scientific revolution' of the seventeenth century. In the eighteenth century, the concept of universalism was established during the western 'age of enlightenment'. From this perspective, western scientific rationality (combined with philosophy in the form of epistemology) was to defend the universal status of western science as the most advanced form of knowledge and rationality. This claim to universality assigned philosophy the role of demonstrating that (western) universal a priori conditions are necessary for the possibility of science. Modern western philosophy-as-epistemology thus functions as a foundational discipline aimed at demonstrating that scientific knowledge can be accounted for in humanistic terms, with man defined as a 'rational being'.

[4] Androcentric: having or regarding man or the male sex as central or primary. *Collins English Dictionary*. London: Harper Collins Publishers.

[5] Underpinning the idea that all men are the same, the awareness grew that cultures differ and exist in different geographical worlds requiring different social strategies for survival. For western science, in the quest to study man, 'race' soon became the marker for different social practices that constitute different cultures. Race therefore became a science 'subject', and racial differences became a cultural 'marker'. The paradoxical result of celebrating differences, respect for pluralism and acknowledgement of identity politics—which became the feature of a liberal—modern democratic outlook—made science a political issue because the science of human differences could only be read in a racial fashion (Malik 2008). This can be referred to as the 'guilt of science'.

processes, products and manufacturing technologies (Raza and du Plessis 2002; Ladyman 2002). The introduction of transdisciplinarity as an approach to manage the complexity of research on these topics (Gibbons and Gummett 1984) took its inspiration from 'science rebels' such as Paul Feyerabend (1975, 1987). There is evidence of links between practices of artifact/craft (product) design, traditional systems of knowledge and understanding of science during the use of technology (Wajcman 1991; Grint and Gill 1995; Gyekye 1997; Godin and Gingras 2000). The epistemological contribution of IKSs is also evident in non-textual communication efforts. In this regard, Isidore Okpewho (1992:3) points to the dominance of oral literature in Africa. He argues that traditional African oral literature (literature delivered by word of mouth) contains a number of different forms in which traditional African philosophical thought is expressed ('orature', 'traditional literature', 'folk literature' and 'folklore', proverbs, sage literature etc.). These terms refer to different forms of communication and indicate the richness of African epistemological options.

To provide a comprehensive understanding of IKSs is a complex undertaking. At present there is much debate, disagreement and skepticism in the attempt to arrive at some universal understanding of what is meant by 'indigenous knowledge systems'. Some efforts have, however, met with limited success in describing the reality of the use and usefulness of traditional knowledge. For example, at the International Symposium on Indigenous Knowledge and Sustainable Development, held in the Philippines in 1992, the participants adopted Michael Warren's definition:

> The term 'indigenous knowledge' is used synonymously with 'traditional/and local knowledge' to differentiate the knowledge developed by a given community from the international knowledge system, sometimes also called the 'Western' system, generated through universities, government research centers and private industry.[6]

An important contribution to furthering the understanding of IKSs has been made by Edward Said (1978, 1994), who indicated that the valorization and defense of IKSs in a world dominated by western science is ultimately about the affirmation and recognition of the self in relation to the 'other'. Said (1994) used the term 'culture' (culture as the 'other') to address conflicting issues of indigenous knowledge in relation to the Eurocentric understanding of IKSs. In this regard, Said (1994) pointed out that research into IKSs has been linked to esthetic theory and practices

---

[6] A further indication of advances in the definition of IKSs was in a document published by the South African Department of Art, Culture, Science and Technology after a visit by a group of South African delegates to India (March 2001). The following definition was provided:

> 'Indigenous Knowledge (IK) is local knowledge generated by people living within a particular community—hence it is unique to a given society or culture. Indigenous knowledge is tacit knowledge and therefore, not easily codifiable. It is dynamic and based on innovation, adaptation, and experimentation, thus codifying IK may lead to the loss of some of its properties. Indigenous knowledge can contribute to a sustainable development strategy that accounts for the potential of the local environment and the experience and wisdom of the indigenous population. Furthermore, IK covers critical issues of primary production, human and animal life, natural resources management, etc.' (Retrieved 24 August 2002 from http://www.dacst.gov.za).

of interpretation, in which the relative autonomy of the esthetic discourse has been separate from, and dogmatically defended against, the economic, social and political discourse. According to Said (1994), IKSs have often been represented, in the Kantian sense, as an esthetic form of judgment, the principal aim of which is to provide esthetic pleasure. This Kantian demarcation of the transcendental form of judgment aimed at esthetic pleasure has contributed significantly to disciplines such as ethnography, historiography, philology, sociology and literary history, where the cultural 'other' has been reduced to the level of providing (exclusively) esthetic (exotic) pleasure for the western observer.

European colonialists initiated the formal academic study of the oral literature of Africa in the mid-nineteenth century. The aim was to assist the colonizers to understand the cultures of the different societies they encountered in their travels in Africa (Serequeberhan 2002). This process was undermined by misinterpretation and misunderstanding, emanating from a racist sense of cultural superiority on the part of many of the European colonialists. Isidore Okpewho (1992:17) mentions a book by British anthropologist Captain R.S. Rattray, *Ashanti proverbs: The primitive ethics of a savage people* (1916), as a prime example of the racial arrogance that characterized the attitude of colonial administrators and field researchers in the quest to understand African cultures.[7]

10.2 Post-Colonial Western Dominance in the Development of Science and Technology in South Africa

Against this background, questions are asked about the impact of the philosophical, epistemological and scientific marginalization of African knowledge systems on the development of science education and science communication during the British colonization of South Africa (Sparg et al. 2001). Saul Dubow (2006:165) describes how scientific and technical agencies—ranging from professional associations, museums, botanical gardens to transport and communication systems—became part of the intellectual and political substructure of the formulation of (white) South Africanism

[7] A number of anthropological terms that were developed by colonial contact with the 'other' still exist today. African traditional cultures are still being humiliated by the categorical classification of 'being primitive'. Christopher Steiner (1994) refers to the rebellion against the so-called 'primitive isolates' followed by earlier anthropologists whose studies followed a 'bounded system' whereby, in a single society, one isolated community within one remote village was studied. Women were mostly not included in the studies and in a certain sense became doubly marginalized. Today, this 'system of investigation' has been revised to contain 'processes of investigation' consisting of the history as well as social changes in the given community. This includes the replacement of old key words such as 'homeostasis, cohesion and balance' with new concepts such as 'pluralism, heterogeneity, crisis, conflict and transformation' (Steiner 1994:1). The African community, therefore, is progressively considered as being part of a global matrix with 'trans-national contacts and macro-scale linkages' (Steiner 1994:1).

from the early 1900s. In the spirit of the European Enlightenment ideal, science was not seen as being overtly 'ideological' and served as an easy conduit for introducing British imperialist ideals during the British colonialization of South Africa. Of interest in the earlier period is the formation in 1877 of the South African Philosophical Society (which gained a royal charter in 1907 and became the South African Royal Society), which served as an overarching organizational body for coordinating scientific activity. Here we find 'philosophy in service to science' in imitation of the European model.

Dubow (2006:174) comments on this notion that science has no political boundaries: because science 'knows no nationality', the British colonialists were sending out a message of political inclusivity (on the basis of continued membership of the imperial fraternity) whereby science would help to break down barriers and prejudices. However, the Afrikaners (white emigrants from the Netherlands, France and Germany, also called 'Boers') soon developed a sense of nationhood that turned upon their British 'benefactors'. The Anglo-Boer War of 1899–1902 created not only a permanent rift between Britain and the new nation state but also led to a truce in the scientific development of South Africa. South African scientists would in future develop science programs on their home ground, aided and abetted by their European colleagues. A new 'problem' that also captivated the attention of Britain and the new nation state appeared at this time: the problem of managing and regulating the majority of the population—the indigenous people of South Africa. This became an area of growing scientific interest in what was referred to as the 'native question' (Dubow 2006:175).

After the Anglo-Boer War, ethnological issues came to the fore (as in most other colonies) and science was progressively considered as a key role player in redefining the South African identity in terms of race. Though the so-called Boers were alleged to be 'backward' and 'degenerate' at that stage, they were of use to enable the British to establish a 'common white future'. 'Being white' had to be linked to 'being racially superior'—hence the 'native question' and the new terminology of (racial) 'segregation'. To address the 'native question' became the work of scientists, and specifically of anthropologists. Publications by Dudley Kidd serve as examples: *The essential kaffir* (1904), *Savage childhood* (1906) and *Kaffir socialism and the dawn of individualism* (1908). Emphasis was placed on the so-called irrational nature of African mentality and belief in magic and sorcery 'in order to argue that Africans were inherently unsuited to the spirit of democratic individualism that defined western civilization' (Dubow 2006:178). Some key aspects in the struggle to establish a new (white-dominated) South African state emphasized ideals of loyalism, patriotism, imperialism, nationalism and progressivism (Dubow 2006:5). Africans were placed in the position of being subjects of scientific investigation with inherent differences from the white race. Africans were 'found' to be pre-scientific and intellectually non-progressive—a prime example of how the universalizing of science can be applied to highly particularized political purposes (Dubow 2006:178). The methodology of science could be invoked to look at the 'native question' and the question of race in a reasonable and disinterested manner. To facilitate these efforts, a large number of organizations in the newly formed Union of South Africa were instituted through the South African Royal Society (1907), ranging from engineering and architecture bodies to the Association for the Advancement of Science.

Two main instruments furthering the imperialist ambition of Britain in South Africa became prominent. The first was the newly found (white) national identity and 'South Africanization' (representing the west 'on behalf' of the empire) of science. The second was the gradual shift towards science in Africa as the property of white South Africans and the quest for the expansion of western scientific understanding 'with the responsibility of disseminating enlightened values of the European civilization to Africa as a whole' (Dubow 2006:212).

10.3 The Apartheid System of Governance (1948–1994)

The battle of the 'politics of knowledge' led to the design of a tool for implementing racial segregation. With the establishment of the apartheid system of governance in the 1950s,[8] a divide of the education system was implemented because 'an ethos of socially prescriptive scientism led to discussions about the educability of Africans and the nature of "Bantu mentality"' (Dubow 2006:229). A divide in the provision of education was therefore deemed appropriate and even necessary. During the early colonial period, special schools that only white children could attend were built, with English as the language of instruction. The overall level of education was initially considered as being of poor quality, and three-quarters of white children were considered to be illiterate by 1875 (Giliomee and Mbenga 2007:190). Education for black and 'colored' (non-white, non-African) children was in the hands of the churches and missionary institutions, and the literacy of those children was never measured. One can safely argue that all children in South Africa from the late 1800s till the early 1900s were educated in a foreign language and had little understanding of the cultural issues related to the English language.[9]

When tracing the historical development of science research and science education in South Africa, it is possible to maintain that the 1950s and 1960s were a nearly cataclysmic period that endeavored to establish a so-called 'notion of nation' that is intimately linked to the ideals of western modernity. South Africa had emerged from a debilitating war against the British (1899–1902), experienced a severe financial depression (after the 1929 Wall Street crash) and was left to its own devices while World Wars I and II were conducted in Europe and elsewhere (1914–1939).[10]

[8] The National Party took over the Union government of General Jan Smuts in 1948.

[9] To prevent complete political exclusion, the more educated section of the Afrikaners, with their burgeoning sense of nationhood, formed the *Genootskap van Regte Afrikaners* ('Association of Real Afrikaners') in 1875. In a similar vein, black South Africans formed the Native Educational Association in 1879.

[10] South African military personnel took part voluntarily in both World Wars.

After these traumatic periods, two events in South Africa proved to have lasting repercussions by retarding the development of local science communication.

The first event was linked to the preparation for the referendum on the establishment of the independent Republic of South Africa (1961) under the leadership of Dr. H.F. Verwoerd. The exclusively 'white' National Party used the 1913 Natives' Land Act that divided South Africa into 'white' and 'black' areas, followed by the setting up of 'native locations' in the 1923 Natives (Urban Areas) Act to facilitate and adapt racial segregation. A raft of newly formalized legislation was also subsequently introduced in the quest for complete racial separation, eventually leading (among other things) to the separation of education along racial lines. The superiority of the 'west' and efforts to establish modernity in Africa were within reach. Science education became a political tool to enforce further dominance (Sparg et al. 2001).

The second event was the armament program South Africa embarked upon. This included the development of nuclear energy, which soon became linked with the design of nuclear weapons. In 1959, the Atomic Energy Board of South Africa established its nuclear energy research program. South Africa became a founding member of the International Atomic Energy Agency in 1957 and established close ties with the United States and Britain, to which it was selling vast quantities of uranium (Venter 2009:59).

However, this 'uranium relationship' soon began to unravel. The Verwoerdian-style institutionalized apartheid system of governance cost South Africa its seat on the board of the International Atomic Energy Agency, and the nation was also forced out of its membership of the Commonwealth of Nations. South Africa's growing isolation, international sanctions and resistance to its internal apartheid policies by neighboring countries such as Namibia, Mozambique and Angola created political pressure, and a decision was made in 1978 to develop a nuclear weapons capability (Steyn et al. 2003:43). A very high proportion of regulated research efforts by government bodies such as the Council for Scientific and Industrial Research and the Armament Corporation of South Africa (Armscor) was channeled into the enrichment of uranium, the production of nuclear fuels and the nuclear explosives program (Steyn et al. 2003:32). Those efforts mainly took place at Pelindaba, a site some 20 km outside of the capital city, Pretoria. Several hundred scientists were involved in the research program, but only a dozen top scientists were responsible for weaponization; all worked under the strictest security (Venter 2009:37).[11] The veil of secrecy that covered the South African armaments industry spilled over into a general silence about almost all government-sponsored scientific and technological research. The public(s)—and the media—were kept in the dark for decades to follow. Science education was geared to produce (white) nuclear scientists, and the absence of efforts to popularize science reflected the apartheid government's low regard for the information needs of the general population.

[11] Towards the end of apartheid, South Africa voluntarily abandoned its nuclear armament program, and in 1993 dismantled its six nuclear bombs (Venter 2009).

10.4 Post-Apartheid Sociopolitical Reconstruction (1994–2010)

With South Africa struggling to overcome the structured 'social engineering' experiment of 'separate development' (as apartheid was euphemistically called by the white National Party), the African National Congress (ANC)[12] government, democratically elected in 1994, engaged in a process of social transformation (McKinley 1997). This process was initially and mainly characterized by the cancelation of a large number of old laws, such as legislation on Bantu education (in which races were separated), various labor-related Acts and even the infamous Immorality Act, which made sex across the racial barrier illegal. The ANC government introduced new policies, which would radically change, for example, the tertiary education landscape and restore land to original owners who were forcefully removed by the apartheid government.

It stands to reason that the legacy of more than 40 years of apartheid will not be erased overnight, and inclusive social change will perhaps take decades to transpire. South Africa is a country of social complexity and contradictions—multiple cultural differences exist in what is often described as a country simultaneously being 'first world' and 'third world'.[13] During the rule of the National Party (1948–1994), the system of apartheid effectively separated so-called 'first world' and 'third world' communities. After 1994, the two worlds are now being joined in a process of social transformation.

In Africa, political independence has contributed greatly to an increase in the acknowledgement of local science practices (IKSs) by scholars who regularly debate the existence of pre-European science on the continent. This is evident in the growing number of publications on IKSs (Jegede 1998; Makhurane 1998). According to Jegede (1998), deliberations include discussions on the link between African design and art and African science and technology. Those deliberations are leading to the incorporation of traditional technological knowledge systems into modern academic science education. South Africa succinctly reflects the complexity involved in the scientific, social and educational development of science communication resulting from merging tradition with modernity. The efforts of the ANC government to promote the public communication of science require the inclusion of traditional knowledge practices and, through policy, embrace the worlds of modern as well as traditional technologies.

---

[12] The ANC was founded in 1912. It survived decades of National Party banning orders and the incarceration of its leadership to become the ruling party in South Africa, which it has been since 1994.

[13] These 'worlds' were part of what was identified during the Cold War as an ideology-based, tripartite structure of a *first world* (western, industrialized, capitalist nations), a *second world* (centralized, command economies in communist countries) and a *third world* (new nations that were previously colonized by the first world). Clear preference was given to the capitalist structure of the developing world, whereby 'the ideological underpinnings of this asymmetric structure politicized the three groups, tainting the transfer of aid and technical assistance with propagandistic overtones' (Margolin 2007:111).

To embark on a more inclusive and efficient post-apartheid research and development (R&D) program in South Africa, a new national R&D strategy was finalized in August 2002. The key objectives were:

- *New strategic considerations for human resources in R&D*—increased investment in human capital and transformation in South Africa's science base
- *Alignment and delivery*—create an effective government science and technology (S&T) system
- *Upgrading of levels of investment and performance in R&D by the South African private sector*—because South Africa currently operates on a finite minerals-based economy, the manufacturing sector will require increased support
- *Intellectual property*—address crucial issues in legislation and legal infrastructure
- *Development of competencies across the system*—universities, research councils and the private sector should play their parts.

To achieve these objectives, examples from newly industrialized South Korea, natural resources oriented Chile and Australia, high-technology fast-follower Malaysia and R&D-intensive Finland were selected to inform the structure of new policies within the various South African government departments. The National System of Innovation was proposed in 1996, indicating a desire to promote the idea of 'innovation pull' rather than 'science push'. Role players in the system include the public and private sectors, the science councils and all local tertiary educational institutions. Long-range initiatives have included the expansion of S&T activities in collaboration with other African countries. To further that aim, the New Partnership for African Development was initiated in 2001 under the leadership of Thabo Mbeki (South African president, 1999–2008), former president Olusegun Obasanjo of Nigeria and President Abdelaziz Bouteflika of Algeria. The partnership operates as an economic development program of the African Union.

Government support for the development of PUS as a discipline is spelled out in the South African *Green paper on science and technology: Preparing for the twenty-first century* (1996), which recommended the launch of campaigns and initiatives to promote public understanding of science and technology (SA Government 1996:84). However, insufficient data is available to assess the outcome of the campaigns that took place. Under 'Point 9: Public Understanding of Science, Engineering and Technology (SET)', the green paper presented further options for the development and promotion of public awareness of S&T:

1. Institutions must be identified that can best respond to disseminating SET information to the public.
2. The kind of information that is required is to be determined by the public who would need to make informed decisions about technology-related issues.
3. The media is identified as the best channel through which SET information can be made accessible to the public.
4. Structures must be established to ensure that the flow of accessible information will actually reach disadvantaged populations, including women and rural populations.

5. Effective S&T awareness initiatives and campaigns must be launched, aimed specifically at politicians (operating at national, provincial and local levels), policymakers and decision-makers in government (SA Government 1996:84).

The South African Government initiated the National System of Innovation and established the National Advisory Council on Innovation in 1998. One of the objectives of the council is to promote the public understanding of S&T, and it is required to play a supportive role in innovation for rural development and social progress. To further those aims, a proposal to establish the Foundation for Technological Innovation was presented to Parliament in June 2007 to bridge the so-called 'innovation chasm' between industry, tertiary educational institutions and government departments. These efforts are ongoing and are yet to show concrete results at the national level.

10.5 Communicating Science in South Africa

According to Bucchi and Trench (2008:57), modern European science communication developed mainly in relation to two broad processes: 'the institutionalisation of research as a profession with higher social status and increasing specialisation; and the growth and spread of the mass media'. Recognizing the obvious intersection between mainstream science and peoples' cultural complexity of thought, new fields of research such as 'science communication' and 'public understanding of science' (PUS) were introduced internationally from the late 1960s.

In recent years (after 1994) in South Africa, science communication and PUS have been developing as separate disciplines. While PUS remains a neglected research field in local academic institutions, science communicators are progressively using the impact of the media and other channels of communication to disseminate science findings—mainly sourced from international information agencies. The focus is on a multimedia *communication process* through, for example, journalistic reporting, science museums, the press, TV and radio. PUS research, according to Martin Bauer (2008:111), covers:

> in the first place a wide field of activities that aim at bringing science closer to the people and promoting PUS in the tradition of a public rhetoric of science. Second it refers to social research that investigates, using empirical methods, what the public's understanding of science might be and how this might vary across time and context. This includes the conceptual analysis of the term 'understanding'.

Dedicated large-scale surveys that were conducted in Europe to establish people's 'science literacy' levels were not replicated in South Africa.

The 'science-in-and-of-society' paradigm (mid-1990s to present) is currently giving recognition to the fact that 'science and technology operate in society and therefore stand relative to other sectors of society' (Bauer 2008:122). This approach opened the door for new opportunities in the further development of PUS and provided, for example, an opportunity for the evaluation of existing perceptions regarding the term 'public understanding of science' (Pitrilli 2003). Initiatives to promote and develop South African S&T education and communication since 1994 have included the following:

- In December 1995, African educators and international scholars under the auspices of the African Forum for Children's Literacy in S&T participated in a conference to reflect on the continent's needs for S&T education and to plan future interventions to promote science education.
- In October 1998, the South African National Research Foundation organized an International Conference on Science and Society in Pretoria.
- In December 2002, the seventh PCST International Conference on Science Communication was held in Cape Town.
- The South African Agency for Science and Technology Advancement, a business unit of the National Research Foundation responsible for a number of community-based science communication activities, hosted the first African Science Communication International Conference in December 2006. A second international conference followed in 2009.

Despite these science communication efforts with a focus on journalism and the media, research in PUS remains undeveloped. A Human Sciences Research Council (HSRC) report, *Science and the publics: A review of public understanding of science studies* (Reddy et al. 2009), commissioned by the South African Agency for Science and Technology Advancement (SAASTA), presented an overview of the South African PUS research position. The report clearly indicates that South Africa does not have a systematic, comprehensive and nuanced assessment of the public's relationship with science. It refers to policies that indicate the need for the transformation of an economy that is resources-based to one that is increasingly knowledge-based—with the expressed aim of harnessing the growth potential of the knowledge economy for socio-economic development. Because South African society is highly stratified, the report recommends the consideration of the relationship with science of the 'public(s)'. There is also (most importantly) recognition that a public's relationship with science is shaped by the culture in which that public is located.

There had been small-scale PUS surveys, focused on one topic and mainly involving the racially classified 'white' population as respondents.[14] This led the HSRC (Reddy et al. 2009) report to reach the following conclusions:

[14] According to the HSRC (Reddy et al. 2009) report, a few surveys in South Africa were conducted in the past by the Foundation for Education, Science and Technology (FEST). FEST later became the South African Agency for Science and Technology Advancement (SAASTA). These surveys included:

- 1991: 1,300 respondents (face-to-face in white suburbs); *Understanding and appreciation of science amongst the public in SA* (Pouris 1991)
- 1993: 400 white and 400 black respondents (face-to-face among teenagers); *Understanding and appreciation of science amongst South African teenagers.* (Pouris 1993)
- 1995: Omnibus survey (HSRC)
- 2003: 7,000 respondents (face-to-face in white suburbs); *Biotechnology* (HSRC)
- 2007: 3,164 respondents; *Climate change* (HSRC) (Reddy et al. 2009).

- Science communication perceptions are still dominated by race perceptions, with a nearly complete absence of a fair demographical representation.
- The public(s)' perceptions exist in theory only.
- The public is still perceived as 'deficient', and scientists were following the by now globally contested 'deficit model' of science communication.
- A fairly recent new focus is on developing a bi-directional relationship between the public and science. Related issues include understanding the communication of messages about S&T, the dynamics of attitude and belief formation regarding S&T and, most importantly, access to information about S&T.

The HSRC (Reddy et al. 2009) report considered the 'scientific literacy' model, originally theorized by Jon Miller (1983, 1998a, b) and developed and adopted in Europe, with its preference for measuring formal science knowledge, to be inadequate for South African needs. In the South African context, there is an a priori reason to focus on practical science literacy. Gauhar Raza's (Raza and du Plessis 2002:57) comment on the complex and heterogeneous nature of society is pertinent here:

> There is an increasing global need to look for alternative models of development which are more compatible with socio-cultural structures prevalent in the so-called third world. The gap between the social, cultural and economic conditions of the west and the developing countries poses numerous problems in implementing developmental strategies as devised by the developed world. The developmental models meant for third world countries often originate in the west. The lack of understanding of culture, which is a decisive force and which inhibits or accelerates the pace of accepting science and technology in a society, introduces distortions in the social fabric. Thus a deeper insight into the cultural complexities of thought that prevail in a society is imperative for suggesting workable solutions to socio-technical problems.

The key findings of the 2009 HSRC (Reddy et al. 2009) report indicate a number of areas in South Africa that are still neglected and need attention:

- Policy in support of PUS is in place due to the efforts of the Department of Science and Technology. Policy commitment has not yet been translated into programs and projects (except for awareness strategies for biotechnology and climate change). The high incidence of 'don't know' responses in surveys on biotechnology and climate change calls for the careful consideration of items and samples to be surveyed.
- South Africa has not undertaken a systematic, comprehensive and nuanced assessment of PUS. There is, however, general agreement on the positive contribution of PUS. The 'science and society' framework is conducive to PUS research. The nation's stratified public(s) need(s) understanding. South Africa needs to develop an appropriate assessment framework.
- There is still uncertainty about the definition of science—currently epitomized by the debates about western science and IKSs.
- More efforts are needed to understand the S&T needs of the school-going population. South Africa has not undertaken large-scale PUS surveys on S&T attitudes among the children attending school.
- South Africa needs to grow an academic understanding of issues related to science communication.

- Policymakers and academia need baseline information describing key indicators. They need to build a record tracking changes over time and the public's input in policy formulation.
- South Africa needs to review conceptual and theoretical frameworks and tools to understand the impact of S&T and science communication on society.

10.6 Conclusion

Political and social transformation takes time. This process is more pronounced when a society emerges from a long history of political subjugation, social oppression and intellectual neglect. This chapter demonstrates that the development of science communication, and more specifically PUS, should be contextualized in an understanding of the specific and unique history of the African continent. The implementation of the 'idea of progress' inherent in the European Enlightenment's communication of science needs to be considered against a background of oppression and marginalization of African philosophical, epistemological and traditional practices. Based on the damaging large-scale effects of colonialism, the social engineering process of the apartheid regime and the efforts after 1994 to institute a democratic and just society in South Africa serve as example of the complexity of facilitating a fair system of science communication.

This chapter tries to illustrate the sociopolitical complexity of governance in South Africa. However, a number of debates still need to happen among science communicators to develop a nuanced and comprehensive theoretical understanding of the complexities involved in the process of communicating science to the public(s). The work of Gouthier (2005) and Greco (2005) provides guidance in this regard. The first aspect that requires detailed reconceptualization is the notion of rationality. Richard Rorty (1980) draws attention to the fact that the relation of the concept of rationality to the so-called 'philosophical dogma of the day' reflects the Kantian endeavor to present a permanent neutral framework for culture, whereby the 'framework is built around a distinction between inquiry into the real—the disciplines which are on the "secure path of a science"—and the rest of culture'. Thus, the Rortian challenge to philosophy: in its distinction between science and non-science, philosophy is endangering all current formulations, endangering philosophy itself and, with it, tampering with the concept of rationality.

Another aspect that needs further probing involves the understanding of the term 'scientific literacy' in its application to further surveys in South Africa. In this regard we need to ask some crucial questions, such as 'Whose science and what knowledge?' To bring African knowledge systems into the global lexicon of philosophical, epistemological and scientific knowledge will be a challenge for the immediate future in order to provide support for a nuanced and informed popularization and communication of science in South Africa.

References

Adorno, T. (1970). *Aesthetic theory* (trans: Hullot-Kentor, R.). Minneapolis: University of Minnesota Press.

Bauer, M. (2008). Survey research and the public understanding of science. In M. Bucchi & B. Trench (Eds.), *Handbook of public communication of science and technology* (pp. 111–129). London: Routledge.

Bucchi, M., & Trench, B. (Eds.). (2008). *Handbook of public communication of science and technology*. London: Routledge.

Cloete, M. (2008). HIV/AIDS and modernity in Africa: A philosophical perspective. *Acta Academica, 40*(2), 53–81.

Descartes, R. (1968; originally published 1637). *Discourse on method and the meditations* (Translated with an introduction by Sutcliffe F. E.). Harmondsworth, Middlesex: Penguin Books.

Descartes, R. (2006; originally published 1644). *A discourse on the method of correctly conducting one's reason and seeking truth in the sciences* (trans: Ian Maclean). Oxford: Oxford University Press.

Dubow, S. (2006). *A commonwealth of knowledge: Science, sensibility and white South Africa 1820–2000*. New York: Oxford University Press.

Eze, E. C. (1997). *Postcolonial African philosophy*. Oxford: Blackwell.

Eze, E. C. (2008). *On reason: Rationality in a world of cultural conflict and racism*. Durham: Duke University Press.

Feyerabend, P. (1975). *Against method: Outline of an anarchistic theory of knowledge*. London: NLB.

Feyerabend, P. (1987). *Farewell to reason*. London: Verso.

Gibbons, M., & Gummett, P. (Eds.). (1984). *Science, technology and society today*. Manchester: Manchester University Press.

Giliomee, H., & Mbenga, B. (2007). *New history of South Africa*. Cape Town: Tafelberg.

Godin, B., & Gingras, Y. (2000). What is scientific and technological culture and how is it measured? A multidimensional model. *Journal of Public Understanding of Science, 9*, 43–58.

Gouthier, D. (2005). Understanding science publics. *Journal of Science Communication, 4*(1).

Greco, P. (2005). What type of science communication best suits emerging countries? *Journal of Science Communication, 4*(3), 1–6.

Grint, K., & Gill, R. (1995). *The gender technology relation*. Bristol: Taylor & Francis.

Gyekye, K. (1997). *Tradition and modernity: Philosophical reflections on the African experience*. New York: Oxford University Press.

Habermas, J. (1987). *The philosophical discourse of modernity*. Oxford: Oxford University Press.

Horkheimer, M. (1939). *The social function of philosophy: Critical theory: Selected essays*. London: Continuum.

Jegede, O. (1998). The knowledge base for learning in science and technology education. In P. Naidoo & M. Savage (Eds.), *African science and technology education into the new millennium: Practice, policy and priorities*. Kenwyn: Juta.

Karp, I., & Masolo, D. (2000). *African philosophy as cultural inquiry*. Bloomington: Indiana University Press.

Ladyman, J. (2002). *Understanding philosophy of science*. London: Routledge.

Lévi-Strauss, C. (1966). *The savage mind*. London: Weidenfeld & Nicholson.

Lévy-Bruhl, L. (1985; originally published 1910). *How natives think*. Princeton: Princeton University Press.

Macamo, E. (Ed.). (2005). *Negotiating modernity*. London/Pretoria/Dakar: Zed Books/UNISA Press/CODESRIA.

Makhurane, P. (1998). The role of science and technology in development. In P. Naidoo & M. Savage (Eds.), *African science and technology education into the new millennium: Practice, policy and priorities*. Kenwyn: Juta.

Malik, K. (2008). *Strange fruit: Why both sides are wrong in the race debate*. Oxford: Oneworld.

Marcuse, H. (1964). *One-dimensional man: Studies in the ideology of advanced industrial society*. Boston: Beacon.

Margolin, V. (2007). Design for development: Towards a history. *Design Studies, 28*(2), 111–115.

Masolo, D. (1994). *African philosophy in search of identity*. Indiana/Edinburgh: Indiana University Press/Edinburgh University Press.

McKinley, D. (1997). *The ANC and the liberation struggle: A critical political biography*. London: Pluto.

Miller, J. (1983). Scientific literacy: A conceptual and empirical review. *Daedalus, 112*(2), 29–48.

Miller, J. (1998). The measurement of civic scientific literacy. *Public Understanding of Science, 7*(3), 203–223.

Mudimbe, V. Y. (1988). *The invention of Africa: Gnosis, philosophy and the order of knowledge*. London: James Curry.

Okpewho, I. (1992). *African oral literature: Backgrounds, character and continuity*. Bloomington/ Indianapolis: Indiana University Press.

Pitrilli, N. (2003). The crisis of the 'public understanding of science' in Great Britain. *Journal of Science Communication, 2*(1).

Pouris, A. (1991). Understanding and appreciation of science by the public in South Africa. *South African Journal of Science, 87*, 358–359.

Pouris, A. (1993). Understanding and appreciation of science amongst South African teenagers. *South African Journal of Science, 89*(2), 68–69.

Raza, G., & du Plessis, H. (2002). *Science, craft and knowledge*. Pretoria: Protea Boekhuis.

Raza, G., & du Plessis, H. (2003). Science and indigenous knowledge in a culturally diverse world. *Africa Insight, 33*(3), 37–42.

Raza, G., & du Plessis, H. (2004). Indigenous culture as a knowledge system. *Tydskrif vir Letterkunde, 41*(2), 85–98.

Reddy, V., Gastrow, M., & Bantwini, B. (2009). *Science and the publics: A review of public understanding of science studies*. Report commissioned by the South African Agency for Science and Technology Advancement.

Riana, D., & Habib, I. (2004). *Domesticating modern science*. New Delhi: Tulika Books.

Rorty, R. (1980). *Philosophy and the mirror of nature*. Princeton: Princeton University Press.

SA (South African) Government. (1996). *Preparing for the 21st century*. Green Paper. Pretoria: South African Government Department of Art, Culture, Science and Technology.

Said, E. (1978). *Orientalism*. London: Penguin.

Said, E. (1994). *Culture and imperialism*. London: Vintage.

Sardar, Z. (1998). *Postmodernism and the other: The new imperialism of western culture*. Washington: Pluto.

Sardar, Z. (2002). *Introducing science*. Cambridge: Icon Books.

Serequeberhan, T. (2002). The critique of Eurocentrism and the practise of African philosophy. In P. H. Coetzee & A. P. J. Roux (Eds.), *Philosophy from Africa: A text with readings* (pp. 64–78). Cape Town: Oxford University Press.

Slater, P. (1977). *Origin and significance of the Frankfurt School: A Marxist perspective*. London: Routledge/Kegan Paul.

Sparg, M., Schreiner, J., & Ansel, G. (Eds.). (2001). *Comrade Jack: The political lectures and diary of Jack Simons, Novo Catengue*. Johannesburg: STE Publishers.

Steiner, C. (1994). *African art in transit*. Cambridge: Cambridge University Press.

Steyn, H., Van der Walt, R., & Van Loggerenberg, J. (2003). *Armament and disarmament: South Africa's nuclear weapons experience*. Pretoria: Network.

Torgovnick, M. (1990). *Gone primitive: Savage intellects, modern lives*. Chicago: University of Chicago Press.

Venter, A. J. (2009). *How South Africa built six atom bombs and then abandoned its nuclear weapons program*. Kyalami Estate: Ashanti.

Wajcman, J. (1991). *Feminism confronts technology*. Cambridge: Polity.

Chapter 11
An Experience of Science Communication in Korea: The Space-Sharing Project with Mass Media

Sook-Kyoung Cho

Abstract Korea's success is attributed to outstanding human resources and highly competitive science & technology (S&T). However, two serious social issues relating to S&T occurred in 2002: youth did not want to study science, technology, engineering and mathematics (STEM) at college level, and scientists had lost their eagerness to do future research. There were many heated debates and serious discussions to find long-term as well as short-term remedies for those problems. The direct and immediate response was the start of Science Korea Movement to enhance public awareness of the importance of S&T. Ten projects were proposed, of which the Space-sharing Project, which published a newspaper science section once a week, was the most successful. During the 18 months of the Project, the science section attracted great attention not only from scientific communities but also from Korean society at large. Many topics relating to S&T which were rarely covered in mass media were dealt with and many of the debates were resolved during or after the Project. Also, project stakeholders gained benefits in one way or another. The Space-sharing Project was unique in that it was strongly government-driven and was based on the social consensus among the Korean people. However, it also had some limitations in the issues of sustainability as well as motivation.

Keywords Science culture in Korea • Space-sharing with newspaper • Science communication • STEM • Science Korea Movement • Students avoid S&T • Public awareness of S&T • Creativity • Involvement • Collaboration • Not-for-profit organizations

S.-K. Cho (✉)
Korea Foundation for the Advancement of Science and Creativity (KOFAC), Seoul, Korea
e-mail: skcho@kofac.re.kr

B. Schiele et al. (eds.), *Science Communication in the World: Practices, Theories and Trends*, DOI 10.1007/978-94-007-4279-6_11,

11.1 Introduction

Korea rose from the ashes of the Korean War to achieve a successful industrialization, known as 'The Miracle on the Han River'. In the 1960s, Korea had a tiny economy; however, by 2009, it ranked 15th in the world in gross domestic product (GDP) and its trading volume was the 9th largest in the world (IMD 2009). Many people say that the great economic success of Korea is attributable to outstanding human resources and highly competitive science & technology (S&T). In 2009, the government allocated 3.37% of GDP to R&D, which is world's fourth highest proportion and much higher than the goal of the European Commission, which aimed for 3% in 2010. More than 80% of Korean high school graduates go to college, which is the highest proportion among the OECD countries (IMD 2010).

Owing to S&T-friendly policies and strong educational motivation and practice, Korea, which used to receive financial assistance from developed countries after the Korean War, has developed to become an aid donor. However, two serious social issues relating to S&T occurred in 2002: first, students no longer wanted to study S&T for their future careers[1]; second, scientific communities had lost their eagerness to do research due to low social morale after the Asian financial crisis of 1997–1998.[2] There were many discussions in different sectors of society to find long-term and short-term remedies, and the Korea Science Foundation (KSF), now the Korea Foundation for the Advancement of Science and Creativity (KOFAC),[3] inaugurated a national campaign called the Science Korea Movement to enhance scientific literacy and awareness among all Koreans.

In this paper, I explain the Science Korea Movement through its objectives and programs. Among those programs, I focus especially on the Space-sharing Project, the aim of which was to publish a newspaper science section once a week in collaboration with a major newspaper company. I discuss the main factors in the success of the project, list some notable outcomes and raise some issues to be considered when a government-driven project or program plans to collaborate with the private sector. Before that, I briefly introduce KOFAC, a non-profit organization (NPO), with its main activities.

11.2 The Functions and Main Activities of KOFAC

In 2008, KOFAC was reborn from the KSF, which was devoted to science popularization during the 40 years from 1967. KOFAC is a non-profit organization fully supported by the Korean Government through the Ministry of Education, Science

---

[1] *Dong-a Daily*, 17 February 2002.

[2] *Yonhapnews*, 27 March 2002. Because of the role of the International Monetary Fund in economic reforms, the Asian financial crisis is referred to in Korea as the 'IMF crisis'.

[3] Korea Foundation for the Advancement of Science and Creativity (http://www.kofac.re.kr).

and Technology. Its aim is to contribute to national development by popularizing S&T and by enhancing public awareness and understanding of science (MOST 2008). The core programs during the 1990s were the running of 'science buses', which were full of experimental kits and books, and the Korea Science Festival, which is the nation's largest hands-on scientific event (KSF 2007).

In 2001, the KSF initiated the Science-Book Start program, in which selected science books donated by the scientific communities and others were sent to every school across the country (KSF 2007). In 2006, the KSF hosted the ninth International Conference on Public Communication of Science and Technology, which was the first time the conference was held in Asia (KSF 2006c). From 2008, KOFAC expanded its functions greatly to cover informal as well as formal science education, targeting not only school students but also the general public. It now deals with science culture, science communication, public understanding of global issues, informal science education for youth, STEM (science, technology, engineering and mathematics) education in schools, and coalitions among the sciences, arts and humanities (Cho 2011). In the following sections, the main activities of KOFAC are introduced under the three headings of involvement, creativity and collaboration.

11.2.1 Involvement

In order to encourage more involvement by academics, professional societies, non-government organizations (NGOs) and individuals, KOFAC runs the Science Culture Grants scheme and supports universities to do research on science culture, science communication and the public understanding of science (Cho 2003). In addition, it jointly runs master's degree courses in science communication at Sogang University[4] and science journalism master's degree courses at the Korea Advanced Institute of Science and Technology.

KOFAC also serves the public by distributing scientific knowledge and information using cyberspace. The internet newspaper *Science Times* and a science portal, ScienceAll.com deal with issues, events, seminars and conferences relating to S&T. For school youths, there are a variety of programs such as Science Ambassadors, the Youth Science Club, the Everyday Science Class and the International Youth Science and Engineering Camp. The Science Ambassadors, consisting of more than 1,000 scientists, travel to schools and give special science lectures related to their research areas. In 2009, a new program of Science Volunteers (Gwa-whal) was started, in which more than 1,000 college students from 80 universities go to underprivileged areas and schools to provide hands-on science classes (Chung 2011).

In 2009, to enhance public understanding of global issues, the RGB campaign was started. R (red) stands for disease issues; G (green) stands for climate change, energy and food shortage issues; and B (blue) stands for water management.

---

[4] Sogang University (http://www.sogang.ac.kr).

Under the RGB campaign, TV documentary programs, books and cartoons on climate change and books on food shortages were developed and distributed in collaboration with the UNESCO (KOFAC et al. 2010a, b, c).

11.2.2 Creativity

To enhance the creativity of society as a whole, KOFAC handles national school curriculums, new types of schools, school science and maths textbooks and teacher training. The Korean Government has developed next-generation science textbooks, which are experiment-oriented narratives full of pictures, for middle and high school students. In 2010, a new type of school was introduced: the 'science core high school', where the science and maths curriculum is taught for 35–45% of classes. Compared to science high schools and science academies for the gifted, science core high schools aim to enhance ordinary students' creativity through science and maths education.

To promote creativity in schools, KOFAC is collecting, developing and distributing world-class science and maths resources, such as STC, Foss (from the United States), BBC Motion Gallery (from the United Kingdom) and La main à la pâte (from France). These educational materials are used for students through leader teacher training by invited world's-best academics and researchers. An education donation movement, in which industrial organizations, universities and research institutes donate their research and industrial facilities and human resources, is very active. Recently, KOFAC introduced the Honours Programme and Undergraduate Research Programme to promote the creativity of the gifted at college level.

11.2.3 Collaboration

Another of KOFAC's important functions is to encourage collaboration among different sectors of the community through a series of activities and forums for scientists, humanists and artists. In the Coalition Café, people from different disciplines gather and discuss specific topics, such as 'Soccer meets science,' 'Brain talks music,' 'Wine tastes science' and 'Science enjoys library'. Scientists, musicians, historians and librarians talk about aspects of soccer, the brain, wine and libraries and find collaborators for future work. Also, at the end of the year, many outcomes from the grants scheme, such as science theater plays, science visualization projects and storytelling projects, are exhibited at a coalition festival. KOFAC has now launched a new paradigm of science education, STEAM, in which science, technology, engineering and maths are combined with the liberal and fine arts.

Table 11.1 Science Korea movement programs

Target group	Programs
Youth	Science class for daily life
Youth	Youth science club
Youth	Science ambassador
Youth/teachers	Scienceall.com (www.scienceall.com)
Children	Science playground
College students	STS curriculum
Opinion leaders	Science for leaders
Opinion leaders	Space-sharing project
General public	Korea science festival
General public	New science museum
General public	Time-sharing project

11.3 Background of the Space-Sharing Project: The Science Korea Movement

The Science Korea Movement, aiming for science for all Koreans, was officially announced by the acting president of Korea on Science Day in 2004, although some projects had been conducted from 2002.[5] The fact that most of the best high school graduate students preferred medical schools to basic research was perceived as a serious problem for the Korean economy. Scientific communities as a whole, after bad experiences during Korea's IMF crisis, lost their eagerness to do research and even objected to their children studying S&T in universities.[6] After much argument and discussion, the Science Korea Movement was proposed as a solution.

The movement was a national initiative to advance Korean society through social innovation on the basis of science culture and science communication. The objectives were promoting the importance of science in every corner of society, laying the foundation for scientific knowledge creation and making people understand the value of creativity, effectiveness and rationality (Cho 2007b). Every sector of society, from government, scientific communities, universities and industries to the mass media and NGOs, actively participated in the movement. KOFAC was designated as the secretariat, and 10 main projects and programs were proposed (KSF 2004). Some were finished and some are still going with great success (see Table 11.1).

Among those, the Space-sharing Project targeting opinion leaders and opinion makers was begun on the basis of national surveys of the public's attitude towards S&T. Every 2 years from 2002, KOFAC surveyed 1,000 members of the public from all parts of the country. In 2002, only 2.9% of newspaper coverage and 8 among 750 regular weekly TV programs dealt with science-related issues. Asked where they obtained science-related information, 50.3% of respondents nominated TV and 20.5% nominated newspapers. Opinion leaders and opinion makers preferred newspapers over TV (KSF 2002).

[5] *Seoul Economic Daily*, 20 April 2004.

[6] *Science*, vol. 295, 15 March 2004.

11.4 The Aims and Method of the Space-Sharing Project

In 2002, the KSF and *Joongang Il-bo Daily*, one of Korea's major newspapers, agreed to produce an S&T section for the paper. Titled 'Science and Future', the eight-page section was published once a week; two pages were handled by the KSF. The project, which was a first for Korea, had two aims. One was to induce youth to study S&T not only by enhancing their parents', their teachers' and society's understanding and awareness of the importance of S&T but also by introducing and praising star scientists as role models. The other was to revitalize scientific communities by encouraging scientists in their research activities and by offering them media space to express their opinions on science-related issues.

The project started from 31 July 2002 and lasted 18 months, until the end of January 2004. *Joongang Il-bo* advertised that 'we produce this paper with scientists. Please send your stories, ideas, research works and cutting-edge science to us by email.[7]' A special committee of professors, researchers, science reporters and KOFAC staff coordinated the content. The committee discussed issues and topics relating to S&T and recommended scientists, engineers and chief technology officers who could be role models for youth. The Science and Future section covered many current and potential science-related issues, agendas and policies that had rarely been featured in the media, such as 'a national innovation system based on S&T', 'science communication in media era', 'women in S&T', 'S&T and politics', 'science meets the arts and humanities' and 'innovating science education'.

One of the topics handled in October 2002 was the importance of science communication and the necessity of cultivating and training science communicators, including science reporters. Under the title of 'Press had a blind spot: scientific information', the section criticized the poor coverage of science in the mass media. It then introduced a successful example of connecting scientific communities and the media—the United Kingdom's Science Media Centre—and compared Korea's mass media science coverage with coverage in advanced European countries. The article strongly suggested the opening of a Korean TV channel devoted wholly to S&T.[8]

Another topic, which in fact received the greatest response from the readers, was a comparison of the political parties' scientific agendas. A national presidential election was due in 2004, and each party promised new scientific policies. There were many demands and opinions from scientific communities about the most urgent S&T issues for the government. The Science and Future section emphasized the role of political leaders in the development of S&T by citing the historic examples of the French Acadèmie des Sciences during the reign of Louis XIV and the École

[7] *Joongang Il-bo*, 31 July 2002.

[8] *Joongang Il-bo*, 24 October 2002.

Fig. 11.1 'Science and Future' section in Korean newspapers (photo © Sook-kyoung Cho 2005)

Polytechnique under Napoleon, and then compared the Korean political parties' policies and priorities for national R&D. The section reported diverse opinions on new policies among scientists from universities, research institutes, science policymakers and NGOs, one of which was to reinforce the functions of the Ministry of Science and Technology in government.[9]

For the project, KOFAC contributed nearly $800,000 and convened the coordinating committee once a week. *Joongang Il-bo* allocated two full-time reporters to cover the topics and items discussed in the committee meetings. Except for the science reporters and KOFAC staff, the committee membership changed from time to time depending on the topics being covered. The committee dealt with research priorities; the social status of scientists; encouraging youth into the STEM area; women in science and technology; the ethical, legal and social implications of S&T; and the coalition of science and art. The members also recommended scientist role models in Korea and overseas and wrote a series of columns expressing their personal opinions.

[9] *Joongang Il-bo*, 5 December 2002.

11.5 Some Outcomes and Legacies of the Project

11.5.1 Project Proposals Realized

One of interesting outcomes of the project is that many proposals raised in the newspaper sections were realized during the life of the project. For example, the need for science communication and for the cultivation and training of science communicators, which was raised in 2002, was partly met by the opening of new university courses. In 2003, a short course called 'Science Culture Academy' and a master's degree course in science communication began with the support of KOFAC. Each 10-week course of Science Culture Academy, which was delivered for four terms a year, was attended by nearly 40 scientists, academicians and public relations people from research institutes, industry and universities. To date, more than 1,000 people have gained certification in science communication through Science Culture Academy. By 2010, 16 students had graduated from the 2-year master's in science communication.[10]

Following this success and responding to more demand, KOFAC instituted a master's degree course in science journalism at the Korea Advanced Institute of Science and Technology in 2009. Compared to science communication courses, the science journalism course is oriented more to cutting-edge scientific knowledge and information.

In addition, Korea's first television science channel, *ScienceTV*, was established by the YTN broadcasting company under the Ministry of Science and Technology, after a great deal of effort from 2002 by KOFAC.[11]

After the argument for reinforcing the functions of the Ministry of Science and Technology was raised in 2002, the newly elected president created a position of presidential adviser on science, technology & IT in his office for the first time. This is a very important and powerful position, providing advice to the president about the appropriateness of national R&D and coordinating the functions of various ministries. The first adviser was one of the committee members working for the Space-sharing Project, and the second adviser was a role model scientist introduced in the newspaper section.[12] In addition, the Ministry of Science and Technology was elevated to the vice prime ministerial level, with the power to coordinate the whole R&D budget, in 2004.[13]

The third and most remarkable example was the many policies made for women scientists and engineers. From 2003, the Ministry of Education and Human Resources urged the universities to accept a quota system for women professors.

---

[10] http://www.kofac.re.kr/academy

[11] *Digital Times*, 17 September 2007.

[12] *Joongang Il-bo*, 2004.

[13] *Hankook Economic Daily*, 1 September 2004.

In 2004, the WISE (Women into Science and Engineering) program was greatly expanded to cover local areas. WIST (Women into Science and Technology) and WATCH 21 (Women's Academy for Technology CHanger in the 21st century) were started by different ministries, and the WIE (Women into Engineering) program began in 2006. In 2011, all the programs for women scientists were integrated under the name of WISET.[14] Korea is now one of the most advanced countries in its policies covering female scientists.[15]

11.5.2 The Visibility of Science and Scientists Greatly Enhanced

Another interesting project outcome is the much greater visibility of science and scientists. The scientists who worked on the project and those introduced in the newspaper section attracted great public attention, becoming well known to scientific communities and then to society as a whole. The newspaper company strengthened its position among the mass media, and KOFAC raised awareness of its own brand (KSF 2006a).

For example, one of the committee members, Dr. Kim (a professor and the director of the course in technology policies at the college of engineering at Seoul National University), was appointed as the first science, technology & IT adviser to the Korean president. He appeared in the paper as a committee member in 2002 and was appointed to the adviser position in 2003.[16] Although it is not easy to measure how much the project contributed to his promotion, it seems clear that the newspaper section enhanced his public visibility. His work as presidential adviser was reported not only in the science section but also in other newspapers and media. As a result, the public's attention to science-related policies and issues was greatly increased.

The visibility of science reporters also received a boost. The leading reporter for *Joongang Il-bo* won a great reputation as a science journalist not only at his own newspaper but also among other journalists. At first he was merely a reporter handling S&T, but he gained the title of 'science specialist reporter' for the first time in Korea, after which he became well known to scientists and the public. As a result, he received awards from the Korea Science Journalists Association (KSJA) and some scientific sectors.[17] In 2008, he won the Korea Science Culture Award from

[14] Korea Advanced Institute of Women in Science, Engineering and Technology (http://www.wiset.re.kr).

[15] http://www.wiset.re.kr.

[16] *Inews24*, 23 February 2003.

[17] *The Kyunghyang shinmun*, 30 November 2003.

the Minister of Science and Technology.[18] He has also twice consecutively elected as the president of the KSJA. Under his leadership, the functions and activities of the KSJA were greatly expanded. In 2011, he won the national medal for science popularization for contributing to the first successful newspaper science section.[19]

11.5.3 Demands from Other Newspaper Companies

Another outcome of the project was a great demand for science sections in other newspapers. When KOFAC first asked the newspaper companies about participating in the Space-sharing Project at the beginning of 2002, they were very reluctant and needed much persuasion. However, as the project went ahead and gained approval from scientists and the public, the other major newspaper companies showed strong interest in publishing their own S&T sections.

After some criticism, KOFAC withdrew the project and revised the way S&T coverage was to be handled in newspapers. Instead of supporting special newspaper sections, KOFAC introduced a new grant scheme for newspaper companies to allocate spaces for S&T-related issues. Under this scheme, 5–8 newspapers have dealt with scientific issues every year since 2005 (KSF 2006b).

11.6 Key Factors for Success and Some Issues

The Space-sharing Project, which went for a relatively short period of 18 months compared to other science communication programs, had great outcomes. Three main factors seem to have been very important in that success. First, this was a strongly government-driven project. Because the Asian countries industrialized very late compared to western nations, they are more oriented to central governments to achieve efficiency and rapid development, particularly Korea. The project was possible because it was handled by KOFAC, which, although it is not a government body, is wholly supported by the government.

Second, there was a social consensus among the Korean people. It was at a time of national crisis, in a sense, that people identified students' reluctance to study S&T for their careers as a serious problem for the future economy. Various sectors of society talked about remedies and agreed to find both short-term and long-term solutions.

Third, the personal eagerness of the people involved drove the project. Scientists, science journalists and KOFAC staff worked hard to find a series of issues and scientists in order to revitalize scientific communities. Science journalists and

[18] *Joongang Il-bo*, 3 November 2008.

[19] *Joonagng Il-bo*, 21 April 2011.

KOFAC staff frequently exchanged ideas not only in committee meetings, but also by email and by mobile phone.

However, the project raised some issues and questions for further consideration. One is sustainability. It was originally expected that the project would build its own momentum, but when economic support was withdrawn, the S&T section paper was replaced with a page only at *Joongang Il-bo*. The section failed to achieve its own momentum or logic to continue. How can we make a close government-driven collaboration with the private sector sustainable? (Cho 2007a).

The other issue is motivation. Scientists and science journalists who were involved directly or indirectly in the project gained great social attention and were rewarded with promotions within Korean society and their communities, but the staff working for KOFAC did not. Is it fair to consider this situation as par for the course? Is it enough to appeal to staff's sense of mission or social responsibility? How can we motivate staff of NPOs to work even harder?

References

Cho, S.-K. (2003). *Science culture: Its meaning and its task.* Science and Technology Policy Institute (STEPI), Korea.

Cho, S.-K. (2007a). *Exploring policy for science culture in Korea through the analysis of core programmes of 'Science in Society' of the European Commission.* Science and Technology Policy Institute (STEPI), Korea.

Cho, S.-K. (2007b). Science communication as practices of science culture. *Journal of Science Technology Studies, 7*, Korea Association of Science Technology Studies.

Cho, S.-K. (2011). *Korean experiences of science culture and STEAM education.* France: IHEST Summer University.

Chung, Y. (2011). *Low carbon green growth and RGB campaign in Korea.* 2011 AAAS Annual Conference, Washington, DC.

IMD (Institute for Management Development). (2009). *IMD world competitiveness yearbook 2009.* IMD.

IMD (Institute for Management Development). (2010). *IMD world competitiveness yearbook 2010.* IMD.

KOFAC, UNESCO Korean office, & KEMCO. (2010a). ESD Series I. *Our planet.*

KOFAC, UNESCO Korean office, & KEMCO. (2010b). ESD Series II. *Energy.*

KOFAC, UNESCO Korean office, & KEMCO. (2010c). ESD Series III. *Climate change.*

KSF (Korea Science Foundation). (2002). *Public attitude toward science and technology.* KSF.

KSF (Korea Science Foundation). (2004). *Action plan for Science Korea Movement.* KSF.

KSF (Korea Science Foundation). (2006a). *Public attitude toward science & technology.* KSF.

KSF (Korea Science Foundation). (2006b). *2006 Annual report.* KSF.

KSF (Korea Science Foundation). (2006c). *The final report of PCST-9 in Seoul.* KSF.

KSF (Korea Science Foundation). (2007). *40 years history of Korea Science Foundation.* KSF.

MOST (Ministry of Science and Technology). (2008). *40 years history of science and technology in Korea.* MOST.

Chapter 12
From Science Popularization to Public Engagement: The History of Science Communication in Korea

Sung Kyum Cho and Ock Tae Kim

Abstract Most Koreans agree that science and technology are critically necessary for national development and individual welfare, and that scientists are important people who work hard to develop science and technology as a foundation of the country's economic growth. However, technological and scientific skills were not always respected in traditional Korean society. The modernization of science and technology was neglected due to successive historical incidents that occurred in Korea. However, science communication has played a role in changing sociocultural attitudes toward science and technology. Over the past few decades, institutions of science and government have adopted three different phases of science communication: popularization of science, public understanding of science and then public engagement in science. In the first phase, popularization of science was a government-led promotion of science to efficiently deploy scientific knowledge from top to bottom. In the second phase, public understanding of science was enhanced by inducing people to participate in scientific events and exhibitions hosted by non-government organizations. Finally, in the third phase, public engagement in science was increased by emphasizing social responsibility and citizen participation in the development of scientific and technological policies. Despite considerable success in diffusing scientific knowledge to the public, Korea is still experiencing

S.K. Cho (✉)
Department of Communication, Chungnam National University,
Daejeon, South Korea
e-mail: skcho@cnu.ac.kr

O.T. Kim
Institute of Communication Research at Seoul National University,
Seoul, South Korea
e-mail: ocktopia@snu.ac.kr

B. Schiele et al. (eds.), *Science Communication in the World: Practices, Theories and Trends*, DOI 10.1007/978-94-007-4279-6_12,

problems resolving some science issues, as differences in judgment make it difficult to reach a consensus on science and technology policies. This implies that scientific knowledge alone cannot resolve the differences, and that the process of judgment needs to be highlighted. The process of judgment can be improved by using scientific methods of thinking and problem solving, rather than just accumulating scientific knowledge.

Keywords Science communication in South Korea • Science and technology policy in South Korea • Science popularization • Public understanding of science • Public engagement in science

12.1 Introduction

A survey of the Korean population (KFASC 2008) showed that 75% of adolescents and adults reported a strong interest in science and technology (S&T). An even greater proportion, 85–90% of respondents, seemed to have a positive attitude towards S&T, saying that it will have a positive effect on our life and work, the next generation, and the nation. The importance of S&T is further reflected in the belief that, as a profession, scientists have the most important role in the development of our country, above executives and educators. Most people agreed that S&T is a critical area necessary for national development and the future of individual welfare, and that scientists are important people working hard to develop S&T as a foundation for the country's economic growth (KFASC 2008).

However, technological and scientific skills were not always respected in traditional Korean society. Although Korea has a history dating back more than 2,000 years of making outstanding scientific and technological innovations, such as the world's first movable metal type, the world's first iron ship and the rain gauge, the Joseon Dynasty, which ruled the Korean peninsula from the fifteenth century to 1910, neglected S&T in its adherence to the fundamental rule of a four-class Confucian society: scholars, farmers, artisans and tradesmen.

The Japanese colonial period from 1910 to 1945 also delayed the modernization and development of S&T. The Korean War from 1950 to 1953 completely destroyed the nation's existing S&T base. In the mid-twentieth century, science and technology were condemned as the hard, dirty work of a lower class of people.

We have to credit the role of science communication in changing the sociocultural attitude towards S&T. Over the past few decades, in an attempt to promote a positive relationship between science and the public, the institutions of science and government have adopted different models of science communication, such as the popularization of science, the public understanding of science and the public engagement in science. The history of Korean science communication can be explained in these three phases.

12.2 Phase 1: The birth of Science Communication (1967~1990): Science Popularization

After the Japanese colonial experience, the pain of war, and political turmoil, Korea started to achieve economic growth through industrial development in the 1960s. In the 1970s, the Korean Government initiated massive investments in the chemical, steel and other heavy industries. Since then, science and technology have been highlighted as the nation's most important means for achieving modernization, promoting economic development and enhancing the quality of life of Koreans.

In particular, 1973 has been regarded as a notable year in the history of science communication in Korea (Kim 2007a). In his New Year press conference, President Park Chung-Hee officially announced the start of the National Movement of Scientification campaign. The campaign had three purposes: it intended to encourage ordinary people to take advantage of scientific knowledge in everyday life; it spurred people to acquaint themselves with at least one scientific skill or ability that could contribute to improving one's life and country; and it intended to spread the government's strategy and policy of investing national resources in S&T.

This campaign was deployed to support a highly industrialized society by increasing the public's scientific thinking and living habits. It was based on the belief that public knowledge of science should be a major element in increasing the quality of life and national prosperity. Thus, government departments, the S&T community, industry and the media were urged to pursue those goals.

The modern history of Korean public science communication stems from the establishment of the Association for Supporting Science and Technology in December 1967, which was restructured as the Korea Foundation for the Advancement of Science and Technology in 1972. The foundation's real activities began with the National Movement of Scientification.

The foundation's first order of business was publishing S&T books, because the scientific book publishing business at that time in South Korea was almost non-existent due to lack of investment in social infrastructure after a long period of historical incidents (Kim 2007b). By 1976, the foundation had supplied a total of 153,000 copies of high-quality scientific books to classrooms across the nation (STEPI 1997). This project has been continued ever since to keep pace with consumers' demand for scientific knowledge.

The Korea Foundation for the Advancement of Science and Technology also introduced films about S&T to the public. A science film library project purchased, translated and dubbed foreign science and science fiction films and screened them at schools and public institutions, particularly in remote rural areas. The foundation also supported junior-school pupils' field trips to science institutions, such as atomic energy research institutes and science technology parks. Lecture tours by prominent scientists across the country, which began in 1972, helped to increase youth's understanding of science and to inspire and motivate them to become scientists (MST 1972).

The Housewives Life Science Course (science lectures targeting housewives, first offered in 1973) educated housewives about how scientific knowledge could be adopted in their household management of food, clothing, housing and health care. In addition, S&T exhibitions were held and science museums were built to promote public awareness of the social contribution of S&T (STEPI 1997).

The Korean Government's approach to popularizing science in this phase can certainly be regarded as a good example of the so-called 'deficit model' of science communication. This approach is usually defined as the assumption that scientists have sufficient scientific knowledge and the public is ignorant about science and lacks appreciation for it (Wynne 1991; Ziman 1991). It also suggests that the public would accept science readily if it knew more basic scientific vocabulary and had some level of understanding of scientific methodology and of scientific (and technological) impacts on society (Miller 1983).

Scientists were asked to act as the main agents of science communication to persuade people of the value of science by becoming more effective at spreading the word. Thus, the flow of scientific information was seen as basically one-way communication from the scientist to the public. The passive public was conceived of as attending with interest to the message, comprehending its content, and subsequently adopting a positive attitude towards science (Kim 2007). Thus, the government focused on providing space, such as S&T museums, and spreading scientific knowledge, which was done by a small number of scientists and science teachers who worked directly with the public (STEPI 1997).

Although it is common in modern western states to regard S&T as a means of civil enlightenment, science communicators in Korea have emphasized the practical application of science. During this time, economic growth was a priority in Korea as part of the nation's attempt to catch up economically with the developed nations. Because science and technology were important factors for economic development, communicators tried to emphasize how they should be used for economic growth. As evidence of this, the government justified its national science policy in the third 5-year plan for economic development (1972–1976), which focused on investing natural resources in economic prosperity rather than social welfare (Arnold 1988).

12.3 Phase 2: The Expansion of Science Communication (1990~2004): Public Understanding of Science

In November 1990, Anmyeondo, a small island off the west coast of Korea, caught the attention of the nation. When the government announced plans to build a low-level radioactive waste disposal facility on Anmyeondo, opinions were divided. Some residents, scientists and local politicians expressed support, but other residents, anti-nuclear NGOs and civil activists opposed the plan. Severe violent confrontations between the two parties made daily news headlines for several weeks.

The Anmyeondo situation meant the failure of two decades of determined effort to find sites for low-level waste disposal since Korea's first nuclear power plant began generating electricity in 1978. The president, National Assembly members, local executives and council members, and provincial governors were unable to counter anti-nuclear demonstrations and opposition from local residents and civic environmental organizations.

During that phase of authoritarian government, the protection of personal gains and collective protests against government science policies were dismissed as opposition to national development. However, the first democratic presidential election in 1987 fundamentally altered the social atmosphere in a way the government had failed to appreciate. The Anmyeondo situation had important implications for science and society. The government now began to understand the necessity of public support for S&T policies (Yim 2002).

This social change resulted in a different kind of science communication in the 1990s. Government science communication activities were mostly conducted through the Korea Foundation for the Advancement of Science and Technology, designated as a dedicated public organization for promoting public understanding of S&T, based on the *Promotion of Science and Technology Act 1991*.[1] The foundation was expanded and restructured as the Korea Science Foundation in 1996 and its formal operation was specified in the Special Law for Science and Technology Development (1999).[2]

Since 1991, the foundation has opened doors for the public to participate in S&T-related events, largely aimed at improving and strengthening public understanding of science. In particular, the foundation attempted to promote personnel and information interchange between science and other fields, such as the arts and humanities. The first Science + Art Exhibition (1991) received considerable acclaim. Subsequently, the South Korea Joint Symposium on Scientific and Cultural Organization, a symposium for women in science popularization, was held (STEPI 1997). The Science and Technology + Politics seminar (1993) was in line with efforts to incorporate S&T into policies by exploring and sharing information with political parties. The Korea Science Festival was first held in 1997 with the theme of 'The New Millennium', along with 'Science and Technology' in Seoul, which 150,000 people attended (STEPI 1997). This annual festival is a public event enabling the public to explore and experience a variety of activities related to S&T, including live experiments, exhibitions, science movies, book readings and tours. It takes place over a week and aims to increase people's interest and support for science and to foster scientific understanding. Since the fourth festival in 2000, the Cyber Science Festival has also been held in order to overcome the limitations of one-time events.

Science fairs for regional students, at which they are encouraged to participate in and enjoy the practical applications of science, have been held since 1998 with the support of local departments of education and other institutions. The fairs have

[1] Korean Law No. 4402. Retrieved 5 February 2011 from www.law.go.kr.

[2] Korean Law No. 5718. Retrieved 7 February 2011 from www.law.go.kr.

grown into some of the largest scientific events in Korea. Science ambassadors, who are opinion leaders in science and engineering, have conducted lectures in schools and community centers all over the nation and have donated books and scientific instruments since 2002.

Also at this time, mass media such as broadcasting and the internet began to be used actively as communication channels. The Korea Science Foundation sponsored the creation of TV programs related to S&T. Science shows such as *Curious Heaven*, *Da Vinci Project*, *Science Café* and the *Quiz Republic of Korea* quiz program have been broadcast on various networks. The nation's first science cable channel, Science TV, first went online in 2007. Since 2001, public service announcements about science and culture have been aired by broadcasting companies.

Due to the emergence of the internet as an important means of communication, especially for the younger generation, the foundation needed to utilize cyberspace as a forum for public relations related to S&T. In order to accomplish this, a comprehensive information network for explaining science, technology and culture, *Science All*,[3] was launched in April 1999; it now has an audience of 30 million internet users. In the beginning, *Science All* consisted of 23 main menu options, including a virtual science laboratory, and 146 submenus. According to the *Science All* homepage, the website received over 20 million hits in 2003, when it had 500,000 members. Established in April 2002, a science webcasting service by *Science All*[4] presented six channels, including The Cyber Science News, Science & Technology Seminar, Click! Scientific Inquiry, Science Show, and Today's Scientific Information. The internet newspaper service *Science Times* has become one of the leading science publications, and contributes to the government's science agenda by providing special feature articles about scientific issues.

During this period, media content relating to science communication increased in quality as well as quantity, and diverse social groups made an effort to provide content for public understanding. Journalists, public relations experts, event and exhibition specialists as well as scientists and science educators were involved, and many of them became adept in using media and communications technologies for science communication. Science communication channels also expanded in this period from the mainstream media to scientific events such as festivals and experiment exhibitions. The diversified formats and media made content delivery much more efficient.

Still, the critical criterion for science communication with the public was the improvement of public understanding of science (or 'scientific literacy') through the direct dissemination of scientific information to the public. People were considered to be passive consumers of humorous and simple representations of science and its knowledge. We can say that the main characteristic of this phase was the expansion of one-way science communication, but also that Korean science communication failed to respond to social and political issues related to S&T.

[3] www.scienceall.com.

[4] www.scienceall.tv.

12.4 Phase 3: Maturity and New Challenges (2004~Present): Public Engagement in Science and More

A new Korean Government launched a plan to build an S&T-oriented society as a major national project in 2003. The Ministry of Science and Technology prepared the Science Korea Movement, which was promoted by the Korea Science Foundation as a specific action plan. In 2004, the President announced that 'the hope of reaching a GDP per capita of $20,000 rests on the shoulders of scientists and engineers', and that 'a priority of science and technology policy would be in inspiring innovation and training engineering talents' (Presidential Secretariat 2004:601). Korea began to emphasize social responsibility and citizen participation in the process of developing S&T policies in the mid-2000s (PACEST 2007).

In 2003, technology assessment was conducted on a trial basis to minimize the potential risks of new S&T, and since then Korea's scientific community has continuously expanded its use. In 2004, a consensus conference for nuclear power policy was held with NGOs and environmental groups, such as Civic Solidarity for Participatory Democracy (PACEST 2007).

In 2005, the national government called a referendum to identify a site for a low-level waste disposal facility. In the same year, life science research ethics and legislation were enacted to educate scientific researchers about ethics after a heated debate about stem cell research (MST 2006). In addition, public understanding of science activities were deployed, for example to foster science culture by pursuing the convergence of science with the arts, humanities and social sciences and by disseminating converged high-quality science and culture content.

One of the hallmarks of this stage is the emphasis on the social responsibility of science. The government made efforts to establish a bi-directional information flow to encourage public engagement in the process of science communication. Another characteristic of this phase is that, although public engagement in science is encouraged, one-way hierarchical science communication is still the main form of communication. In science and culture projects, S&T experts unilaterally provide scientific knowledge. There were attempts to reflect public opinion on S&T policies through public hearings or seminars, but the main participants are professionals and scientists rather than ordinary people (Kim 2005a).

Government agencies, including the Korea Science Foundation, are still the most strongly represented sources of science policy and science culture activities. NGO activities in S&T are increasing, but the number of groups and activities are still relatively small.

It has been observed that resources for science communication have often been weighted towards youth-oriented events, that mass media are poorly used, and that there is not enough good scientific content (KFASC 2008). The proportion of resources allocated to science in print media, broadcast media and the internet is relatively low. Korea is struggling to develop quality content mainly due to a lack of professional staff and lack of funding. In the area of policy, the targets of public policy engagement have been limited only to scientific issues that could have a negative impact on the daily lives of citizens (H.S. Kim 2005a).

12.5 Lessons from the History of Science Communication in Korea

Over the past 50 years, South Korea has undergone remarkable economic development and integration into the high-tech modern world. Although there are other factors favoring Korea's growth, such as an extremely competitive education system and a highly skilled and motivated workforce, some researchers have pointed out that the national policy towards S&T has been one of the major factors contributing to growth and development (Choe 2001; Song 2009).

Since technological innovation is affected to a certain extent by public awareness of S&T, the promotion of public understanding of S&T has become a key policy. Such promotion has been necessary because investing a significant amount of money to support S&T requires the consent of the people.

In 2008, the Korea Science Foundation was renamed the Korea Foundation for the Advancement of Science and Creativity. It continues to work to enhance the public understanding of science, develop creative human resources and create an integrated culture.

Government-supported programs aim to bring about a society that will understand and value S&T, respect the key concepts and principles of S&T as social tools, and learn to use S&T knowledge in ways that enhance personal, social, economic and community development. One type of science communication has included institutional (government) initiatives to persuade the public of the benefits of science and to justify government investment in S&T. Another type has included government-supported initiatives to disseminate science information about daily life conditions among the population (Kim 2005b).

On the evidence available, the promotion of public understanding of S&T has been generally successful, as was demonstrated in the 2008 national survey (KFASC 2008). It can be said that the major cognitive effect of science communication in Korea is a good 'impression' of science. An impression is different from knowledge, attitude and image, which may neglect people's self-informing or self-instructing capabilities. Since impressions are often spontaneous, consequential and situational, good impressions may elicit or direct actual behaviors based on the utilization of scientific knowledge.

There are some probable reasons for this success in the public understanding of science. It could be due to efficient government-led promotion of science and communication in the early phase. It can also be said that appropriate content strategy has contributed to the success of communication efforts. From the early campaigns concerning S&T, the value and uses of science and scientific knowledge were deeply connected not only with concrete matters of daily life and public health, but also with more abstract and fundamental issues, such as rationalism, freedom and equality of people and society. In other words, S&T, national development and quality of life have been combined as a goal of successful science communication.

Although cultural and economic circumstances have changed continuously, science communication has met the needs of the times and increased the social acceptance of huge investments in S&T. For example, in the 1980s, the government and electronics companies promoted personal computers as educational tools for children and adolescents. The broad penetration of PCs across the nation may have been an important factor in Korean information technology innovation in the 1990s.

However, the government's nearly two decade long effort to build a low- and mid-grade nuclear waste storage facility repeatedly ended in failure, mainly due to the prevailing 'not-in-my-backyard' syndrome. Therefore, the government is striving to find a solution in a more transparent and democratic manner through public engagement with problems or issues relating to science. This new strategy reflects the notion that it is difficult to improve public understanding of science or scientific literacy through the direct dissemination of scientific information to the public. However, recent scientific issues, such as the safety of genetically modified food, the use of stem cell research and the selection of the radioactive waste disposal facility site, raise questions about the effectiveness of public engagement and more transparent government processes as the solution to these problems. Improved knowledge of science and public engagement are imperfect solutions. Rather, public relations in science policy needs to adopt two-way communication to negotiate with the public to resolve conflict and promote mutual understanding and respect between the organization and the public.

According to Webster (1991), groups that seek to influence the direction of science have differing effects depending on the political culture within which the S&T debate is located. For example, the more pluralistic and open the political culture, the more possible it is for interest groups to participate in a debate about science, even though this can mean that the debate is adversarial in style and may be resolved through litigation in courts. Whatever the particular political culture, there is also the question of how scientists, particularly the elite or senior members of the principal professional organizations and elite research institutions, respond to the call for a wider public involvement in the direction and prioritizing of science.

Koreans generally have a good attitude about science, but people who simply have a good attitude without specific scientific knowledge may become indifferent to science-related issues, such as the low-level waste faculty, if it is not to be located in their own neighborhood. However, when they hear the news about events such as public demonstrations, they tend to condemn the demonstrators as being selfish and thinking only on a local or regional scale. It can be said that a positive impression of S&T, without knowledge and understanding, has limits in helping to resolve problems (Lee 2007).

In the 1990s, there was a mutual distrust between policymakers and the public. Korea's government agencies and management experts did not believe in the ability of the public and resisted the transfer of policymaking functions to the public, and the public distrusted the intentions of the policymakers. Frequently, one hears the argument that the public is not sufficiently informed about science to be able to make rational judgments about areas of research. For its part, the public might in fact accept the wisdom of science and seek only to be informed in order to comment on areas of individual or regional concern. The sort of public concern that is expressed over S&T clearly depends on the image and understanding of S&T among the public. However, there must be a distinction between the public's understanding of scientific knowledge and its understanding of science knowledge as a social institution.

As Collins (1987) commented, while it is undoubtedly important that people understand more about science in the technical sense, it would be wrong to presume that they will therefore be able to enjoy a more 'objective' and 'authoritative' position in regard to science. Especially in areas of controversy, Collins observed that 'even among the experts themselves, who have been trained to many levels above what can be expected of the public's understanding, radically different opinions are to be found.' He concludes, therefore, that 'it is dangerously misleading to pretend that the citizen can judge between the competing views of technical experts when the experts themselves cannot agree' (Collins 1987:691). Nelkin (1979) also said that there is not much evidence that technical arguments can change anyone's mind. In the disputes over fetal research and in the various controversies about selecting a site for storing nuclear waste, no amount of data could resolve differences in values. Each side used technical information mainly to legitimize a position based on its own priorities. Ultimately, dramatic events or significant political changes had more effect than experts. We can see from this discussion that scientific expertise certainly has a role to play in policymaking, but it is one that will be limited by wider sociopolitical factors.

Although people have their own purposes and ideological reasons for using science, scientific knowledge alone is not enough to overcome their differences. This is because coming to a consensus on S&T policies such as nuclear power plants and environmental issues requires a process of judgment. Therefore, the differences can be resolved by using scientific methods of thinking and problem solving (Kim 2002). In other words, it seems more relevant to adopt a scientific way of thinking and problem solving to come to a scientific consensus, rather than just accumulating scientific knowledge. Therefore, both have become important.

References

Arnold, W. (1988). Science and technology development in Taiwan and South Korea. *Asian Survey, 28*(4), 437–450.

Choe, Y. H. (2001). Reflection and suggestions on the last 30 years of science and technology: Specification of science-led national development strategy. *Science and Technology, 34*(6), 82–93.

Collins, H. M. (1987). Certainty and the public understanding of science: Science on television. *Social Studies of Science, 17*, 689–713.

PACEST (Presidential Advisory Council on Education, Science and Technology) (2007). *Report on improving the interactions between the public and science and technology policy*. PACEST.

KFASC (Korea Foundation for Advancement of Science and Creativity) (2008). *Survey of public understanding of science in Korea*. Korea Foundation for Advancement of Science & Creativity.

Kim, Y. S. (2002). *Study on improving science and technology policy-making process in Korea: Focused on increasing public participation*. Science and Technology Policy Institute.

Kim, H. S. (2005a). *Understanding of science communication in South Korea*. Seoul: Iljin.

Kim, H. S. (2005b). *A new approach to science communication.* Seoul: Iljin.
Kim, H. S. (2007). A new model for communicative effectiveness of science. *Science Communication, 28*(3), 287–313.
Kim, H. J. (2007a). Setting goals for science popularization. *Science Times*, 31 January. Retrieved March 15, 2011 from www.sciencetimes.co.kr
Kim, H. J. (2007b). Beginning of the Korea Foundation for the Advancement of Science and Technology and the national movement of scientification. *Science Times*, 31 July. Retrieved March 15, 2011 from www.sciencetimes.co.kr
Lee, J. J. (2007). *Report on improving the interactions between public and science and technology policy*. Presidential Advisory Council on Education, Science and Technology.
Miller, J. D. (1983). Scientific literacy: A conceptual and empirical review. *Daedalus, 112*(2), 29–48.
MST (Ministry of Science and Technology). (1972). *Science and technology annual.* Seoul: MST.
MST (Ministry of Science and Technology). (2006). *Science and technology annual.* Seoul: MST.
Nelkin, D. (1979). *Controversy: Politics of technical decisions*. New York: Sage.
Presidential Secretariat (2004). *A collection of president Noh Moo-Hyun's speeches*, vol. 1. Presidential Secretariat.
Song, J. G. (2009). Research on future science and technology based national development. Science and Technology Policy Institute.
STEPI (Science and Technology Policy Institute) (1997). *Fifty years of Korea's science and technology policies*. STEPI.
Webster, A. (1991). *Science, technology, and society: New directions*. New Jersey: Rutgers University Press.
Wynne, B. (1991). Knowledge in context. *Science, Technology, and Human Values, 16*(1), 111–121.
Yim, S. J. (2002). Environment, political, and economic problems of nuclear plants: Criticism on Korea's nuclear plant development policy, and new directions for energy policies. *Politics and Information Studies, 5*(2), 223–253.
Ziman, J. (1991). Public understanding of science. *Science, Technology, and Human Values, 16*(1), 99–105.

Chapter 13
Spanish PCST and the European Science in Society Strategy

Vladimir de Semir

Abstract The dissemination of scientific culture and the public communication of science and technology have come of age in the academic, social, cultural and political fields. However, there is still a long way to go. Europe aims to strengthen a specific model in the 'Science in Society' strategy, while all the sectors involved are undergoing enormous changes that result from adapting to the communication and knowledge society. These changes are fundamentally linked to the development of information and communication technologies, but involve a radical economic, social and cultural adjustment. The internet marks a turning point—as did the printing press in its day—in the way things are done. The spread of science in this context is easier and more complex at the same time, although this appears paradoxical. The way has been opened for the public communication of science to be part of cultural programs at all levels, as occurred with the genesis and evolution of environmental concerns and movements. However, considerable political will at all levels (states, regions, cities) is still needed to make the necessary qualitative leap. In this regard, Spain has provided a good laboratory over the past 30 years for initiatives and experiences in the public communication of science and technology.

Keywords Science communication • Scientific culture • Science journalism • Science museums • Science centers • Public understanding of science • Science in society • Public communication of science and technology

V. de Semir (✉)
Science Communication Observatory, Pompeu Fabra University, Barcelona, Spain
e-mail: vladimir.semir@upf.edu

B. Schiele et al. (eds.), *Science Communication in the World: Practices, Theories and Trends*, DOI 10.1007/978-94-007-4279-6_13,

13.1 The Transition to the Knowledge Society

Our society is in the throes of evolution. It is in transition from an industrial society, based on our ability to apply manufacturing technology to the transformation of raw materials, to a new society that we have agreed to call the 'knowledge' society, based on our ability to use information and communications technology for the creative transformation of ideas and the generation of new employment opportunities characterized by high added value and social and economic impact. Or, in other words: until now, we have based our progress on our ability to transform the tangible (molecules); from now on, we add our ability to transform the intangible (bits).

The development of the new knowledge society and economy (which does not replace but is superimposed on skills developed during the Industrial Revolution, which has blazed the trail for the past two centuries) is principally founded on the rapid incorporation of scientific innovation and its technological application to our professional circumstances and day-to-day lives. Thus, our individual and collective ability to assimilate those changes is essential to the success of the continuous learning and constant adaptation that characterize the new society. The transformation is inherently associated with a conceptual evolution that must provide us with a better understanding of and ability to manage the innovations that mark the rapid consolidation of the knowledge economy and society.

We must not forget that the transformations that accompany and condition this evolution are not just economic, but also (and especially) cultural and social. For that reason, it is increasingly necessary to reinforce strategies aimed at disseminating new knowledge and its technical applications in the context of the ethical, cultural, social, economic and political questions they pose. The end goal is, then, for our society to better adapt to this new structure of life and work that will become consolidated over the course of the twenty-first century.

The importance of educating and training citizens in science and technology to ensure the development of the knowledge society has been highlighted by experts such as Richard V. Knight, a highly influential academic in the study of the knowledge-based economy. Knight (1989) assures us that this development requires certain essential conditions:

- Knowledge resources must be thought of in regional terms.
- Cities must provide incentives for knowledge-intensive economic activity and promote their centers of excellence.
- Knowledge must be defined and perceived by society as a form of wealth.
- The general public (society as a whole) must understand and assimilate the nature and origin of knowledge resources.
- Development based on knowledge activities means improving human and organizational skills and creating an environment that leads to innovation, learning, creativity and change.

The generalization of ideas and opportunities that derive from the application of scientific knowledge and its associated technology is, therefore, an unarguable necessity for achieving a cohesive society. This socialization of knowledge must ensure the ability of citizens to adapt in order to avoid the risks of creating a greater social divide between those who know and those who do not, with the aim of facilitating access to the new, high-added-value professions generated by the knowledge economy.

Those objectives must be part of the social and cultural policies of the government of any community, whether it is a city, a region, a country or a state. Consequently, for Spain to play a leading role in this global context of change, its political strategy must include the promotion of the culture of science and technology. This is necessary to ensure that citizens do not miss the train to the new knowledge society and, in addition, to ensure that society understands, accepts and feels complicit in the process of change and so supports the large public investment in education, research and technology that is necessary.

13.2 The Spanish Scenario in the European Context

It is not enough, as has been the case up to now, for political will to drive (with greater or lesser conviction) research, development, innovation and general access to information and communication technology. The adaptation of the necessary infrastructure and of the budgets that are essential for developing these activities must be accompanied by a decided political commitment to the dissemination and generalization of an appropriate and effective effort to lift the educational and cultural level of the population. The aim is for citizens to be able to access new opportunities for personal development (essentially occupational) and to be in a position to take part with sufficiently critical spirit in the social, ethical and political debate produced by most scientific and technical advances. We should not forget that the major social and economic transformations that are taking place will have a direct effect on the quality of life and work options of all citizens. Factors such as excessive specialization in services or delocalization of industry and traditional economic activity may be—indeed, already are—a serious danger for communities that have not understood that cultural diversity is a potential source of added value and that have not been able to develop their ability to adapt to the new social and economic reality. Unfortunately, Spain is a conspicuous example of this very problem, which manifested in an unemployment rate double the European average at the outset of the global financial crisis.

It is not just a matter of achieving better public understanding of science in the cultural context, as has been the traditional aim since C.P. Snow's famous discourse, *The two cultures*, in the middle of the last century (Snow 1959), but of citizens having the opportunity to better understand the world we live in through scientific knowledge—an understanding that will allow us to develop a critical capacity for furthering the democratic construction of our society.

For this reason, the public communication of science has taken on strategic importance in recent years. With the changes that are taking place in the transition towards a knowledge-based economy, the now traditional formula of RD&I (research, development and innovation) has become essential for reaching a competent place in the new economy. However, it seems evident that the sum of R + D + I ignores another important factor if this socioeconomic chain reaction is to occur and turn us into a society at the forefront of the twenty-first century, with citizens who are prepared, who understand and, consequently, who are able to participate in the necessary and unavoidable adaptation to the new economic, social and even cultural model that is the knowledge society. It is clear that we must add a variable to our traditional RD&I formula for it to be really effective. This essential variable that, like a catalyst, will make the reaction work properly is the *C* of science communication, scientific culture and creative citizenship.

Thus, the resulting formula that is essential for establishing this strategic scenario, which we must promote using the right policies for a successful transition to the knowledge society, is RDI&C, where:

R is the necessary training in basic and applied scientific research
D is the sufficient level of social and economic development
I is the decisive drive for individual and collective innovation and creation
C is the essential strengthening of public communication of science.

The right value of *C* will allow us to reach an appropriate level of scientific culture to configure a citizenship able to participate in and influence democratic decision-making that will configure and consolidate our future.

Spain started with a clear disadvantage in this process, given the accumulated lag at the time of the nation's historic transition from dictatorship to democracy in the 1970s. Scientific education and research had to adapt over the last third of the twentieth century, hampered by the fact that the important process of political, social and cultural modernization over the previous 40 years had tended to prioritize many other factors and infrastructures. It has therefore been difficult, year after year, for education, universities and scientific research to find the money needed to progress rapidly enough to make up for lost time. It thus comes as no surprise that Spanish R&D as a proportion of gross domestic product (GDP) is far below that of other states (and the average) of the European Union. Today, at 1.4%, we have finally managed to exceed 1% of GDP, but we are still a long way from the European level of 3%, established by the Lisbon Summit in 2000 (Lisbon European 2000). The summit laid the foundations and set out the guidelines for Europe to become a competent knowledge society. Those aims were ratified in 2010 in the updated 'Europe 2020: A strategy for creating smart, sustainable and inclusive growth' project (EC 2010).

Spain has certainly made considerable progress in scientific research in recent years and, in some specific fields, such as biomedicine, it has managed to achieve an honorable mention, while international publications by our researchers have increased significantly. However, private initiative needs to really get on the science bandwagon, as its contribution to R&D continues to be extremely modest, and the public sector should not let its guard down and must continue to invest in new educational and scientific infrastructure, as it has done in recent years. And a stable professionalization of scientific research as a career must be established once and for all—a task still pending in our society.

Nor should we be surprised that all the reports evaluating the knowledge of our students at different levels of education also place Spain in an embarrassing position in European rankings. Successive PISA reports and the *Report on the state of science teaching* presented in the Spanish senate on 13 May 2003[1] highlighted a drop in the level of scientific knowledge among secondary school students. To solve this problem, the 2003 report states that public authorities must promote initiatives that ensure that all students of arts and sciences have basic theoretical and practical scientific knowledge. The report recommended:

- Overcoming the traditional separation between science and arts, and keeping the humanistic connection in the teaching of science by considering scientific knowledge as an essential part of the history of humankind
- Promoting the creation of science infrastructure and scientific culture (science parks, natural parks, botanic gardens, museums and science houses)
- Promoting science outreach through the media and through public and private institutions.

The traditional divorce between the arts and the sciences continues, therefore, to be a reality and has even become exacerbated in Spain, although it is largely a general feature of Europe.

13.3 International Influences

In 1999, a world conference—'Science for the twenty-first century: A new commitment'—was held in Budapest, Hungary, with the sponsorship of UNESCO and the International Council for Science.[2] The conference produced the 'Declaration on Science and the Use of Scientific Knowledge', which included a commitment to:

> promote dialog between the scientific community and society, and to act ethically and in the spirit of cooperation in its areas of responsibility in order to consolidate scientific culture and its application for specific purposes in the world.[3]

Since then, studies and actions aimed at increasing general public awareness in favor of the interaction between science and society have become widespread.

Clearly, this object is realized largely through the intervention of the mass media. 'Society's relationship with science is in a critical phase' is the opening sentence of *Science and technology*, a wide-ranging British parliamentary committee report published in March 2000 (SCST 2000). The report indicates that there is a crisis of confidence in science and that many values are being questioned because, among other reasons, there is a fundamental reticence on the part of the public about scientific

---

[1] Boletín Oficial de las Cortes Generales, no. 660, 22 May 2003 (http://www.rsme.es/comis/educ/senado/I0660.pdf).

[2] International Council for Science (www.icsu.org).

[3] Declaration on Science and the Use of Scientific Knowledge (http://www.unesco.org/science/wcs/eng/declaration_e.htm).

authority and because most of the information that citizens receive after school is determined by the creation of a reality that is distorted by the media (mainly television). Indeed, the media are fingered as being one of the main entities responsible for the trivialization of cultural messages. 'As well as the negative image of real science,' the report states, 'the media provide an exotic range of material that goes beyond scientific respectability that tends to debilitate the mind.' As we can see, those arguments reflect a profound crisis of values that undoubtedly forms part of the direction in which the global information system is heading and which compromises the cultural level of our society (De Semir 2008)—a factor that, in Spain, is added to the poor levels of education with which we are still encumbered.

While the 2000 report's conclusions refer to the social and cultural environment of Britain, they coincide fully with those of many other studies carried out in recent years in Europe and, particularly, in Spain. We only have to take a close look at the Eurobarometers *Europeans, science and technology* (EC 2005) and *Scientific research in the media* (EC 2007a), or the surveys carried out every two years by the Spanish Foundation for Science and Technology on the social perception of science in Spain (SFST 2008), to obtain a clear cultural snapshot of the intersection of Spanish citizens with scientific knowledge. In summary:

- A broad majority of Spanish citizens surveyed (2 out of 3) acknowledge that they received a low or very low level of scientific and technical education while in school
- Spanish society has a positive image of science and technology, built on a certain (though limited) curiosity, that does not correspond to the education and information that citizens believe they possess and receive in this area.

This not only results in a cultural problem, but has negative implications in many other social and economic areas. It even plays a role in the considerable reduction in scientific vocations and also, to a large extent, technological vocations, thus posing a threat to Spanish society and compromising its future welfare and competence. Without good scientists, Spain may end up becoming the Florida of Europe (with tourists, retirees, residential activity etc.), rather than the California of Europe (with knowledge-intensive activities, creativity, research, innovation, patents etc.). The latter model ensures a solid but diversified economy with the ability to adapt to change and, therefore, able to support a balanced and strongly cohesive society.

We should not be surprised that the need to promote scientific culture has already entered the political arena and that the European Commission has set in motion the Science and Society Action Plan (EC 2002), or that this is one of the lines of work in the seventh Framework Programme of the European Union (2007–2013), which establishes specific grants to highlight the role of science in society and to promote science communication and the perception of science by the public (EC 2007b).

This is the context when we analyze the interrelation between science, knowledge, culture and society in Spain. At this point, we must inevitably ask: Where do we come from? How have we got here?

With the boost given to science in Spain to make up for our structural and skills deficits, the dissemination and promotion of scientific culture came of age in the last third of the twentieth century. Based on a solid tradition of cultural popularization of the sciences that was consolidated in the nineteenth century, as in the rest of Europe, and developed intensively and in parallel with the great scientific and technological advances that took place in the twentieth century (De Semir et al. 2008), the promotion of scientific culture has taken on the features of political strategy.

To some extent, it began with the sociocultural consequences of the great, sometimes brutal, demonstrations of human technological capability arising after World War II. For example, a definitive consolidation of science journalism took place in the midst of the Cold War, with the space race as the communications battlefield between Americans and Soviets. John N. Wilford, one of the founding journalists of the 'Science Times' section first published on 14 November 1978 in *The New York Times,* expressed this clearly when he stated, 'I am a science journalist thanks to Sputnik' (Wilford 2004). From the atomic bomb (1945) and Sputnik, the first man-made satellite (1957), to the Moon landing (1969), the economic and technological rivalry of the United States and the Soviet Union was largely played out in the world of the public communication of science and technology. Science journalism became a strategic tool for explaining to the world what was happening in the science and technology race between the two superpowers. The goal was to show, in each case, that the particular superpower's political and economic models were more efficient at dealing with the new world that had arisen from World War II.

In that context, the first survey of public understanding of science was launched in the United States by the National Association of Science Writers in 1957. It found that 'Americans have a generally positive attitude to science, although their knowledge of the subject is poor.' Coinciding with the beeping signal of the Sputnik, which overflew American territory several times a day, concern arose that the population should have sufficient science education, thus ensuring a competent country that could lead on the world stage. It is no coincidence that in the same year (1969) that a man landed on the Moon (a spectacular achievement by the United States to counteract the unquestionable initial successes of the Soviet Union in the space race: the first artificial satellite, the first animal in orbit and the first human in space), the world's first interactive center for the popularization of science was created: the Exploratorium in San Francisco, California.[4] Its founder was Frank Oppenheimer, an atomic physicist who worked on the Manhattan Project (directed by his brother Robert at Los Alamos) to design the atomic bomb.

The first phase of this strategy, born in 1957 and based on the promotion of science popularization, journalism and, in general, public 'literacy' in science to consolidate a model of society, culminated in 1985 when the American Association for the Advancement of Science set in motion Project 2061 to help all Americans achieve sufficient education in the sciences, mathematics and technology.[5] Naturally, this American educational and cultural convergence for the integration and promotion of scientific knowledge translated to Europe and, specifically, to Spain.

[4] Exploratorium: the museum of science, art and human perception (http://www.exploratorium.edu/).

[5] Project 2061 is so called because 2061 is when Halley's comet will return (its last appearance was in 1985), and that span of time is considered to be necessary for American society to achieve the quality of education necessary to be able to deal with scientific and technological changes (http://www.project2061.org/about/default.htm).

13.4 Science Museums and Science Centers

On 2 June 1981, the Science Museum of the La Caixa Foundation opened to the public in Barcelona. It was clearly inspired by the interactive ideas of Oppenheimer, and based on the importance of sensory perceptions as a starting point for the exhibitions in order to awaken interest in the sciences (Oppenheimer 1968). Today, the museum, renamed CosmoCaixa in a clear show of marketing nous, receives an average of 800,000 visitors each year,[6] after considerable remodeling and expansion was completed in 2004.

In mid-1985, the Casa de las Ciencias opened in A Coruña as the first publicly owned interactive museum in Spain to be sponsored by a city council. Today, the Galician city has other popular-science centers, including Domus, dedicated to the human species (1995), and Aquarium Finesterrae (1999), dedicated to the ocean ecosystem. These three centers make up the Coruña Science Museums,[7] which will be joined in the near future by the National Museum of Science and Technology, which the Spanish Government has decided to locate in A Coruña in an unprecedented decentralizing political decision.

This multiplication of science centers (although most are not exactly museums but rather centers of science animation, interpretation and participation) has also occurred in other Spanish capitals due to public initiatives, such as the Science Park in Granada (inaugurated in May 1995 and expanded in 2008 to become one of the most important science parks in Spain).[8] And we must not forget the many local museums that have opened in recent years in small towns and that have an important impact on local cultural tourism. In this regard, we should mention the network of museums throughout the territory belonging to the National Museum of Science and Technology of Catalonia,[9] which includes specific themed museums that are the legacy of the industrial activity in each county, such as the Cercs Mining Museum in El Berguedà.

13.5 Media Channels

The Spanish media have also evolved, as the media in the United States did before them. On 10 October 1982, the Barcelona newspaper *La Vanguardia* published the first weekly insert specializing in science in the contemporary Spanish press; in 1987, the inserts became science and medicine supplements and references in

---

[6] Department of Science, Research and Environment, Obra Social 'La Caixa'.

[7] Museos Científicos Coruñeses (http://www.casaciencias.org/).

[8] Parque de las Ciencias de Granada (http://www.parqueciencias.com/).

[9] Museo Nacional de la Ciencia y de la Técnica de Cataluña (http://www.mnactec.cat).

modern science journalism, which survived in various formats until 1997.[10] Other newspapers have followed this line and, today, scientific information is fully present in the press, on radio and on television, with greater or lesser intensity and quality. A popular science magazine, *Muy Interesante*, is the biggest seller among Spanish monthly magazines.

In 1985, several collections of popular science books appeared. Biblioteca Científica de Editorial Salvat—which published 100 volumes—and Biblioteca de Divulgación Científica 'Muy Interesante' from Ediciones Orbis, have both published translations of well-known popular-science writers from all over the world and have achieved impressive sales in bookshops and kiosks. Today, literary popular science is in good health, as witnessed by the many collections that can be found in bookshops (Metatemas-Tusquets, Drakontos-Crítica etc.), although cultural surveys continue to show that 50% of the Spanish population never buys or reads a book!

In the political sphere, the Catalan Government put forward a pioneering proposal in 1988 when the then Catalan Minister for Culture, Joan Guitart, created a committee to stimulate scientific culture; for several years, the committee developed an innovative and fertile program to encourage the integration of science in the world of culture (CASC 1989), until the minister was replaced in 1996. This initiative was cut short, although it was continued to a small extent by the Catalan Research Foundation, particularly in coordination with the Science Week held each November—an initiative sponsored by the European Commission that has become widespread throughout Spain through the Spanish Foundation for Science and Technology. Other Spanish autonomous communities also have their own programs, such as the important 'Madrid es Ciencia' fair, which has been held since 2000. The fair received more than 150,000 visitors in 2008 in the 4 days that it was open to the public.[11]

13.6 Political and Administrative Actions

Barcelona City Council is possibly the first city council to have overcome the timidity with which scientific culture has so far been included in the political action of municipal governments, if we exclude the above-mentioned case of A Coruña under the stewardship of Mayor Francisco Vázquez. In 1999, Barcelona created a City of Knowledge department, whose 1999–2003 political action program (taken on board by the municipal government as a whole) included the promotion of scientific and technological culture. In its day, this initiative was rated by the European Commission as best political practice in the field of scientific culture. In this case, it was also due

[10] Morales, Pol 'Vladimir de Semir: 25 años de periodismo científico (1982–2007)'— Cuadernos de la Fundación Dr. Antonio Esteve, no. 11, 2007 (http://www.upf.edu/pcstacademy/_docs/200710_desemir_intro.pdf).

[11] http://www.madrimasd.org/cienciaysociedad/feria/.

to the personal sensitivity and will of a politician—Mayor Joan Clos. Before becoming mayor, when he was first deputy mayor and was preparing to replace the Olympics mayor, Pasqual Maragall, in one of his first interviews as a candidate for the office he responded to the question of whether he would emphasize a scientific Barcelona by saying that:

> [T]oday, there is not such a big difference between the concepts of the scientific and the humanistic. But it is true that I would like to stimulate the scientific role of Barcelona. I am concerned that the city may miss the boat in terms of biotechnology or telecommunications, and I would be delighted to contribute, too, to the popularization of science. In truth, I would be delighted to contribute to reconciling the public with knowledge, which seems to me to be one of the most urgent challenges of our age.[12]

During the following term (2003–2007), the Mayor's Committee for the Promotion and Dissemination of Scientific Culture was created within the Barcelona Institute for Culture. The committee sponsored the strategic plan, 'La Ciudad por la Ciencia' (2004–2007), on instructions from the then councilor for culture, Ferran Mascarell, based on the following pragmatic decalogue:

1. Promotion of a more active citizenship by increasing their critical spirit and decision-making ability in the face of the new scientific, medical and technological challenges.
2. Consolidation of the public image of research and innovation as activities that generate wealth and, therefore, as key elements of development.
3. Improvement of options for accessing new opportunities, thereby strengthening cohesion and reducing the educational and digital divides.
4. Promotion of the process of transformation of Barcelona's science centers and museums to the new Natural History Museum of Catalonia, while continuing with the task of promoting the centers and their activities as elements that articulate the scientific culture of the city.
5. Reinforcement of the city–university–enterprise triangle by encouraging synergies between the three sectors. This triangle should also serve to promote dialog between the production sectors of Barcelona (especially those with a strong scientific and technological base) and the citizens.
6. Identification and promotion of strategic areas in the areas of science, medicine and technology, which are key to the development of Barcelona in the knowledge society.
7. Encouragement of knowledge and dissemination of local scientific, medical and technological values, whether generated in our universities, small and medium enterprises, or new businesses in the sector.
8. Resolving the crisis in scientific vocations among young people, thereby ensuring the future competitiveness of the city in terms of research and development and, at the same time, promoting a solid scientific culture in future generations.

[12] *El País*, 21 September 1997.

9. Correction of the gender inequality in the area of science and technology by promoting greater visibility of women, thereby favoring a less biased image of this activity.
10. International projection of the values of Barcelona as a nursery city that generates ideas, projects and opportunities by seeking alliances and strategic cooperation with the rest of Spain, with Europe, the Mediterranean and the rest of the world. (ICB 2004)

It was in this context that a commitment arose in the plenary municipal session of Barcelona City Council to dedicate 2007 to 'Science and Scientific Culture', thanks to the 'The City for Science' government measure drafted by the Committee for Scientific Culture and presented by Councilor for Culture, Ferran Mascarell, the prime architect of the strategy to include science and culture in municipal policy. As occurred in 2005 with the Year of the Book and Reading, this initiative was also promoted throughout Spain. In this case, this was basically due to the decisive action of the Secretary of State for Universities and Research, Miguel Ángel Quintanilla,[13] who successfully brought the initiative to the prime minister's office from the Spanish Ministry of Education and Science. The Spanish Parliament gave its full support to a proposal to declare 2007 the Year of Science, and the Spanish Government promoted the year to the fullest extent with a royal decree and the appointment of Deputy Prime Minister María Teresa Fernández de la Vega as chair of a statewide coordinating committee. The Spanish Foundation for Science and Technology was responsible for coordinating the initiative throughout Spain.

The results? Some 4,000 events in all Spain and 1,000 events in Barcelona alone.[14] Barcelona involved a million people throughout the year in a science outreach program implemented in a 100 different places in the city. A special role was played by the public libraries in the different municipal districts. More than 250 entities of all types formed part of the promoting council that drove the initiative from the Barcelona Institute for Culture.[15]

Science outreach or popularization was established as one of the ten basic program lines of the Strategic Plan for the Cultural Sector of Barcelona,[16] and the dissemination of science has been consolidated as a substantial part of the city's innovative cultural policy. Few European cities have included science in their cultural programs—a fact acknowledged by the European Commission, which

[13] 'Éste es el Año de la Ciencia para los ciudadanos', interview with Miguel Ángel Quintanilla, *El País*, 7 February 2007 (http://www.elpais.com/articulo/futuro/Ano/Ciencia/ciudadanos/elpepusocfut/20070207elpepifut_2/Tes).

[14] Informe Final Comisión Nacional del Año de la Ciencia 2007 (http://www.upf.edu/pcstacademy/_docs/InformeAxoCiencia.pdf).

[15] *Barcelona Ciencia 2007 Report* (http://www.bcn.cat/ciencia2007/cat/MemoriaBcnCiencia2007.pdf).

[16] Barcelona Strategic Plan for Culture (http://www.bcn.es/plaestrategicdecultura/english/index.html).

supported the Science and the City project born in Barcelona.[17] As a result, the commission decided to call an important European science journalism forum at Pompeu Fabra University as one of the activities of Barcelona Science 2007.[18]

The Spanish Government closed the Year of Science on 7 February 2008 with the presentation of a report[19] and a barometer[20] evaluating the activities and initiatives carried out in 2007. Deputy Prime Minister María Teresa Fernández de la Vega stated that:

> [A]s Newton said, science is like the earth: only a small part of it can be possessed. Its value resides in the fact that scientific activity is a collective work. An activity that involves the participation of scientists and researchers on the front lines, but that must have the support of society behind it.[21]

Mercedes Cabrera, the Minister for Education and Science, highlighted the fact that:

> the Year of Science did not finish in 2007 and, indeed, today we are not shutting down an initiative, but are giving a new impulse to a new mentality: science for the citizens and dialog on science with the citizens.

To extend this dialog between science and the citizens, four stable continuing structures were created as part of the Year of Science:

- *The Science Information and News System (SINC)*: an internet portal where institutions and scientists can post all kinds of science news. This content can also be accessed by journalists dedicated to the popularization of Spanish science and technology and by the general public.[22]
- *Scientific culture units*: providing research centers with experts in science communication, who also provide the SINC with more relevant news on science and technology.
- *A network of local scientific culture agents*: technical personnel to develop science outreach activities in small municipalities of between 15,000 and 75,000 inhabitants, where there is a demand for popular science but generally little activity.
- *The National Network of Science and Technology Museums*: a collaborative network for the better use of resources among some 30 popular science museums in different institutions or administrations.The dissemination of scientific culture

[17] *Europe, science and the city: Promoting scientific culture at local level*, www.escity.org. The project has a continuity in the seventh Framework Programme with the new Places project: Platform of local authorities and communicators engaged in science (www.openplaces.eu).

[18] European Forum on Science Journalism (http://ec.europa.eu/research/conferences/2007/bcn2007/index_en.htm).

[19] *Final report of the National Science Year Committee*, 7 February 2008 (www.upf.edu/pcstacademy/_docs/InformeAxoCiencia.pdf).

[20] Barometer evaluating and monitoring the results of the Year of Science (www.upf.edu/pcstacademy/_docs/BarometroAxoCiencia.pdf).

[21] Speech by the Deputy Prime Minister at the closing act of the Year of Science (www.la-moncloa.es/NR/exeres/885A3CC0-54F1-4851-B7CB-E36FCFA776E,frameless.htm?NRMODE=Published).

[22] SINC (www.plataformasinc.es).

has come of age in the political sphere, and the way has opened for it to be part of cultural programs at all levels. A similar phenomenon previously happened with the evolution of concerns for the environment and the emergence of the environmental movement, which was to some degree facilitated by observations of the fragility of Earth from the Moon and from man-made satellites.

So, in May 2011, a new Spanish law devoted to science included an article on scientific and technological culture in these terms[23]:

1. The government shall encourage activities leading to improving scientific and technological culture in society through education, training and outreach, and properly recognize activities of agents of the Spanish System of Science, Technology and Innovation in this field.
2. The Scientific and Technical Research and Innovation Plan will include measures to achieve the following objectives:

    (a) Improve scientific training and the innovative society, so that everyone can at any time form their own opinion on changes that occur in their natural environment and technology
    (b) Promote the popularization of science, technology and innovation
    (c) Support institutions involved in the development of science and technological culture through the promotion and incentives for activities of museums, planetariums and science centers
    (d) Foster scientific and innovative communication by agents implementing the Spanish System of Science, Technology and Innovation
    (e) Protect the historic scientific and technological heritage
    (f) Include the culture of science, technology and innovation across the entire education system.

We will see in the future whether this fine proposal will become a reality.

13.7 Duty and Right

Disseminating science in a way that is useful and valuable both for science and for society continues to be a challenge, not only because the deficit model that underlies the public understanding of science is still strongly rooted among (some) scientists, political leaders and media. The solution lies not in providing more information about science, but in more effective communication and dialog (EC 2009).

Despite many statements to the contrary, the practice of science communication is still linked to the 'transmission model'. The public understanding of science concept dominates broad sectors of science communication, and is concerned more with informing the public than with capturing the public. The relatively hidden

---

[23] See Article 38 in http://www.congreso.es/public_oficiales/L9/SEN/BOCG/2011/BOCG_D_09_59_402.PDF.

objective is, above all, to generate acceptance of and fascination for science, making public understanding of science a type of marketing dominated by economic interests and innovation. The underlying ideology may be expressed simply: society must accept science, technology and innovation and needs more engineers and scientists. In this way, science and society do not communicate (communication is a two-way process): rather, science addresses society. In this context, we identify five challenges for achieving a successful interaction between science and society.

First, it is necessary to do away with the myth of a unique public. There are many audiences (scientists, financial organizations, politicians, journalists and NGOs), many reasons for becoming involved (education, entertainment, deliberation/dialog) and, as a result, multiple voices (layperson, expert, experimental and codified) and different types of intermediaries (journalists, teachers, civil-society organizations etc.). The challenge is to use different mechanisms at different moments and with different training, both for the providers and for the users of the information, thereby allowing them to choose the most appropriate medium (or media). Scientists receive many requests to communicate, including internal communication with their colleagues, external communication for purposes of responsibility, and much broader communication with the general public. The complex processes of communication are related to all stages of research, including planning, funding, production, use and dissemination. Each one involves many actors, and a one-way (from science to society) and one-dimensional view of the public will not produce results.

Second, scientists often view society as an enormous, irrational, unknown monster with an aversion to risk, which sometimes behaves unpredictably. Scientists may be very negative about the functions of the media, of primary and secondary education and of politicians. Ethical research and technological evaluation, and activities with ethical, legal and social implications, are sometimes viewed by scientists as obstacles to scientific progress, or even as dangerous to science because they may awaken the 'monster'. This opinion has doubtless been exacerbated by experiences of severe scientific and technological conflicts. Nevertheless, recent advances in the social debate, for example on nanotechnology, suggest that positive changes are possible. For a number of years, nanoscientists, political leaders and funding agencies have been concerned about the public perception of nanotechnology. In the first few years of the twenty-first century, there was a widespread opinion that nanotechnology (after nuclear energy and genetics) would become the next science–society communications disaster. Proposals by NGOs for a moratorium on the use and promotion of nanoparticles feed such expectations and fears.[24]

A third obstacle is the strong dependence on science journals and on the press releases they generate. Scientific reports in other media involve little more than extracting the information from professional publications such as *Nature*, *Science*, *The Lancet* and *The New England Journal of Medicine*. The rigorous system of

---

[24] 'Enough talk already: Governments should act on researchers' attempts to engage the public over nanotechnology', editorial, *Nature*, 5 July 2007, 448, 1–2 (http://www.nature.com/nature/journal/v448/n7149/full/448001b.html).

evaluation used by those journals gives general reporters confidence that they are sources of reliable and thoroughly researched information. Nevertheless, particularly in the case of medical research, professional publications may not be so reliable and neutral, especially when pharmaceutical companies find ways to use them to publish their own results. All the parties require information, albeit of different types. Journalists must understand how scientific knowledge is produced and what its limits are. Scientists must increase their skills and their understanding of the possibilities and limitations of the different media for communicating to different audiences. And audiences must be versed in both the media and science.

A fourth challenge affects the rights and responsibilities of both science and society. Science communication has become a 'duty' for scientists and a 'right' for the public (a right to know and a right to participate), but the duty is not always welcome and the rights are not always exercised with enthusiasm. These rights and obligations have emerged with the proliferation of public involvement in Science (with a capital S) and of two-way communication. Various interactions have been generated between the audiences and the actors involved in the new sciences and technologies, but there are increasing doubts about the true value of those interactions. This is partly a matter of communication and partly a matter of government. The group of experts supervising scientific activity in society (the MASIS Project of the European Union[25]) suggests focusing on the involvement of the public in science from the perspective of communication, with clearly defined responsibilities for the actors. For this to be effective, there must be a greater understanding by all the parties of the nature of science as an evolving activity. There are many forums that allow us to applaud great scientists and celebrate surprising discoveries, but for public involvement to be effective more attention should be paid to the decisions that must be made, the resources that must be assigned, and the work and methods of individual scientists and research organizations.

Fifth, while scientific knowledge has shown a notable ability to transcend the barriers of politics and language, science communication is highly culturally specific. In Europe, there are very different traditions and regulations covering journalistic and scientific autonomy. Those differences have consequences: they produce different modalities (which should not be ignored or underestimated) for disseminating science in each country and transnationally.

Finally, scientists should play a more active role in encouraging useful communication of science in the popular media. This is even more crucial, given that there is currently a greater demand for transparency in scientific information at the same time that newspaper science sections are suffering from cutbacks due to the crisis in the print media, forcing researchers to deal with less experienced science reporters. Scientists can and must help to ensure that information on science continues to be documented and accurate.[26]

---

[25] MASIS: http://www.masis.eu/. See also *Science, economy and society: Research policy* (2009), European Research Area (http://www.researchprofessional.com/media/pdf/Highlights3397.pdf).

[26] 'Getting the word out', editorial, *Nature Neuroscience*, March 2009, 12, 235 (http://www.nature.com/neuro/journal/v12/n3/pdf/nn0309-235.pdf).

13.8 The Way Ahead

Today, the different administrations and institutions must be able to guide adaptation to the knowledge culture, as not all citizens have the same ability to make use of the opportunities provided by new technologies or to apply scientific and technological knowledge. In each revolution in the history of humanity, from the time the first signs of change were detected until the changes become fully implemented in society, at least one generation was unable to adapt to the changes. Continuous education and cultural training are essential for adaptation, as is the ability to transmit the values of the knowledge society that we already possess. Scientists and technologists play a decisive role in this society and therefore have a special responsibility not only to carry out their work but to explain it to society, as it is essential that we are all able to assimilate and participate in the rapid innovation inherent in the knowledge society.

'Democracy is necessary, but not sufficient,' said the currently somewhat forgotten philosopher Bertrand Russell (1952). If the essential critical spirit of a society is not developed (which, we insist, requires emphasis on education and culture), our democracy will be incomplete. Albert Einstein made explicit the risk we run if we do not do this: 'The restriction of knowledge to an elite group destroys the spirit of society and leads to its intellectual impoverishment' (Einstein 1948). We must therefore actively fight to prevent the consolidation of the *third culture* that would be imposed upon us—a non-culture based on the superficial and on the mediocre uniformity of the circular circulation of ideas rooted in the trivial and directed *pensée unique*. It is therefore more essential than ever that the two cultures come together in a single culture, *the culture*, based on critical thinking, which allows us to become genuine protagonists responsible for our evolution in order to become competent citizens in a cohesive and more just society.

References

CASC (Commission for the Advancement of Scientific Culture) (1989). *Science is culture: Statement of purpose*. Retrieved from http://www.upf.edu/pcstacademy/_docs/CoEstimulen.pdf

ICB (Instituto de Cultura de Barcelona) (2004). *La Ciudad por la Ciencia: Plan Estratégico 2004–2007*. Retrieved from http://www.upf.edu/pcstacademy/_docs/plan_estrategico_2004_2007_noviembre.pdf

De Semir, V. (2008). El declive de las noticias, incluidas las científicas. *Periodística* 11, Barcelona. Retrieved from http://84.88.10.30/index.php/Periodistica/article/view/178083/230453

EC (European Commission) (2002). *Science and Society Action Plan*. Retrieved from http://ec.europa.eu/research/science-society/pdf/ss_ap_en.pdf

EC (European Commission) (2005). *Europeans, science and technology*. Retrieved from http://ec.europa.eu/public_opinion/archives/ebs/ebs_224_report_en.pdf

EC (European Commission) (2007a). *Scientific research in the media*. Retrieved from http://ec.europa.eu/public_opinion/archives/ebs/ebs_282_en.pdf

EC (European Commission) (2007b). *Work programme 2008, Capacities, Part 5, Science in society*, C(2007) 5759 of 29 November. Retrieved from http://www.upf.edu/pcstacademy/_docs/FP7SciencexSociety.pdf

EC (European Commission) (2009). Science communication. In *Challenging the future of science in society*, MASIS report from Directorate General for Research—European Commission, September. Retrieved from http://ec.europa.eu/research/science-society/document_library/pdf_06/the-masis-report_en.pdf

EC (European Commission) (2010). *Europe 2020: A European strategy for smart, sustainable and inclusive growth*. Retrieved from http://ec.europa.eu/commission_2010-2014/president/news/documents/pdf/20100303_1_en.pdf

Einstein, A. (1948). Foreword. In L. Barnett (Ed.), *The universe and Doctor Einstein*. New York: Harper & Brothers.

Knight, R. V. (1989). *Cities in a global society* (Urban Affairs Annual Reviews). Newbury Park: Sage.

Lisbon European Council (2000). *Presidency conclusions*. Retrieved from http://www.europarl.europa.eu/summits/lis1_en.htm.

Oppenheimer, F. (1968). A rationale for a science museum. *Curator*, 11. Retrieved from http://www.upf.edu/pcstacademy/_docs/Oppenheimer1.pdf

Russell, B. (1952). *The impact of science on society*. London: Unwin Hyman.

SCST (House of Lords Science and Technology Committee) (2000). *Science and technology: Third report*. Retrieved from http://www.publications.parliament.uk/pa/ld199900/ldselect/ldsctech/38/3801.htm

SFST (Spanish Foundation for Science and Technology) (2008). *Cuarta encuesta nacional sobre Percepción de la Ciencia y de la Tecnología en España*. Retrieved from http://www.fecyt.es/fecyt/docs/tmp/-1406741440.ppt

Snow, C. P. (1959). *The two cultures*. Cambridge: Cambridge University Press.

Wilford, J. N. (2004). Homenaje al *Science Times* en su 25 aniversario. *Quark*, 34. Retrieved from www.prbb.org/quark/34/default.htm

Chapter 14
Science Museums and Cultural Images of Modernity: Scientific Communication, New Identities and Sociopolitical Constraints on Science Museums in Spain

Xavier Roigé

Abstract This paper studies the role of science museums in Spain and their contribution to the public communication of science and technology (PCST). In particular, it analyzes the social and political contexts in which science museums have developed in Spain over recent decades and evaluates how sociopolitical circumstances have conditioned both the content of the museums and the way in which the museum projects have been executed. An analysis of the institutional context in which these museums have been created in Spain suggests that there is an interrelation between scientific dissemination and the institutional and sociopolitical context in which it took place. The proliferation of museums and science centers has brought about greater dissemination of science in Spain, but their existence is not simply a response to the desire for scientific communication, as the museums are not merely places for the transmission of scientific knowledge, or places where science is consumed. They are also scenarios and symbols, institutions used to construct new discourses of an identity based on the idea of modernity, and are used politically to locate the local, regional and national in a globalized context.

Keywords Science museums • Science centers • National museums • Science in society • Scientific culture • Museums in Spain

The proliferation of museums and science centers in Spain, as Vladimir de Semir points out in Chapter 13 of this volume, has been one of the most significant elements in the advance of the public communication of science in this country. If for many years the dissemination of scientific culture was found in books, magazines and audio-visual media, today museums have taken on an important, almost vital, role in this

X. Roigé (✉)
Department of Social Anthropology, Research Project CSO 2011-29413,
University of Barcelona, Barcelona, Spain
e-mail: roige@ub.edu

B. Schiele et al. (eds.), *Science Communication in the World: Practices, Theories and Trends*, DOI 10.1007/978-94-007-4279-6_14,

educational endeavor. As de Semir demonstrates (2012), a variety of reports drawn up by experts in scientific communication and education recommended promoting the creation of infrastructure dedicated to the dissemination of science, such as museums and science centers, in order to overcome the problems of communicating science both to the general public and, more specifically, to schoolchildren. Museums and science centers, by allowing the visitor to 'manipulate' science, make a great impact on public opinion (Renvillard 2005:8), improving the public image of science by linking it to pleasant moments, contexts and sensations (Nuñez 2003). This has been particularly significant on the Spanish scene, where, according to social and school indicators, interest in science and culture is lower than in other countries in the area.[1]

The rapid creation in Spain of a large number of museums and science centers in recent decades has not been due only to plans to communicate science or to the international influence of interactive science museology. This paper studies the social and political contexts in which science museums in Spain have developed over recent decades, and evaluates how those circumstances have conditioned both the content of the museums and the way in which museum projects have been executed. Our hypothesis is that the communication of science to the public, particularly in museums, cannot be analyzed without taking into account the sociopolitical circumstances involved in the creation of scientific heritage and its *musealization*.[2] We consider that in Spain the science museums have been a key element in the construction of new *cultural landscapes*, as part of public policies to create new urban images. The proliferation of a large number of museums and science centers is, to a great extent, the result of national, regional and local public policies to create new identities based on icons of modernity.[3]

All museums, in one way or another, play a decisive role in defining and redefining national and local identities through the use of heritage as an element of identity. The very existence of museums is inseparable from the desire to democratize heritage and scientific knowledge, but museums are also institutions which create identity—they are places where heritage is upheld as a standard-bearer in the struggle against the tendency towards cultural standardization. Each museum is both an institution which creates identity and a symbolic element of identity—in other words, a producer and a product of identity. The appropriation of and use of museum heritage is never neutral (Prats 1997). Museum institutions, through their discourse, their exhibitions and their choice of content, act as agents that produce and feature certain elements of heritage (and not others) and are a fundamental element in the creation of social discourses.[4]

[1] See, particularly, 'Informe de la Ponencia sobre la situación de las enseñanzas científicas en la educación secundaria', *Boletín Oficial de las Cortes Generales Senado*, 22 May 2003.

[2] The term 'musealization' is a neologism much used in French and Spanish museological literature to define the process by which a cultural element is converted into an element of heritage, and used in museums (Roigé and Frigolé 2011).

[3] Not only do we refer to the fact that the construction of new buildings generates new city symbols, but also to the fact that science is used to create new images of 'modern cities'.

[4] This happens in all museums, but in some (such as history or ethnology) it is more obvious, while in others (such as art, technology or natural history) it is less explicit.

Science museums are often presented as instruments for the dissemination or communication of a universal science, as places for the discussion and spreading of knowledge. But they are not free from social and political interests, and their existence and content are also conditioned by public policies aimed at creating new identities and new images of society. For this reason, science museums are not only a means of disseminating scientific culture; they also convey a discourse about society and its history, ideology and political structure. Beyond their apparently neutral character, museums of science and technology are not only centers for the communication of scientific culture, but also institutions which reproduce political and cultural discourses (Roigé and Arrieta 2010). Like other museums, science museums select some elements of scientific and technological heritage while omitting others, depending on criteria related to identity, politics, economic perspectives, and even the effects of cultural tourism. Museums of science and technology thus collect 'artefacts of national identity' (Freitas 2010).[5]

14.1 The Emergence of Science Museums in Spain and Their Sociopolitical Context

In Spain there are 59 large museums devoted to science and technology, which between them receive more than 5.1 million visitors a year.[6] They are the third most popular type of museum, just behind art and contemporary art museums. The majority of them have been created in the past 25 years as part of a museum restructuring process carried out by the Spanish Government and in particular by the governments of the 17 Spanish autonomous regions.

To understand how these museums have been created, it is necessary to understand the context in which they were created. As Reuben Holo (1999) pointed out, the way in which museums were set up in Spain by the new democratic system following the death of Franco in 1975 was the result of the nation's new territorial organization based on autonomous communities, and this meant a move from a centralized and unifying structure to a more decentralized system in which responsibilities for cultural matters lay with the regional powers. Every autonomous region has tried to equip itself with a network of museums which reflects its idiosyncrasies. The autonomies and also the large cities have had to design (and in many cases invent) a new identity. In many cases, rather than going back to the past, this has developed from ideas of modernity and progress aimed at overcoming old images of traditional and economically backward regions (Holo 1999; Roigé and Arrieta 2010).

[5] Interesting questions are raised with regard to museums and science centers. How can we reconcile the universality of science with local characteristics? How can we avoid the risk of 'cloning' science centers and put more emphasis on local culture? (Renvillard 2005:7–8).

[6] Source: Spanish National Institute of Statistics, 2008.

This has entailed the creation of a large number of museums, providing a supply which has not always been welcomed by the public (Díaz Balerdi 2007), but which is key in proposing new cultural images and new elements of social relevance.

This tension between the capital of Spain (Madrid) and the various autonomous regions (with important cities such as Barcelona, Valencia and Bilbao) helps one to understand the institutional context which has completely changed the museum scene in Spain. On the one hand, the various autonomous regions and large cities have created large museums which correspond to the idea of an updated modernity, and to the desire to establish new urban icons. On the other hand, the Spanish Government has tried to rebuild a new conception of the state and this has led to the creation of the large Spanish state heritage museums, mainly located in Madrid, with projects intended to turn the city into a large cultural capital and international showcase.

Paradoxically, this process of museum planning has focused little on creating museums which reflect any specific identity using history or ethnology (Roigé and Arrieta 2010). In most cases, the autonomous communities have opened museums of modern art and science museums, which have been key elements in the creation of new identities using new architectural symbols and new discourses based on the idea of modernity (Layuno 2002). As Pryterch and Huntoon (2005:41) claim, this cultural creation has been carried out as a type of 'entrepreneurial regionalism', with planning which creates an interrelation between economic globalization and the restructuring of state and cultural policy in a new 'Europe of Regions'. In this way, we could say that these 'new museums' are, in some way, the result of the cultural policies of 'new regionalism' of the various regions which have constructed large new museums to define a new identity (Holo 1999).

Science museums based on the interactive model are a good reflection of this museum creation process and of the search for new cultural images. Thus we can see two broad categories of museum: the private museums promoted by financial entities, and the public museums which are a result of town council or regional initiatives, generally created to demonstrate their modernity and new identity as science museums.

14.1.1 The Importance of Private Museums

The importance of private museums can be explained by the strong presence of savings banks (now banks), which are obliged by law to devote a substantial amount of their budget to cultural and social activities. This was the origin of the first science museum in Spain, inaugurated in Barcelona and promoted by La Caixa Foundation.[7] This museum was a pioneer in many respects and can, to a certain extent, be described as the inspiration for science museums in Spain. In 2004, the museum reinvented

[7] Foundation of the savings bank 'La Caixa', today the biggest bank in Catalonia and the third largest in Spain.

itself, both with a new building extending its surface area to 50,000m^2 and by rebranding itself and renewing its vision. In a clear marketing operation, the museum renamed itself *Cosmocaixa*, in clear reference to the financial institution ('la Caixa'). This museum and its counterpart *Caixaforum* (devoted to large-scale art exhibitions) are the two flagships of the institution's cultural policy. The museum has another building in Madrid, which was part of the bank's drive to expand in that city.

The renovation of the museum is based on a new museological concept, and the institution is defined as a new type of museum based on 'total museology'. According to Jorge Wagensberg, its director, with the new museum:

> we have broken the frontiers between content types. We offer a complete synthesis, an interdisciplinary museum ... this is not a theme park, with entertainment for the sake of entertainment. We offer entertainment for intelligence. With this we want to change our visitors' lives in some way. We want them to talk about what they have seen and to leave with more questions than they came with.[8]

The originality of the museum consists in combining the Oppenheimer interactive museum model with original collections and an approach in which the most important thing is not to teach, nor to protect heritage, nor even to educate or inform, but to provoke the visitors, to generate more questions than answers—in short, to stimulate curiosity in science (Wagensberg 2000).

The museum certainly projects a very attractive image which has had a very favorable response from the public and has gained international recognition, receiving the Prize for the Best European Museum in 2005. In Barcelona, Cosmocaixa has become in a very few years the second most visited museum in the town, with almost 2.6 million visitors a year, while in Madrid there are nearly 1.2 million. But, while it receives more visitors than most of the museums in the town, its public has very different characteristics from those of other museums, being mainly families and from a different geographical area. Tourists are in the minority, 27% of the museum's visitors are from schools, and the museum attracts people who are not regular museum visitors. The museum has therefore been an excellent image for the bank, which has demonstrated its modernity. In addition, by promoting itself via its network of banks (and giving free or discounted entry to its customers), it has managed to connect the museum's image to the corporate image of the bank, as they use the same logo and web designs. Based on the public's evaluation during the first 5 years of the museum's existence, the bank which owns the museum considers scientific dissemination as one of the priorities of its 'Social Programme', and for that reason its strategic plan aims to increase the number of visitors using new formats and strategic alliances with leading science communicators.[9]

Although on a smaller scale, the *Eureka! Zientzia Museoa* (which changed its name to *KutxaEspacio de Ciencia* in 2011 and is promoted by the Kutxa Basque Country Savings Bank) is similar. With 138,264 annual visitors[10] (32% schoolchildren,

---

[8] *La Vanguardia*, 23 September 2004, pp. 40–41.

[9] Fundació La Caixa (2010). *El nuevo plan estratégico de Divulgación Científica (2010–2016).*

[10] Source: Eustat, 2009.

52% families and 9% organized groups) it has become one of most visited museums in the Basque country, exceeded in visitor numbers only by the Guggenheim and the Bilbao Fine Art Museum.

Among other justifications, financial entities' interest in creating these museums is in creating a public brand image which is modern and attractive—a type of heritage which is forward rather than backward looking. But, in addition, it attracts a public not found in other museums, as it aims to be a space for entertainment. Thus, in addition to Eureka! Zientzia Museoa's new educational and dissemination objectives, we find:

> making scientific and technological applications more comprehensible … an extension of cultural leisure through new spaces of simulation, continuing the renovation of the museum, and being a holiday resource for families.[11]

As in the case of Cosmocaixa, the recreational element of the museum makes this a museum which is aimed mainly at families.

14.1.2 New Regional and Local Images of Modernity

In addition to the private museums, there has been a rapid increase in the number of interactive science museums opened in various Spanish cities. Generally, the science museums have been designed as part of a campaign to project a new image of these towns, using the dissemination of scientific knowledge as an element of 'modernity'.

The case of the City of the Arts and Sciences in Valencia is paradigmatic. As Prytherch (2003) pointed out, the construction of this large architectural complex is consistent with the idea of creating a new cultural landscape which is committed to the idea of 'modernity'. The government of the Valencia autonomous region built an 'ideological landscape' (Olwig 2002): the new 'town of Valencia' emerges from a new regionalist discourse based on global competitiveness and obsessed with modernity, presenting the town of Valencia as both tourism-oriented and focused on international business. Although the project originated from a suggestion by the region's Socialist president, Juan Lerma, in 1989, it was developed with the support of the conservative People's Party from 1995 onwards, and was finally opened in 2000. For the then regional president, the city would convert 'our community into a global benchmark for science, leisure and technology on the eve of the third millennium' so that 'scientific innovation and regional authority would work together to contribute to the cohesion of the Valencia autonomous community'. The museum, for the Valencia president, was 'an opportunity to get ahead of the future' (Prytherch 2006:204). For his part, the museum architect stated in a press interview that 'I am the essence of Valencia's drive for modernity.'[12] The City of the Arts and the Sciences

[11] 'KutxaEspacio cambia su nombre por "Eureka! Zientzia Museoa" e inicia nueva etapa' (http://www.euskalmuseoak.com).

[12] *La Vanguardia*, 6 March 2005, p. 49.

in Valencia receives 4.2 million visitors a year; 2.4 million visit the science museum, and the rest the aquarium and the IMAX cinema.

The construction of this large museum has implied a certain 'theming' of the city. As Montaner (2003:22) has pointed out:

> the most themed city in Spain is Valencia: it is the city which has put together the most simple and commercial image, focused on the City of the Arts and the Sciences and on the Mediterranean coast, generating disproportionate growth whilst abandoning the city center and hoping that its decline will lead to it being completely transformed in the future.

The large-scale City of the Arts and Sciences has emerged as a new and powerful urban image, but its plans for science were far from clearly defined at the outset and had no clear objectives with regards to scientific dissemination. In a newspaper article, Ten Ros (1999) explained:

> Bit by bit, the City of Science ceased to be a great educational project and became a kind of postcard from Valencia ... Calatrava [the architect responsible for the project] put spectacle before functionality, rather than joining the two, and the initial, rather modest, investment soared to stratospheric heights. The initial project for contents, drawn up by 52 internationally recognized scientists, was forgotten and what replaced it was limited to a few vague items.

The *Casa de las Ciencias* (the House of Sciences), opened in 1983 in La Coruña, was the first publicly owned interactive science museum in Spain (Nuñez 2003). The museum was also intended to be an instrument of urban renewal—a way to create a new identity and project local modernity. The House of Sciences was finished later, in 1995, with an aquarium, and the House of Man (Domus) was added to it in order to create the complex which is now called the La Coruña Science Museums. As the local newspaper, *La Voz de Galicia,* pointed out on the occasion of the twenty-fifth anniversary celebrations of the museum, 'the city would be inconceivable today without its House of Sciences, the Domus and the Aquarium.' In his farewell speech, the mayor, who was the driving force behind the project, maintained that 'The House of the Sciences brought out our pride in being from La Coruña. We all realized that we could be pioneers, that we could be different, that La Coruña could be the equal of any European city.'[13] He also indicated that 'of all the things that we Socialists have done, the science museums are closest to my heart. They are my fondest creations.'[14]

These two cases are not the only examples. We could also note the Museum of Science and Water in Murcia, the Park of the Sciences in Granada, the Museum of the Sciences in Castilla–La Mancha, the Museum of Science in Valladolid, the House of Sciences in Logroño, the Elder Museum in Las Palmas in Gran Canaria, the Museum of Science and the Cosmos in Tenerife, the Acciona Museum in Alcobendas, and others. And yet more projects are underway. In all cases, they define themselves as museums for the dissemination of science, but, using science as a pretext, they also prove to be important instruments for the creation of an idea

[13] *La Voz de Galicia*, 25 May 2008.

[14] *La Voz de Galicia*, 5 June 2010.

of modernity in the cities where they are built. As Prytcherch and Huntoon (2005:41) point out, when one comes to consider the process of regional museum creation in Spain, one must not forget the overwhelming importance of the autonomous regions' policies, which they consider to be 'entrepreneurial regionalism', in creating new elements of identity and policies of differentiation which are no longer based so much on historical references as on a discourse of 'modernity'. The creation of new urban icons through the creation of museums has been a fundamental element of these policies.

14.2 How Do Museums Disseminate Science? Museographical Constraints and Socioscientific Problems

All science museums in Spain, as in other countries, define themselves as institutions devoted to the 'democratization' of science using the interactive model of Oppenheimer to exhibit content and transmit messages. Communication becomes the most important element, and at the same time interactivity is seen as the means, *par excellence,* of enabling the visitor to come to terms with the contents. All these museums' projects highlight their role as communicators—their mission to connect with society, based on the dissemination of scientific knowledge.

The objectives of informing citizens, enabling scientists to form relationships with society and democratizing science are present in the definitions of all these museums, which have almost identical discourses and missions. The museums present themselves as public places or as forums for the discussion of science (Einsiedel and Einsiedel 2004); they are places (Durant 2004) which enable a vision that can be shared between the different members involved in science (scientists, businesses, society in general, scientific institutions, researchers etc.). When defining their objectives, most also aim to overcome the traditional vision of museums and become public forums which foment an active dialog between scientists and museums, to promote critical reflection and prompt citizens to participate in taking decisions, following a communications model which seeks to make knowledge more widely available to all (Schiele 1994, 2001).

If one analyzes the museums more closely, however, certain interesting questions are raised. We have to ask ourselves, as Gonzalez, Gil and Vilchez (2002) have done, 'to what extent are science museums contributing to a better understanding of the problems which confront humanity today and of the measures needed to overcome them?' Or, in other words, to what extent do science museums actually fulfill the role of transmitters of knowledge, or are they visited only for their role as providers of entertainment? As Davallon (1992:116) suggests, the constraints of the public can lead to noble aims being downgraded, with the greatest number of visitors being satisfied with the same product. In the 'new' science museums in Spain, developed in the institutional context outlined above, we can observe these problems in both the temporary and permanent exhibitions, and we can divide them broadly into two types: museographical and sociopolitical difficulties.

14.2.1 Does Museography Condition Contents?

In all the museums, one can observe great efforts being made to make museographically attractive products, and this can sometimes lead to exhibitions being more concerned about form than substance (Soichot 2011:76–77). But they all face the same challenge of selecting the most appropriate way of transmitting highly specialized knowledge and practices to general and extremely diverse audiences (Massarani and de Castro 2004:32).

All the museums have broadly similar exhibition resources. Thus, elements such as Foucault's pendulum, children's play areas, an abundance of multimedia and a planetarium are present in almost all of them. Generally, the designs are based on bright colors and an esthetic which reminds one more of a shopping mall than a museum. But, beyond the design, many centers have been conceived without an appropriate museological plan, leaving the museography in the hands of companies specializing in making exhibitions which make spaces that are visually attractive but sometimes lacking in content. Thus we find disorganized and confused discourses (Meyer 2010) in which the visitor gets lost, wanders randomly around the different spaces, and becomes a compulsive consumer, impatient to press as many buttons and play with as many interactive displays as possible. In addition to this, as has been pointed out in other countries, there are difficulties in renewing content, in bringing exhibitions up to date and above all in showing science as it is carried out in reality (Soichot 2011:76).

These new museums reveal a change of paradigm; they move from 'science culture' to 'science entertainment', seeking to raise awareness of science and to turn it into a show (Belaën 2005:104). As centers of experience, museums aim to reach a public with highly diverse sensitivities, and for that reason one of their main aims is to create a museography which is as attractive as possible. Multimedia resources become attractions for the fascinated spectator and resources to attract visitors (Deloche 2005)—resources which must also contain attractions that go beyond the simple museography of the object, and at the same time are much more spectacular than the multimedia facilities found in the homes of potential visitors. Thus, for example, the exhibitions make reference to immersion, to leisure and to fun, as can be seen in some of the exhibition publicity:

- 'Visit Zero Gravity in the Museum of Science, and be wrapped up in the stars' (Museum of Science, Valencia)[15]
- 'The Theatre of Electricity, an experience which will make your hair stand on end' (Museum of Science, Valencia)
- '"The Secrets of Butterflies" will have spectacular photography, magnifying glasses and microscopes, mechanical interactive displays which you will be able

[15] Ciudad de las Artes y las Ciencias, Valencia (http://www.cac.es/).

to manipulate and spaces for children with children's games' (House of Sciences, Logroño)[16]
- 'We invite you to go on a fun journey from the inside of your body up to the stars' (Eureka! Zientzia Museoa, San Sebastián).[17]

As we have seen, many of the science museum projects, as is the case in the City of Arts and Sciences in Valencia, originated with the idea of being powerful symbols of new identities, and buildings have even been designed before the actual content of the museums was thought of. For this reason, as Prytherch points out (2006:209), behind the monumental and astonishing façades of the Valencia museums:

> we find content which is rather banal and mediocre, and although the museum was defined initially by its contents, the museum proclaims modernity without really articulating what this really means.

To a certain extent, the formal constraints (architectural and museographic) have conditioned the content and affected the way science is disseminated.

14.2.2 Conceptual Difficulties and Sociopolitical Determinants

Despite the museums' objectives, which are held up as being mainly those of the dissemination and democratization of science, the contents exhibited often go far beyond simple communication (Durant 1996). If we take a look at the content and the exhibitions of the science museums in Spain, we can see how the choice of content is conditioned more by sociopolitical factors than by any kind of scientific neutrality. Although the museums are conceived as educational and neutral, the choice of content is often a response to debates specific to the area in which they are located. Thus, for example, aspects related to sustainable development, pollution or the degradation of ecosystems are some of the most frequent themes in the Spanish science museums (González et al. 2002), while other more social aspects are rarer or are treated indirectly (social imbalances, human rights, the social role of science). Thus it is no surprise that in Murcia, a region short of water for irrigation, and which wants water to be diverted from neighboring regions which refuse to do so, there is a museum which has water as its main theme. Its exhibitions 'Drop by drop (Murcia)' and 'The culture of water' are therefore related to the problems of drought in the region. In Granada, the project indicated that 'this important investment will contribute, amongst other things, towards covering one of the areas of science which generates most demand amongst citizens, the area of health and everything related to life.' It is also usual to incorporate current affairs topics into the exhibitions in these museums. Thus, for example, several days after the disaster involving the oil tanker, *Prestige* (which sank in 2002 off the Galician coast, causing an oil spill that

[16] Casa de las Ciencias, Logroño (http://www.logro-o.org/casadelasciencias/).

[17] Eureka! Zientzia Museoa, San Sebastián (http://www.eurekamuseoa.es/).

caused one of the largest ecological catastrophes in the history of the country and sparking a large political debate over the unwise decisions of the government in trying to contain the pollution), the House of the Sciences in La Coruña prepared teaching guides and workshops about fuel and its nature, origin and effects on the habitat. The then KutxaEspacio Science museum in the Basque country also prepared a week of activities explaining the origin of fuel, its composition, and the reasons why the oil slick was bound to end up in the Basque country, as indeed happened later.

This sociopolitical context means that scientific dissemination is caught up in a web of business, economic, political and economic interests. Thus, for example, in the relaunching of the Zientzia Museoa in San Sebastian, one of the main objectives of the museum was:

> the conversion of the center into a showcase for businesses, where science of the highest level is shown to the general public in a meeting point where they can display their plans.

Science museums do not always act with the necessary freedom to turn their exhibitions into real forums or 'agora' for debate and discussion, despite the confidence which the public has in them and in their legitimacy as centers for the mediation of scientific culture (Natali 2007). As Wagensberg (2005) points out, scientific questions do not affect only the scientific community. Scientific data (apparently objective, intelligible and based on logic) can be interpreted in different or even contradictory ways, turning museums into spaces for the creation of public opinion. But the culture of science and technology must not be conceived in isolation, but rather within the political and ideological context in which it is created. Hence, most exhibitions offer a didactic discourse and are far from the idea of the museum as a space where ideas can be set against each other and debated. As Soichot points out (2011:74):

> museums are far from having adopted contextualized approaches which accentuate socio-scientific problems in their multiple facets, promoting debate and the empowering of the public.

14.3 Conclusions: The New Museums in the Context of the Dissemination of Scientific Culture

The new science museums in Spain are, to a degree, a reflection of the situation of scientific culture in this country. They were conceived as forums for knowledge, or places of exchange, but we must ask ourselves how much of an interaction with the public there is, and to what extent do they follow a certain 'educational governance' which is based on the idea of an education that aims to make up for the lack of scientific knowledge, making it possible for citizens to understand the experts' discourse (Healey 2005). Or rather, as Magro (2008:151) suggests, we could position these museums within a new paradigm in which the important thing is to 'get citizens hooked on the wonderful world of science' so that 'science would be just another product and the place where you get your supply would be little different from a theme park or a shopping mall.' In this way, the public would become a type of science consumer. But this author goes even further, in considering that in some way this way of seeing the public as consumers is present even in the definition of scientific culture in the Spanish National Research and Development Plan (2008–2011),[18] which is based on the profitability of science and its dissemination as factors of progress and 'is more appropriate to the ideas of the Universal Exhibitions than to the concepts of today's citizens and the participatory society' (Magro 2008:152)

PCST in Spain is thus confronted with a paradox. Various public studies confirm that citizens consider science very important and value scientists very positively, but on the other hand they have a low or very low level of science education and knowledge (Castellanos 2008:198).[19] Given their characteristics, the science museums should be places which bring scientists and citizens together, not only so that knowledge can be transmitted comprehensibly, but also as places in which citizens can question and contribute to science. But are they? An analysis of the institutional context in which these museums have been created in Spain suggests two questions about the interrelation between scientific dissemination and the institutional and sociopolitical context in which it takes place. First, museums have indeed become fundamental institutions for the transmission of scientific knowledge and for bringing science to society. That said, rather than being democratizing institutions of science, museums act as new cultural intermediaries (Bourdieu 1991), presenting science ready for consumption. Second, as I have pointed out, the new museums are also scenarios and symbols; they are institutions used to construct new discourses of an identity based on the idea of modernity, used politically to locate the local, regional and national in a globalized context.

As early as 1993, in a special report titled 'The big bang of the science museums', the *La Vanguardia*[20] newspaper pointed to the proliferation of these interactive science museums in Spain and predicted spectacular growth in their number, as has turned out to be the case. All these museums, despite all the issues raised, have succeeded in bringing scientific knowledge to the citizens and have become educational spaces which bring the public closer to the scientific arena. The museums have taken on an important, almost vital, role in this educational endeavor. As we have seen in this paper, these institutions are at the center of social debates and contradictions, and constitute a privileged scenario through which to understand the interrelation of science and society.

---

[18] VI Plan Nacional de Investigación Científica, Desarrollo e Innovación Tecnológica 2008–2011 (http://www.micinn.es/portal/site/MICINN/menuitem).

[19] Castellanos based his 2008 work on studies carried out by the Sociedad General de Autores, by the Spanish Foundation for Science and Technology and by Eurobarometer: *Europeans, science and technology*.

[20] 'Suplemento Ciencia y tecnologia', *La Vanguardia*, 8 May 1993.

References

Belaën, F. (2005). L'immersion dans les musées de science médiation ou seduction? *Culture & Musées, 5*, 91–110.

Bourdieu, P. (1991). *The love of art: European art museums and their public*. Stanford: Stanford University Press.

Castellanos, P. (2008). *Los museos de ciencias y el consume cultural*. Barcelona: Editorial UOC.

Davallon, J. (1992). Le musée est-il vraiment un média? *Publics et musées, 2*, 99–123.

de Semir, V. (2012). Spanish PSCT and the European Science in Society Strategy.

Deloche, B. (2005). ¿Es el museo virtual un competidor real para el museo institucional? *Mus-A. Revista de los museos de Andalucía, 5*, 16–23.

Díaz Balerdi, I. (2007). Museos y normalización política en la España postfranquista. *Cuadernos de Museología, 28*, 23–37.

Durant, J. (1996). Science museums, or just museums of sciences? In S. Pearce (Ed.), *Exploring science in museums* (pp. 184–186). London: Atlantic Highlands.

Durant, J. (2004). The challenge and the opportunity of presenting 'unfinished science'. In D. Chittenden, G. Farmelo, & B. V. Lewenstein (Eds.), *Creating connections: Museums and the public understanding of current research* (pp. 47–60). Walnut Creek: AltaMira.

Einsiedel, A. A., & Einsiedel, E. F. (2004). Museums as agora: Diversifying approaches to engaging publics in research. In D. Chittenden, G. Farmelo, & B. V. Lewenstein (Eds.), *Creating connections: Museums and the public understanding of current research* (pp. 73–86). Walnut Creek: Altamira.

Freitas, J. (2010). Artefactos da identidade nacional: O caso do sextante de Coutinho. In C. del Mármol, J. Frigolé, & S. Narotzky (Eds.), *Los lindes del patrimonio. Consumo y valores del pasado* (pp. 335–370). Barcelona: Icaria/Institut Català d'Antropologia.

González, M., Gil, D., & Vilchez, A. (2002). Los museos de ciencia como instrumentos de reflexión sobre los problemas del planeta. *Revista Tecne, Epísteme y Dídaxis, 12*, 98–112.

Healey, P. (2005). *Science, technology and governance in Europe (STAGE): Challenges of public engagement: Project report*. HPSE-CT2001-50003. London.

Holo, S. R. (1999). *Beyond the Prado: Museums and identity in democratic Spain*. Washington: Smithsonian Institution.

Layuno, M. A. (2002). *Los nuevos museos en España*. Madrid: Edilupa.

Magro, C. (2008). Plan nacional y cultura científica. *Revista Madri+d, 21*, 142–154.

Massarani, L., & de Castro, I. (2004). Divulgación de la ciencia: Perspectivas históricas y dilemas permanentes. *Quark, 32*, 30–35.

Meyer, M. (2010). From 'cold' science to 'hot' research: The texture of controversy. In F. Cameron & L. Kelly (Eds.), *Hot topics, public culture, museums* (pp. 129–149). Cambridge: Cambridge Scholar.

Montaner, J. M. (2003). La ciudad, ¿museo o parque temático? *La Vanguardia*, 19 November, Suplemento Culturas, 22.

Natali, J.-P. (2007). Le rôle des scientifiques dans les productions muséales: Légitimité, validité et pertinence d'enonciation dans la mise en culture de la science. *Culture et musées, 10*, 37–61.

Nuñez, R. (2003). El papel de los nuevos Museos en la educación Científica. *Informe de la Ponenca sobre la situación de las enseñanzas científicas en la educación secundaria, Boletín Oficial de las Cortes Generales,* 22 May, 85–87.

Olwig, K. (2002). *Landscape, nature and the body politic*. Madison: University of Wisconsin Press.

Prats, L. (1997). *Antropología y patrimonio*. Madrid: Ariel.

Prytcherch, D. L. (2006). Reconstruir el paisatge per a reconstruir el Regionalisme? L'Horta, la Ciutat de les Ciències i la política ideològia de la modernitat valenciana. *Treballs de la Societat Catalana de Geografia, 61–62*, 89–213.

Prytherch, D. L. (2003). Urban planning and a Europe transformed: The landscape politics of scale in Valencia. *Cities, 20*(6), 421–428.

Prytherch, D. L., & Huntoon, L. (2005). Entrepreneurial regionalist planning in a rescaled Spain: The cases of Bilbao and Valencia. *GeoJournal, 62*(1–2), 41–50.

Renvillard, M. (2005). ECSITE 2004 Barcelona. *Quark, 35*, 7–11.

Roigé, X., & Arrieta, I. (2010). Construcción de identidades en los museos de Cataluña y País Vasco: entre lo local, nacional y global. *Pasos, 8*(4), 539–555.

Roigé, X., & Frigolé, J. (Eds.). (2011). *Constructing cultural and natural heritage: Parks, museums and rural heritage*. Gerona: ICRPC.

Schiele, B. (Ed.). (1994). *When science becomes culture: World survey of scientific culture*. Ottawa: University of Ottawa Press/Éditions MultiMondes.

Schiele, B. (2001). *Le musée de sciences: Montée du modèle communicationnel et recomposition du champ muséal*. Paris: L'Harmattan.

Soichot, M. (2011). *Les musées et centres de sciences face au changement climatique. Quelles médiations muséales pour un problème socioscientifique?* Doctoral thesis, Museum National d'Histoire Naturelle, Paris.

Ten Ros, A. E. (1999). La Ciudad de las Ciencias, entre la educación y el turismo. Crónica de cincuenta mil millones sin rumbo. *Levante*, 14 March.

Wagensberg, J. (2000). Principios fundamentals de la museología científica moderna. *Alambique, 26*, 15–19.

Wagensberg, J. (2005). Los museos de la ciencia: espacios para la creación de opinión pública. In F. J. Rubia (coord.) & I. Fuentes, Isabel (dir.), *Percepción social de la ciencia* (pp. 251–262). Madrid: Academia Europea de Ciencias y Artes, UNED Ediciones.

Part II
Horizontal Issues

Chapter 15
Slowly But Surely: How the European Union Promotes Science Communication

Michel Claessens

Abstract The chapter reviews the initiatives taken by the European Union to stimulate and promote science communication through transnational research projects. The story starts in the 1980s with European Science and Technology Week. Through its successive Framework Programmes, the EU has supported transnational projects related to raising public awareness of science and technology and to the public communication of science and technology. In addition, the European Commission decided to include specific provisions regarding public information and communication in the contracts signed by the beneficiaries of its Framework Programmes. In November 2005, the commission organized a major conference, Communicating European Research 2005. It was the first ever conference on science communication organized by the commission and probably one of the biggest ones on that subject organized for scientists. It showed the growing importance and recognition of public communication of science and technology. All these initiatives show that the European Commission draws the attention of participants in EU-funded projects to the fact that they can no longer ignore the 'public communication' dimension of their activity. The commission has also set up specific tools to disseminate good practices in science communication. In addition, it monitors public opinion in Europe about science and technology through major opinion surveys (the 'Eurobarometers').

Keywords Public communication of science and technology • European Union • Policy • Framework Programme • Media • Eurobarometer • Communication • Public perception

---

M. Claessens (✉)
ITER Organization, Route de Vinon sur Verdon, F-13115 St Paul-lez-Durance, France
e-mail: michel.claessens@iter.org

B. Schiele et al. (eds.), *Science Communication in the World: Practices, Theories and Trends*, DOI 10.1007/978-94-007-4279-6_15,

Since 1984, the European Union (EU) has been implementing a genuine research policy through successive so-called 'Framework Programmes' (FPs), which set out priorities in several fields for 5-year (now 7-year) periods. Research teams made up of partners from a minimum of three Member States or associated countries bid for funding in competitive calls for proposals. The Seventh Framework Programme (FP7) covers the period from 2007 to 2013 and has a budget of €53 billion, which is a 63% increase compared to FP6 (2002–2006). Still, EU R&D expenditures remain, in relative terms, quite low: the 27 Member States, even though when taken together they are the world's largest economy, devote only an average of 2.01% of their gross domestic product (GDP) to R&D compared to 2.77% and 3.44%[1] for the US and Japan, respectively.

As a result of the increase of the FPs' budgets, Europe's role in research has become increasingly visible inside Europe as well as outside it. In addition, the EU has made important efforts to study and develop the science–society interface and to improve communication between scientists and the European public in order to ensure that public awareness is keeping pace with rapid scientific and technological development.

15.1 Framework Programme Initiatives

In 1993, under FP3 (1990–1994), the European Week for Scientific Culture was launched by the then Commissioner for Research, Antonio Ruberti. The aim of the week, which ran each year until 2006, was twofold: first, to support awareness-raising activities on the European dimension of science; second, to provide a European focus for public understanding activities already going on in the EU Member States. Thus, for more than 10 years, the European Commission has supported a range of initiatives that involve, in particular, young people from different Member States. These cooperative projects aimed at showing, rather than telling, (young) Europeans how science and technology affects them, from the simplest gadgets to the most sophisticated satellite technology, through transnational and coordinated projects. As a result, the week provided a framework for demonstrating 'European science' to the public through a diverse range of EU-funded activities, including competitions, exhibitions, interactive internet debates, school projects and videos.

In 1998, the Week for Scientific Culture adopted 'Sea and Space' as a theme. Funding was provided to four activities, which culminated in an exhibition in parallel with Expo'98 in Lisbon in August 1998. But the thematic approach was not universally welcomed and encouraged lobbying from the 'big science' community, often based around established large-scale facilities.

During 1997–1998, it was decided that the week would formally become part of FP5, and it was grouped with three new activities—networks, round tables and

[1] 2009 data.

information services—aimed at raising public awareness of science and technology under the action line 'Promoting Scientific and Technological Excellence' of the Human Potential program. At the same time, the week became the 'European Week for Scientific and Technological Culture', to highlight the fact that many projects were not about 'blue sky' research but concrete and technological applications.

Actually, the European Commission was at that time trying to relaunch the week and give it another chance, as the scale of activity remained very modest and the impact marginal. In order to increase the participation base, introduce competition between ideas for projects and ensure that the selection of projects was based on equality of treatment and transparency, it was decided that proposals would be solicited on the basis of annual calls for proposals. However, because FP5 was adopted in December 1998 there was insufficient time to organize a Science Week in 1999.

The week continued to be marked until 2006, with on average a dozen proposals funded each year for a total budget which, by EU standards, remained very modest (typically €1–2 million).

Following an external assessment of these activities, the European Commission decided to interrupt the week in 2007. The main reason was that the week funded lots of very small local activities that had only marginal impacts on the research community and the public. Very few large-scale projects were proposed and funded by the week, and most of them were not perceived as being part of a larger initiative despite the fact that individually they 'delivered the goods' (such as the eponymous 'Science on Stage'). Clearly, the EU budget was too low to fund a critical mass of such projects. In consequence, the European Commission abandoned the week and sought to fund activities that would introduce a European dimension into national science weeks, but there again success was limited.

The commission then tried to coordinate science weeks across Europe, hoping that this would trigger a bigger impact and a genuine European dimension. The underlying objective (which did not gain wide support across the scientific community) was to have a single event, a giant 'European' science week across the whole continent, on the assumption that this would generate huge awareness, visibility and impact among the public. To that end, the commission set up an ad hoc group, which met several times. But there were many good arguments against such coordination: the situation that prevailed before then gave more flexibility to national coordinators and allowed for exchanges of materials. Despite several trials, the commission's attempts to encourage the consolidation of national science weeks around one or two dates in the calendar clearly failed: they remain uncoordinated and take place at different dates in the year.

The latest attempt along this line took place in the PLACES project (under FP7). The aim is to mobilize science museum networks, science event organizers and cities in projects promoting the science and society dialog (see Table 15.1).

Under FP5 (1998–2002), and therefore in parallel to the science week, the EU supported Raising Public Awareness projects. The rationale was that, in addition to or complementing national activities, EU-supported projects could catalyze the exchange of good practices and achieve a critical mass of research (and researchers). It was also felt that these activities deserved a special budget because they were seen

Table 15.1 Some major initiatives supported by the EU's Framework Programmes to stimulate and promote science communication and education through transnational research projects (Source: CORDIS (www.cordis.lu))

FP	Project title	Start date	End date	Objective	EU funding (€ million)
FP5	AlphaGalileo	01/04/2001	31/03/2003	Provide an internet-based bridge from S&T practitioners to the media. The aim is to stimulate European media to cover European science and technology, and to ensure that it gets the recognition that it deserves	0.787
FP5	Physics on stage	01/02/2000	31/12/2000	Inform the public and physics teachers in particular about innovative ways to convey information about physics	0.432
FP5	Enscot—European science communication teachers network	01/02/2000	01/07/2003	Establish a European network of science communication teachers to exchange best practice and act as a center for further expansion and continuation of the network	0.420
FP6	ESCW: the European science communication workshops	01/07/2005	31/08/2008	Bring together science communication teachers across Europe to exchange information and raise key issues in the training of young scientists in communication with the media, policymakers and the 'general public'	0.324
FP6	Pollen	01/01/2006	30/06/2009	Help to develop a model for the renewal of science education in primary schools based on an inquiry approach that allowed 12 interacting seed cities, after 3.5 years, to consolidate quality science teaching in their schools	1.750
FP7	Wonders	01/01/2006	28/02/2007	Promote greater awareness of science through major European networks and institutions in the field of science communication such as science weeks, science festivals, EUSCEA, ECSITE and EUSJA	0.800
FP7	ESConet trainers	01/01/2009	31/07/2011	Train EC-funded scientists in science communication skills, so that they can engage European citizens in dialog about science-in-society issues	0.544
FP7	PLACES	01/06/2010	31/05/2014	PLACES partners (science museums, science events, cities) develop a common platform to promote cooperation in their activities at the city or regional levels	5.190

as tackling two major European issues: young people's decreasing interest in science and the aging population (which means an additional need for scientists).

In total, 54 projects were funded under FP5, involving thousands of researchers from more than 30 European countries who shared an overall budget of €16 million. The projects aimed to exchange best practices between science communicators, to encourage dialog between scientists and the public, and to support the flow of science-related information, in particular through electronic networks. More precisely, projects were selected in the following four categories:

- *Targeting the general public*: 23 projects provided valuable insights into methods of reaching the general public with science information (such as 'Composites on tour', 'Maths in action', 'Science on the buses' and 'Ocean as a link between research and citizens' concerns')
- *Targeting teachers and students*: 14 projects addressed new teaching methods and young people's awareness (including 'Physics on stage', 'European project on the sun' and 'Kids and science')
- *Targeting Europe's media*: 3 projects focused on issues related to science journalism and the media ('AlphaGalileo—Media resource service for European S&T', 'Communication of S&T through television and European drama' and 'Associating science and society in European new drama')
- *Targeting science communicators*: 14 projects developed networks of science communicators to foster the sharing of ideas and encourage the trickle down of new techniques in the field (for example, 'Enscot—European science communication teachers network', 'Opus—Optimising public understanding of science and technology in Europe' and 'Maths alive').

These activities were quite limited in FP5 and, although important, were not central to the program. Nevertheless, interest in them increased dramatically through the 4 years of the program, and the final call for proposals was 16 times oversubscribed. As the then Director of the Science and Society Directorate in the European Commission, Rainer Gerold, put it: 'We have to be realistic, and recognize that activities to raise *awareness* should primarily be at local or regional and national levels.' However, some activities initiated under FP5 are still up and running, such as 'Science on stage' (the successor of 'Physics on stage') and 'AlphaGalileo'.

In FP6 (2002–2006) and FP7 (2007–2013), the European Commission followed more or less the same approach, but with two major differences. The first is that raising awareness and public communication activities are not only parts of a specific program ('Science and Society', which has been renamed 'Science in Society' under FP7), but are also central to the Framework Programme as all project participants since FP6 have had a contractual obligation to communicate their results to the public. The second difference is that the European Commission is encouraging voluntary coordination between Member States through a group set up to exchange best practice and network at government level.

In FP7, some funded projects focus on science communication. One example is ESConet Trainers, organized by the European Science Communication Network, which provides science communication training to scientists involved in FP projects.[2] The training aims to improve their ability to interact with national and international media, for example by teaching them how to present their work on TV and radio. It also includes training on how to use 'new' media, with the objective of establishing a better dialog with society. A considerable number of other FP5, FP6 and FP7 projects have funded public communication training courses for scientists at all levels and for other interested professionals. Other projects, such as My Science (My Science European Program for Young Journalists) and RELATE (Research Labs for Teaching Journalists), aim to bridge relations between the media and research laboratories involved in EU-funded activities.

15.2 Contractual Obligations

Publicly funded support for research in Europe increasingly encourages and even obliges the beneficiaries to engage with the public and the media. For example, most of the United Kingdom's research councils promote scientific outreach activities and provide the training needed by scientists to allow them to carry out those activities effectively. In particular, grant-holders have to develop a public communication strategy, and are able to get help from council's information and press staff.

In the EU's Framework Programmes, dissemination of results has been a contractual obligation for participation in research initiatives since FP6. What are commonly referred to as the 'participation rules[3]' lay down not only the rules for participation in FPs but also the rules for the dissemination of research results. The rules specify in particular that the quality of the planned dissemination is assessed at the proposal evaluation stage, and Article 20 states that the grant agreement requires 'the submission to the Commission of a plan for the use and dissemination of foreground [generated results]'.

Thus, beneficiaries of EU funding are bound to develop public communication activities. With a view to enhancing the impact of research funded by the EU, and to foster dialog and debate, the FP7 grant agreement requires project participants to communicate and engage with actors beyond the research community.[4] Plans for the outreach activities should be outlined at proposal stage, and are taken into account

[2] http://www.esconet.org/ESConet/Welcome.html

[3] Regulation (EC) No. 1906/2006 of the European Parliament and of the Council of 18 December 2006 laying down the rules for the participation of undertakings, research centers and universities in actions under the Seventh Framework Programme and for the dissemination of research results.

[4] General Conditions, II.2, Organization of the *consortium* and role of *coordinator*:

> 'Beneficiaries shall fulfill the following obligations as a consortium: ... engage, whenever appropriate, with actors beyond the research community and with the public in order to foster dialog and debate on the research agenda, on research results and on related scientific issues with policy makers and civil society; create synergies with education at all levels and conduct activities promoting the socioeconomic impact of the research.'

during the evaluation process. However, the commission does not provide any indication about the funding level that should be allocated to those activities.

The specific aims of this provision are to promote knowledge sharing, greater public awareness, transparency and education. Consortia of researchers are required to provide tangible proof that collaborative research not only exists, but also pays dividends in academic excellence, industrial competitiveness, employment opportunities, environmental improvements and enhanced quality of life for all.

We are halfway through FP7 at the time of writing, and the experience so far shows that engaging with the public and the media brings benefits to project management, such as increasing public visibility and awareness of the science, achieving successful integration with stakeholders, promoting internal communication, networking and marketing the consortium, and disseminating research results. Some participants also argue that having a good communication/dissemination plan increases the success rate of a proposal. At a higher (policy) level, the activities also contribute to bridging the gap between scientists and the public and making European research more attractive.

15.3 In Practical Terms

The European Commission's Research Directorate-General is actively involved in communicating the results of EU-funded research to the media and the general public. Support and help are provided to assist research project coordinators and team leaders to generate an effective flow of information and publicity about the objectives and results of their work, the contributions made to European knowledge and scientific excellence, the value of collaboration on a Europe-wide scale, and the benefits to EU citizens in general.

Some other initiatives have been taken by the European Commission to improve communication, outreach and dissemination of results from EU-funded research projects, and to facilitate the work of project contractors in that area. Researchers involved in FP-funded projects are encouraged to contact the Research Directorate-General's Communication Unit to cooperate to produce online news, press releases, video clips and so on.

The commission also manages three major websites to provide information on EU programs and initiatives, including projects funded and results obtained:

General Conditions, II.12., Information and communication:

> '1. The *beneficiaries* shall, throughout the duration of the *project*, take appropriate measures to engage with the public and the media about the *project* and to highlight the *Community* financial support. Unless the *Commission* requests otherwise, any publicity, including at a conference or seminar or any type of information or promotional material (brochure, leaflet, poster, presentation etc.), must specify that the *project* has received *Community* research funding and display the European emblem ….'

- EUROPA, the official EU website,[5] currently has some 30,000 pages devoted to EU research that are looked at by more than 300,000 separate visitors each month, as well as pages of historical interest (for example, pages on FP5 and FP4), which are visited less often. EUROPA provides up-to-date information on the latest decisions and latest advances in European research. There have been nearly 9.5 million visits to this site, leading to 18 million page views.
- CORDIS,[6] the Community Research and Development Information Service, is run separately and is designed primarily for current and potential participants in the Framework Programmes. In addition to providing information on FP7, CORDIS is intended to enhance the exploitation of research results and to promote the dissemination of knowledge.
- The so-called 'Participant Portal[7]' groups all the information necessary to participate in FPs (calls for proposals, a guide for proposers, participation rules etc.). The portal has been created to become a single entry point for the participants in the research programs implemented by the European Commission. It aims to facilitate the monitoring and the management of proposals and EU-funded projects throughout their lifecycles.

The European Commission does not impose any predetermined format or structure for the communication activities undertaken by FP7 projects. Rather, it draws the attention of participants to the fact that they can no longer ignore the 'public communication' dimension of their activity. Exposing non-specialists to the results of research work helps to improve their understanding of scientific and technological developments and stimulate public discussion of important issues, which not only meets a very real social need but also contributes to the success of a research policy. As Schiele (1983) put it: 'A science policy relies first and foremost on a science communication policy.'

At the same time, the communication of results and the announcement of exploitable developments are of direct value to the participants themselves. Suitably framed messages can help by drawing the attention of national governments, regional authorities and other public and private funding sources to the needs and eventual benefits of the research; by attracting the interest of potential partners and/or correspondents; by encouraging talented students and scientists to join the partner institutes and enterprises; and by generating market demand for the developed products or services.

The European Commission's Research Directorate-General published two practical guides (EC 2006, 2007) and a dedicated website[8] to help project coordinators and team leaders generate an effective flow of information and publicity. Guidelines and best practices are provided to help project participants in communicating and disseminating their research results, with the aim that the results will be presented and discussed.

---

[5] http://ec.europa.eu/research/index.cfm

[6] http://cordis.lu

[7] http://ec.europa.eu/research/participants/portal/page/home

[8] http://ec.europa.eu/research/science-society/science-communication/index_en.htm

The guides available on the communication website particularly address relationships with the mass media (TV, radio and print), the workings of which are less familiar to scientific or academic partners. They also cover websites and other internally generated support, such as print publications, CDs and video. They provide good practices that can be employed in defining key messages, establishing target audiences, selecting the appropriate modes of communication, tailoring information to the intended outlets, building good relationships with the media, evaluating results, maximizing the exposure of messages, and tapping useful European Commission and other external resources.

As well as providing sound advice on how best to proceed, the content includes examples of successful approaches that have been used to date (see also Christensen 2007).

The European Commission also organized the Communicating European Research 2005 conference for beneficiaries of EU research funding. The conference took place in Brussels on 14 and 15 November 2005, and showed the growing importance and recognition of public communication of science and technology.[9] In opening the conference, EU Commissioner Janez Potočnik, in charge of science and research at that time, said Claessens (2007):

> 'The objective of this conference is to explore how and why science needs to reach out and touch a wider audience … It is quite an achievement to bring together close to 3,000 scientists, journalists and policy makers under the same roof, all of whom face the same challenge. This challenge is twofold: on the one hand we need to improve the ways in which we communicate research and on the other hand, we need to improve the image of science in society … Therefore, communicating research and engaging with the public is more than a priority. It is an obligation.'

The conference, which was the first ever organized by the commission on this issue, was a major success, as illustrated by the number of participants and the feedback received from them.[10] During 2 days, participants (including project coordinators, journalists and other communication professionals, press officers and representatives from research organizations) met to promote mutual understanding of their respective roles, to share best practices and to define strategies to improve communication, outreach and dissemination of research results to the public and the press at a European level.

In addition to communication and outreach activities undertaken at the level of the projects, the European Commission has also designed its own specific communication strategy for research, which sets out the following main objectives:

> To communicate Europe as a leading and innovative place for doing and investing in research; show the results and benefits of European research to European citizens and hence foster understanding of research as a driver for European integration and for uniting people beyond the EU and provide first-class information on possibilities under FP7.[11]

[9] More information on the conference, including the program and speakers' presentations, is available on the website (http://ec.europa.eu/research/conferences/2005/cer2005/).

[10] According to internal commission reports, participants who attended in order to exchange best communication practice considered the event overall to be of high standard. The conference was rated as very good across all the different components.

[11] European Commission, internal document.

15.4 The European Scene

When dealing with science communication, it is important to take into account the European background on science and technology to understand the role and the European specificities of these activities.

In 2010, the European Commission published a new Eurobarometer[12] public opinion survey that aimed to evaluate European citizens' general attitudes towards science and technology (EC 2010). In January and February 2010, 26,671 people in 32 European[13] countries were interviewed. The 2010 Eurobarometer followed previous surveys organized in 1989, 1992, 2001 and 2005; some relevant items reached back to 1977.

According to the 2010 results, European citizens appear to have a clear and positive view about the image of science and technology. However, Europeans have a less clear insight into the work of scientists, what scientists actually do, or the structure of the scientific community. They feel slightly less informed about science and technology than they were in 2005 and feel less well informed than their level of interest deserves. There is no clear view on the effectiveness of European research. The perception of the current level of investment in scientific research at the EU level is also not clear. However, for a large majority of Europeans, collaborative research is more creative and efficient than national research, and an expansion of EU-funded research in the future would be a beneficial development.

Across the 27 EU countries, 61% of people are very or moderately interested in scientific discoveries and technological developments.[14] There is obviously a large variation between countries; Luxembourg and France are the most informed (79% and 77%, respectively) and Romania and Bulgaria the least (35%). Fifty-seven percent of EU citizens agree that scientists do not put enough effort into informing the public about new developments in science and technology.

Significantly, the majority of European citizens (63% of all respondents) feel that scientists working at universities or government laboratories are the best qualified to explain scientific and technological developments (they are the highest rated of all groups). The perceived role of newspaper journalists has diminished from 25% in 2005 to 16% in 2010, and television journalists also have a reduced role, declining from 32% to 20%. The role of consumer organizations has increased from 16% to 23%, and that of environmental protection organizations from 21% to 24%.

The survey shows that people are generally not publicly active in science and technology. Europeans are most active in donating money to fundraising campaigns for medical research, such as research into cancer (39% of respondents did so).

[12] Since 1973, the European Commission has organized regular public opinion surveys ('Eurobarometers') to monitor the evolution of public opinion in the EU Member States on major topics concerning European citizenship: enlargement, social situation, health, culture, information technology, environment, the Euro, defense, science and technology etc. This helps the European Commission in the preparation of texts, decision-making and the evaluation of its work.

[13] The 27 EU Member States, plus Iceland, Croatia, Norway, Switzerland and Turkey.

[14] See pages 8–18 of the report.

Only 13% of respondents engage in signing petitions or street demonstrations on matters of nuclear power, biotechnology or the environment; 86% of respondent never did so. Even fewer Europeans (9%) attend public meetings or debates about science and technology (91% of respondents have never or hardly ever done so).

On the behavior of scientists and the integrity of science, one in two respondents supported the view that private funding of scientific and technological research limits our ability to understand things fully. Fifty-eight percent of Europeans agree that scientists cannot be trusted to tell the truth about controversial scientific and technological issues because they increasingly depend on money from industry.[15]

Overall, the latest survey shows that European citizens are fairly optimistic about science and technology, but there has been a slight shift towards skepticism compared to the 2005 survey. Although science and technology may bring benefits, Europeans do not have very high hopes that they can solve all the worlds' problems. A clear majority of 54% believe that science and technology can play a role in improving the environment, but very few (22% of EU27 respondents) think that science can solve all problems.

On the interaction between science and society, most Europeans feel that decisions about science and technology should be made by scientists, engineers and politicians, and the public should be informed about those decisions (36%).[16] Only 29% want a more active role for citizens, such as being consulted when decisions about science and technology are being made.

Europeans have a positive view of the effect of involvement with science on young people but feel that governments are not doing enough to stimulate wide interest.[17] More efforts by governments to encourage women to be involved with science are seen as necessary and would, if successful, have a positive effect on the development of the sciences in Europe.

The Eurobarometer survey also shows that nearly 60% of Europeans think that scientists should put more effort into communicating about their work. In 2007, EURAB, the European Research Advisory Board of the European Commission, published a report that warned the scientific community about not paying enough attention to the dialog with society:

> European publics are not questioning the scientific information as much as they are actually questioning the institutions generating it (a loss of confidence in business, government and academia). Research is seen to be good when it solves problems and is relevant to people's lives—when research is useful to society, and not just in an economic sense. Too often though, researchers are perceived to be addressing issues that the public may not necessarily consider as beneficial to society. Researchers work in systems that are rational and instrumental, and have a tendency to assume that society behaves likewise. But society does not always behave rationally, and in certain sensitive areas, researchers should keep in mind that their systems operate in a public context. (EURAB 2007)

[15] See pages 24–27 of the report.

[16] See pages 85–87 of the report.

[17] See pages 95–106 of the report.

To avoid lost opportunities and suspicion about R&D in the future, the report urged more societal engagement and open dialog on emerging research fields, such as nanotechnology and therapeutic food additives:

> In Europe, GMOs, nuclear energy and crop protection science are examples where all research elements were in place but the societal concerns were misrepresented or not adequately considered, leading to a loss of public trust that has been detrimental to the innovation process.

'Europe is a real mess,' Charles de Gaulle once said. To some extent, that also applies to research, where European diversity adds to the difficulty of science communication. There are indeed some Europe-specific challenges to science communication. As the European Research Area's linking of national efforts becomes a reality, Europe is sorely lacking a genuine mechanism enabling it to draw full benefit from its 'home-grown' research activities. At present, there is no structured mechanism for informing the media in one Member State of scientific activities going on in another and giving the highest possible profile to European research. A survey carried out by ESO (the European Southern Observatory) showed that most articles published in Germany on space and astronomy concerned United States research. Although Europe has leading facilities in that field, American research still tends to dominate European media.[18]

15.5 The Assessment so Far

There is not much information available about the public and media engagement activities undertaken by FP6 and FP7 projects. According to the European Commission's internal data, 93% of FP6 projects did have a project website. The second FP7 monitoring report[19] published by the commission in 2008 covers the implementation of FP7 and provides some data on communication and dissemination of results following a survey by National Contact Points, which helps the commission in providing information and assistance related to participation in FP7. Of the respondents, 32.9% found the communication and dissemination of FP7 project findings by the project consortia to be satisfactory (and 24.9% good); 33.2% found communication and dissemination by the commission to be satisfactory (and 32.2% good).

Those who commented acknowledged that it was still very early in the program to make definitive judgments about the dissemination of project findings, but there was some agreement that it had improved in FP7 and that knowledge transfer remained a challenge to research funding agencies across the board.

---

[18] Claus Madsen, European Southern Observatory, personal communication.

[19] http://ec.europa.eu/research/reports/2009/pdf/second_fp7_monitoring_en.pdf#view=fit&pagemode=none

The most interesting result is that scientific knowledge, as measured by the questions included in the Eurobarometer surveys,[20] seems to have increased substantially in most European countries between 1992 and 2005. Over 15% increases in correct answers have been observed in Luxembourg, Belgium, Greece, the Netherlands and Germany; among the new EU Member States, the Czech Republic and Slovenia showed a 10% increase in only 3 years. Sweden achieved the highest rates of correct answers. Further analysis of the Eurobarometer data confirmed the overall trend towards higher scientific literacy in all European countries (Claessens 2008).

Interpretation of this trend remains unclear, but it is likely that the increase of science festivals, science museums and science centers in Europe is part of the answer. Also, one may argue that media coverage of recent crises and controversies in Europe (climate change, nuclear energy, genetically modified organisms, avian flu, swine flu, mad cow disease, contaminated blood etc.) has brought many scientific and technological concepts and issues onto the public radar and has subsequently raised the overall public understanding of science in the EU countries. This is supported by the analysis of Shimizu (2007), who argues that the 1995 Kobe earthquake contributed to the public understanding of plate tectonics (Shimizu 2007).

However, despite a growing number of public communication of science activities in Europe, and increasing support from the public authorities for them, there is still a large gap between science and society. Science is seen as a 'closed shop', with the public having no say in its development. Rarely do discussions and public debates accompany decisions about research issues and priorities. Europe still needs a genuine science communication culture. Without it, the public is not in a position to anticipate scientific and technological crises or deal with future developments. We need only look at three examples—nuclear energy, genetically modified organisms and cloning—and think of the public concern to see that this is the case. In short, we are not there yet: issues such as scientific research and what should be done at the EU level versus the national level remain, in the strictest sense of the word, 'uncommunicated'.

This situation badly handicaps science–society relationships and the public acceptance of advancements in science and technology (Royal Society 2006). Furthermore, Europeans want to be consulted and involved in shaping the course of 'progress' and the decision-making process. How do we build public trust? How do we improve the dialog between science and society?

[20] The 1992, 2001, 2002 and 2005 Eurobarometers included the following questions on science and technology. 'Here is a little quiz. For each of the following statements, please tell me if it is true or false. If you don't know, say so, and we will go on to the next one. The Sun goes around the Earth; The center of the Earth is very hot; The oxygen we breathe comes from plants; Radioactive milk can be made safe by boiling it; Electrons are smaller than atoms; The continents on which we live have been moving for millions of years and will continue to move in the future; It is the mother's genes that decide whether the baby is a boy or a girl; The earliest humans lived at the same time as the dinosaurs; Antibiotics kill viruses as well as bacteria; Lasers work by focusing sound waves; All radioactivity is man-made; Human beings, as we know them today, developed from earlier species of animals; It takes 1 month for the Earth to go around the Sun.'

Scientists are encouraged or even obliged to inform audiences about what they are doing, but they also have an imperative to listen (Claessens 2011). Researchers these days must understand the social context within which they operate: what people worry about, what they expect or need from science, what they do not want in their lives (Cheng et al. 2008). In short, the ivory tower is no longer an option.

One proposed solution is the systematic and even perhaps institutionalized organization of public consensus conferences. These scientific 'grand juries' could stimulate communication and political decision-making in scientifically controversial areas. Used judiciously, they could offer a realistic answer to our society's inability to control and appreciate the development and the application of science and technology (Claessens 2010). Communicating is truly an imperative in a democracy, and this applies also to scientific research if one is to build trust and legitimacy for activities funded in great part by the public.

References

Cheng, D., Claessens, M., Gascoigne, T., Metcalfe, J., Schiele, B., & Shi, S. (2008). *Communicating science in social contexts*. New York: Springer.

Christensen, L. L. (2007). *The hands-on guide for science communicators*. New York: Springer.

Claessens, M. (2007). *Communicating European research 2005*. Dordrecht: Springer.

Claessens, M. (2008). European trends in science communication. In D. Cheng, M. Claessens, T. Gascoigne, J. Metcalfe, B. Schiele, & S. Shi (Eds.), *Communicating science in social contexts: New models, new practices*. New York: Springer.

Claessens, M. (2010). *Science et communication: pour le meilleur ou pour le pire?* Paris: Quae.

Claessens, M. (2011). *Allo la science?* Paris: Hermann.

EC (European Commission). (2006). *Communicating science: A scientist's survival kit*, Brussels: European Commission. Retrieved from http://ec.europa.eu/research/science-society/pdf/communicating-science_en.pdf

EC (European Commission). (2007). *European research in the media: The researcher's point of view*. Brussels: EC. Retrieved from http://ec.europa.eu/research/conferences/2007/bcn2007/researchers_en.pdf

EC (European Commission). (2010). *Europeans, science and technology*, Eurobarometer 340. Brussels: EC. Retrieved from http://ec.europa.eu/public_opinion/archives/ebs/ebs_340_en.pdf

EURAB (European Research Advisory Board). (2007). *Research and societal engagement: Final report*, 07.013. Brussels: EURAB.

Royal Society. (2006). *Science communication: Survey of factors affecting science communication by scientists and engineers*. Retrieved from http://royalsociety.org/uploadedFiles/Royal_Society_Content/Influencing_Policy/Themes_and_Projects/Themes/Governance/Final_Report_-_on_website_-_and_amended_by_SK.pdf

Schiele, B. (1983). Enjeux cachés de la vulgarisation scientifique. *Communication–Information, V*(2–3), 157–859.

Shimizu, K. (2007). *Japanese survey of the public understanding of science and technology: Review of results, impact and recent secondary analysis*. Communication at the International Indicators of Science and the Public meeting organized by the Royal Society, 5–6 Nov 2007, London.

Chapter 16
Vital and Vulnerable: Science Communication as a University Subject

Brian Trench

Abstract Over nearly three decades, science communication has become established as a subject of teaching and research in universities across the world. Its standing as an academic discipline continues to be debated, but graduate degree programs and doctoral research in the field are increasing. Partly reflecting its inherent multi- and interdisciplinary content, science communication is embedded in different institutions in different ways. These developments have been driven mainly by individual champions, but in some cases also by institutional and government policies. The diversity of science communication programs reflects in part the various histories and institutional affiliations of the programs. The diversity can be seen as a sign of the subject's vitality but it is also a condition of its vulnerability. Many science communication teaching programs have given rise to consultancies, applied research, publishing and, perhaps most notably, doctoral research, but information from the promoters of science communication programs indicates that some programs are particularly exposed to the rationalization affecting higher education institutions in many countries. Science communication's position between and across disciplines and departments may mean it is not always well equipped to defend itself just when this need is most apparent.

Keywords Science communication • Institutional support • Interdisciplinarity • Master's programs • Ph.D. research • Economic conditions

---

B. Trench (✉)
School of Communications, Dublin City University, Dublin 9, Ireland
e-mail: brian.trench@dcu.ie

B. Schiele et al. (eds.), *Science Communication in the World: Practices, Theories and Trends*, DOI 10.1007/978-94-007-4279-6_16,

16.1 Introduction

Over the past quarter of a century, a new subject has emerged in universities across the world. In many countries, on several continents, at widely differing institutions, science communication has become a recognized subject of individual courses in broader science programs or of denominated degree programs at bachelor's and (mainly) master's levels. The spread of these programs since the mid-1980s parallels the staging of international conferences on the public communication of science and technology and the foundation and repurposing of academic journals to cover this field. Publication of handbooks, textbooks and other collected volumes, which has intensified since the late 2000s, both reflects the growth of taught programs and promotes the field (see, for example, Bennett and Jennings 2011; Bucchi and Trench 2008; Brake and Weitkamp 2010; Cheng et al. 2008; Holliman et al. 2009a, b; Kahlor and Stout 2009). The increasing research activity, reflected in the number of doctorates in this field—also spread across several continents and many countries—represents the consolidation of science communication as a university subject.

Several published reviews and discussions have addressed the status of science communication as a discipline or as a distinct field of study. Others have examined its underlying assumptions and models or explored its possible research agendas. Several reviews of taught university programs in science communication have been published. All of this activity is also to a large extent a reflection of the growing and significant number of people and units in higher education who are engaged with science communication.

Mellor et al. (2008) defended the theoretical content and academic validity of science communication in the face of criticism from practitioners in the field. Miller (2008) also examined that divide through a survey of practitioners that indicated they were in large part unaware of the possible contributions of theoretical work to their professional activity. Priest (2010) reflected on science communication's 'coming of age' and particularly on its hybrid status as both interdisciplinary and multidisciplinary (that is, partly integrated between established disciplines and partly based on multiple inputs from various disciplines). Trench and Bucchi (2010) considered science communication 'an emerging discipline' that met some recognized criteria of a discipline but remained weak in terms of theoretical development and the clear definition of its boundaries with cognate areas. Gascoigne et al. (2010) argued that science communication deserves 'special attention' because it 'contributes powerfully to pressing questions the modern world faces' and can derive benefits from not being a full discipline 'because it allows science communicators to plunder all disciplines and fields of study to conduct their work most effectively'.

There have been many other examples of such discussions, including in closely related areas. The status of journalism as an academic subject remains contested but is 'quintessentially cross-disciplinary' (de Burgh 2003). After 40 years of publication of the journal of the same name, social studies of science could be seen as 'still emerging' and are classified as such by the US National Research Council; 'the field has settled into a shape that is akin to that of a discipline, though still often prefaced with "inter" and "trans"' (Lynch 2011). The discussion of science communication in terms of the relationship between theory and practice and as a discipline and a field of study may be seen not as a weakness but as a sign of the subject's vitality. Increasingly, however, the terms of the discussion refer to institutional policies and economic conditions, pointing more to the subject's vulnerability.

Science communication has spread and diversified as a subject in a phase of economic growth and of general optimism about economic and social prospects and the contribution of science to those prospects. The worldwide discussion of and commitment to the knowledge economy gave strong emphasis to the role of scientific research in driving the economy and encouraged science communication initiatives in many countries. While very many social, cultural and other factors have also contributed to science communication's growth, the international economic crisis, particularly as it affects third level education in developed countries, may now be a brake on its further expansion.

Universities across the developed world in particular face accountability pressures and viability audits, some of which have identified science communication as no longer sustainable. The relative novelty of the subject and its uncertain status as a discipline are factors in its possible vulnerability. The subject's inherent interdisciplinarity is a primary source of intellectual stimulation but also a cause of institutional difficulties for those directly involved. In a curious dialectic, the indications of the subject's vitality and of its vulnerability come from the same sources. From the evidence and examples adduced below we will see that science communication faces a challenging future as a university subject. However, just as its advance has been uneven, with surges of parallel growth but also lapses of many years between similar countries, its retreat—or consolidation, as it might be seen more optimistically—is also patchy and contradictory: openings and closures are happening side by side.

In this chapter, I review the short history of science communication programs, consider their common and differentiating characteristics, outline their current challenges and opportunities, and reflect on the subject's prospects. This chapter does not offer a formal comparative study but looks at an international topic in an international perspective, drawing on publications, websites and correspondents' contributions from around the world.

16.2 The Global Spread of Science Communication Programs

Science communication programs at universities did not spring fully formed from the imagination of their individual champions. From the mid-twentieth century, a couple of decades before these programs emerged, science communication and related subjects have been taught to science students in institutions that required them to have some liberal arts instruction. Reflecting different higher education cultures, this element has accounted for a very minor and/or optional part of science degree programs. Single modules or part-modules in science writing, presentation or other aspects of science communication have been taught either by scientists who have taken up this activity (and sometimes, as advocates, this cause) or, less frequently, by specialists in communication providing a tailored teaching service outside their own home departments.[1]

Two other trends are also part of the background to science communication's emergence as a recognized university subject: the provision of short training courses for professional scientists and the inclusion of science writing or science journalism modules within broader communication and journalism programs.

The factors influencing the growth within scientific communities of 'science literacy' or 'public understanding of science' initiatives have been widely discussed elsewhere. One of the manifestations of that interest, broadly shared across the leading industrialized countries and then spreading rapidly from there, was the demand from scientists for communication training. Professional societies were often the hosts of such short courses; university teachers were often the providers.

Science journalism established itself as a recognized specialism in the United States from the 1950s onwards and in Europe and elsewhere somewhat later. The accelerating growth of specialist newspaper sections and broadcast programs from the 1970s onwards prompted the development of specialist modules within journalism bachelor's and master's programs. Here too, the earliest initiatives were taken in the United States.

While these several contributory developments are all important, we date the start of science communication as a university subject to the late 1980s, when the first taught postgraduate programs denominated as 'science communication' appeared in several countries. Those programs are distinguished in several ways from what went before, not only by the breadth of the topics covered but also by their ambition to provide a professional qualification in an emerging area of work. The Australian National University (ANU) in Canberra was one of the first to establish such a program in 1987; the program was linked very directly to a professional outlet in informal science education, specifically a traveling science center. The identification of this 'science circus' in the description of the university program, which is supported by corporate sponsorship maintained for a quarter-century, underlines the highly applied character of the training.

Other science communication programs started in the years immediately after the ANU innovation tended to be less specifically tied to particular professional outlets. Many were oriented to informal or formal education or to media, reflecting their origins, the backgrounds of their champions and their home departments. But some aimed to provide a general university education, equivalent to a degree in other humanities or social sciences subjects.

Over the decade after the ANU program started, other postgraduate programs in science communication—awarding either diplomas or master's degrees—began in

[1] The terminology varies from country to country, and from institution to institution within countries; I use 'program' to refer to studies over 1 year or more leading to a degree or similar award; I use 'module' to refer to an element of a program, typically lasting a semester. I use 'subject' to refer to the name and content of a program. I have avoided 'course', as it can refer to either a single element or a combination of elements.

Australia, Britain, Spain, Italy, France and Ireland. From 1989, some of those involved were meeting through the biennial PCST (public communication of science and technology) conferences. In the same period, science communication sections were started (though not maintained) in the International Communication Association and the International Association of Mass Communication Research, both of which assemble university-based communication researchers in large annual gatherings. But the initiatives to start the taught programs were more or less independent of each other. Their circumstances varied considerably: in some cases, the programs were delivered by small teams also teaching and researching in natural sciences; in other cases, they were the outgrowth of longer established programs in communication, journalism or social studies of science.

Turney (1994) reported on the growth of such programs and courses in the United Kingdom and proposed a simple typology that distinguished programs and courses focused on communications skills from those providing 'skills with added theory' or presenting 'the big picture'. Whereas the first type referred to short courses and the second to individual courses within broader undergraduate programs, only the third type fits with our present interest. Turney reflected on the difficulties of achieving balance between the theoretical and practical elements of such programs.

Science communication programs were started in Netherlands, Mexico and elsewhere over the following decade, and the diffusion of programs across the globe continued in the present century. By the early 2000s, India had several postgraduate diploma and degree programs in science communication, largely on the basis of active guidance and scholarship funding from government. In 2005, the Korea Foundation for the Advancement of Science and Creativity supported the establishment of a science communication master's degree program at Sogang University and in 2009 of a science journalism master's degree course at the Korea Advanced Institute of Science and Technology. In Japan, Hokkaido University established a Master's in Science Communication in 2006. In New Zealand, Otago University added a Master's of Science Communication to its innovative postgraduate program in science and natural history film-making. In Brazil, a Master's in Scientific and Cultural Communication was added in 2007 to the existing offering in science journalism at the University of Campinas (Vogt et al. 2009). At the National Autonomous University of Mexico, the program in science popularization, which started in 1996 through a close association with a science museum, was broadened from 2008 and linked to longer established studies in the philosophy of science (Haynes 2009).

Laurentian University, Ontario, Canada, describes its Graduate Diploma in Science Communication, a joint initiative with the Science North science center, as 'North America's first and only comprehensive Science Communication program'. The relatively weak representation of North America in this story is notable. United States universities have often been to the fore in driving the professionalization of new sectors of employment through career-oriented programs, most notably, in the present context, journalism and public relations. Long established in graduate and undergraduate education in science writing and science, health and environmental journalism, US universities have not adopted the 'science communication' rubric for taught programs as widely as have universities in many other countries.

The Directory of Science Communication Courses and Programs maintained at the University of Wisconsin–Madison covers many types of program with a science communication dimension,[2] but it offers very few examples of self-contained graduate or undergraduate degree programs in science communication of the kind referred to above. Science communication tends to be represented more frequently as the subject of single modules within sciences or humanities degree programs. Relatively rare exceptions are the Master of Science in Communication, with a specialization in science communication, at Drexel University, Philadelphia, and the Master's Track in Science/Health Communication at the University of Florida.

A discussion between representatives of science communication programs at the 2006 international PCST conference led to a survey and report (Mulder et al. 2008). The survey covered science communication programs in 19 universities in 10 countries. Based on the responses, the authors defined science communication programs in terms of four 'areas of study'—science, education studies, social studies of science and communication studies. Supports for teaching science communication and content for such programs come in various blends from those four sectors. In a more detailed examination of seven postgraduate programs, however, Mulder et al. found more variation: direct engagement with science or education studies was not part of several of the programs. The authors proposed the establishment of a core framework to which science communication programs would subscribe.

More recently, Hong and Wehrmann (2010), in their review of 20 science communication programs based on curriculum information available online, found that that a fifth included science content and a quarter covered education studies. They were seeking to establish whether the programs prepared students to work as science communication professionals, and looked at the presence or absence of an internship and the fit of curriculum and program objectives with the profiles of science communication professionals. They found the fit was generally poor and recommended that programs 'provide clear objectives and profiles of science communicators' as well as indicating how modules in the four areas of study outlined by Mulder et al. (2008) fit into the professional profiles.

A series of articles published in the *Journal of Science Communication* (vol. 8, no. 1, 2009) offered an overview of master's programs in science communication, with contributions from the coordinators of six of those programs in six countries.[3] The coordinators had varying views of the scope of science communication as a subject for study and reflection. For de Semir (2009), it concerns 'the process of public transmission and diffusion of scientific knowledge'; for Trench (2009), it encompasses 'the relations between the organizations and institutions of science and those of society (including politics, education and media)'; and for Greco (2009), 'it is a complex dynamic system that functions on many intercommunicating levels and involves not only the mind, but also the body and the spirit.' However, the engagement between disciplines of natural sciences, social sciences and humanities

[2] Posted at http://dsc.journalism.wisc.edu/index.html.

[3] Posted at http://jcom.sissa.it/archive/08/01.

is a common thread of these self-reviews, as is the engagement between the theory and practice of science communication.

As a science film-maker and science communication lecturer, Davis (2010) takes a different view, arguing that science communication practice and theory (or research) should be separated. In his view:

> [W]e should be treating the *practice of science communication* as a separate and recognizable academic entity that draws its theoretical and research wing from those studying narrative, writing, filmmaking, design and digital communication … [R]esearch and theory that currently falls under the auspices of science communication … is a perfectly valid and valuable area of study but, for the sake of clarity, we should be distinguishing it as a separate academic area from those aspects of science communication to do with its practice.

The several published listings of science communication programs tend to reinforce this image of diversity, and specifically this unresolved tension between mainly professional and mainly academic orientations. In 2007, the European Commission commissioned a guide to science journalism training, updated most recently in 2010 (EC 2010), that also includes more broadly based science communication programs not specifically geared to journalism. An expert group advising the British Government on issues in science and the media included a listing of three undergraduate and eight postgraduate programs in science communication as part of the relevant training provision (Science and the Media Expert Group 2010). The EU listing for the United Kingdom and the expert group listing do not completely overlap, although the information was gathered for both at roughly the same time.

The expansion of science communication as a university subject has been uneven in substance and in space and time. There have been program closures as there have been openings. In two of the countries with the earliest and strongest presence in the field, Britain and Australia, some programs have stopped. In Britain, undergraduate programs at Sheffield Hallam University, the University of Bristol and the University of Western England were discontinued, as was the Master's in Communicating Science at the University of Glamorgan. Imperial College London added a Master's in Science Media Production to its pioneering Master's in Science Communication, while it suspended the associated Master's in Creative Non-Fiction Writing after 2 years of operation. At the University of New South Wales in Australia, a bachelor's science communication program with two degree tracks that included significant science content as well as communication and other humanities and social sciences subjects closed in 2010 after a decade of operation. By the end of the 2000s, indeed, there were several signs that the tide was turning or, at least, that expansion had slowed and existing programs faced more demanding requirements.

16.3 Increasing Challenges, Remaining Opportunities

This was confirmed, though by no means uniformly, in email correspondence (plus one direct interview) with senior lecturing staff responsible for the coordination of science communication programs. Many expressed concerns arising from the generally difficult financial conditions facing universities in many countries. Their comments included references to the impact of increased student fees and the difficulties of replacing staff who leave. There were also reassurances from some correspondents, however, that staffing was adequate and that vacancies had been filled. The correspondents also identified challenges and opportunities that are more particularly linked to the character of science communication programs.

Fourteen responses were received from 11 countries. In order to illustrate some of the common or related features, selected points from the responses are presented below. They are identified by country; where there was more than one response from an individual country and the responses differed in substance, the responses are given discrete numbers.

- Challenges
  - Suspension of programs due to recruits falling below viability thresholds (Germany, Thailand, Australia 1)
  - Increased pressure to meet financial reporting targets (UK 1, Netherlands 1)
  - Not seen as core business and therefore vulnerable to cutbacks (Netherlands 1, Italy)
  - Insufficient academic staff to handle increased project and thus restraint on taking in additional students (UK 2)
  - Small specialist teams running programs politically weaker than department-size units (Australia)
  - Research effort affected by overload of education and coordination duties (Netherlands 1)
  - Different affiliations within the university and different evaluation frameworks for teaching and for research (UK 2, Netherlands 1)
  - Strategy adopted of ‘hiding’ in a larger structure that recruits larger numbers of students (France 2)
  - Change in program’s character through relocation to another department (Mexico)
  - Continuing need to explain or justify science communication and science communication research in a natural sciences institutional setting (UK 2, Netherlands 1)
  - Imbalance of professional education openings and professional positions (Finland, Spain).
- Opportunities
  - Start of a new program with strands in journalism and museums (Hungary)
  - Start of a new program in Latin America based on an existing, long-established program (Spain/Argentina)
  - New master’s program in science communication developed in association with existing programs in related fields (Australia 2)
  - Recognition of science communication team’s distinctive contribution to the university’s public profile (Spain, UK 1, UK 2, Italy)

- Encouragement to expand program and strengthen its international dimension (Italy)
- Protection of program through good reputation and association with university's strategic aims in science and society (France 2)
- External support from institutions promoting science-and-society initiatives (Spain)
- Satisfactory recruitment levels without major promotional effort (Spain, UK, Finland, France 1, France 2)
- Satisfactory institutional support, staffing and prospects for growth (Finland, Netherlands 2)
- Improved protection through relocation from a natural sciences to a humanities department (France 1, Mexico)
- Addition of a program and short courses providing continuing professional development for those in relevant employment (UK 2)
- Increased recruitment of Ph.D. students in recent years (UK 2)
- Demand for delivery of courses in science communication for other programs' students (Spain, UK 1, UK 2).

These summarized responses illustrate the uncertain status of a still-emerging university subject. The balances of challenges and opportunities reported were broadly similar across the countries sampled. However, some demands seen as a challenge or difficulty in one case were seen in another as an opportunity. This may be taken as reflecting different institutional and economic conditions, independent of the particular subject of science communication.

My own direct experience of founding and leading a science communication master's program over 15 years also gives evidence of similar challenges and opportunities and illustrates some of the issues facing programs of this kind. The Master's in Science Communication established in 1996 at Dublin City University has been seen as contributing to the university's reputation for innovation and the university's international reach, both in attracting students from several countries and in functioning as a base for participation in international projects. The program has been maintained and a retiring lecturer/coordinator replaced, although its student recruitment has come close to the (shifting) viability thresholds. However, the program has probably fallen short of its original ambitions for interdisciplinary collaboration. It started in a collaboration between two departments in different disciplines in two institutions in two jurisdictions. For the first 7 years of its life, the program was delivered jointly by Dublin City University (Ireland) and Queen's University Belfast (Northern Ireland, United Kingdom). Its coordinators were based in the School of Communications in the first university and the Department of Physics in the second. The lecturers who contributed to the program included medical scientists, archeologists, historians of science, a psychologist, journalists, a philosopher, chemists, biologists, media analysts, communication theorists and others. However, there has been limited interaction between the lecturers in natural sciences on the one hand and humanities and social sciences on the other.

Queen's University Belfast withdrew from the program in 2003 for strategic reasons of the kind already touched on in the correspondents' comments summarized above: the program was not a core activity of any department and it did not fit well with the increasing emphasis on quantitatively measured research output in natural sciences departments. The withdrawal of Queen's reduced somewhat the diversity of the inputs to the program; a single module among six that are specific to the program (there are six further modules, some optional, shared with other programs) has significant natural-sciences content and is delivered from the Faculty of Science and Health.

In other ways, however, the interdisciplinary aspect is prominent. The discourses and systems of natural sciences and social sciences are both examined critically. Most students have backgrounds in natural sciences, though each annual cohort includes some with qualifications in media studies, literature, languages or other humanities or social sciences subjects. The cross-disciplinary experience of the classroom has been a major feature of the program. The inclusion of students from various backgrounds is not universal in science communication programs. Some master's programs specify that entrants must have science degrees; in this way, they are set up as conversion courses. The lecturers include individuals with primary degrees in biological sciences and higher degrees (master's and Ph.D.s) in communication or business. Natural scientists have often reported that they found teaching in a more reflective manner and broader context especially stimulating, and the social scientists and humanists reported that they relished the challenge of engaging students with backgrounds in the natural sciences with the methods and logics of the humanities and social sciences.

Having lecturers who have moved from natural sciences to communication either through additional qualification or research is one of the most usual expressions of interdisciplinarity in the delivery of science communication programs. An unusual expression of interdisciplinarity is found in SISSA (the International School for Advanced Studies), Italy, where classes in science content (mainly neuroscience and physics, reflecting the special interests of the home institution) are delivered by research scientists with a communication specialist alongside to explore the media and other social dimensions.

More critically, being interdisciplinary tends to mean looking two or more ways simultaneously and being rooted in neither one recognized institutional setting nor the other. The difficulties of negotiating these relations with representatives of various disciplines and with institutional leaders representing various approaches are reflected in some of the observations above on the current situation and are expanded in the comments of several program coordinators:

> Science communication is in the tricky position of having to set itself up as an academic subject, while at the same time attending to its relationship with the scientific establishment. Science communication has a base in the universities, but in the bulk it's an 'out-of-doors' activity. History and Philosophy of Science can turn all its attention to the academic corridors in a way that science communication academics may not feel completely sure about. It's possible that the problems science communication faces in the academy could be turned into virtues—its intellectual agenda, its courses—and its 'impact' on society.
>
> —Stephen Webster, Imperial College London, pers. comm., January 2011

> I'm not sure the scientists understand completely what we do and they could have some problems with some of it. We are in a strange balance. They understand that we are useful. It depends on different boundary conditions: we could become a kind of outreach department or a research department, though this is less likely. Mostly, the scientists in our institute have in mind a popularization model for science communication.
>
> —Nico Pitrelli, SISSA, Italy, pers. comm., November 2010

> We would not be able to survive very long on our own because we are too small. The program's good reputation and, especially, the fact that the university has structures dedicated to science communication protect us. The university established a vice-presidency for science and society 2 years ago.
>
> —Elsa Poupardin, University of Strasbourg, pers. comm., January 2011

> The status of science communication programs is likely to reflect the political insight of the director, the program's ties to influential actors in the university and outside, and the educational context (which faculty the program sits in, whether students are willing to pay for postgraduate study, the competition from other programs, cost etc.). Having powerful allies is essential as is having redundancy in place, where possible, for when key allies leave. That said, when budgets are cut, the interdisciplinary program seems more likely to go first.
>
> —Will Rifkin, University of New South Wales, Australia, pers. comm., March 2011

> We were very fragile when we depended directly on science departments but our relocation to Letters seems to protect us. Our Masters is really a professional Masters and the departments in Letters don't have professionally oriented programs. The literary people want to keep us because of the professional dimension to our education and we are able to participate actively in the life of the department.
>
> —Baudouin Jurdant, University of Paris 7, pers. comm., January 2011

16.4 Conditions for Sustainability

From the survey of existing programs, the correspondence and comments of program coordinators, and personal experience in the field, we can identify several key criteria for differentiating science communication programs. Using those criteria, it may be possible to extract factors for the success and sustainability of some programs, although political and economic factors in individual countries and their higher education sectors need to be included to provide a full picture.

As the clear majority of the programs surveyed are postgraduate (master's and diplomas) and the rate of attrition of undergraduate programs has been significantly higher, these key criteria apply mainly to postgraduate programs:

- Breadth of student recruitment—open to science graduates only or open to all graduates with interest in science?
- Institutional setting—based in natural sciences faculty or in humanities or social sciences?
- Mode of delivery—full-time, part-time, or both?

- Target markets—new graduates, those already in related employment, or both?
- Balance of content—mainly science content, social and communication studies or professional skills?
- Disciplinary connections—linked mainly to natural sciences, science education, media/communication, other humanities or social sciences?
- Institutional strategies—support for interdisciplinary collaborations, for science-and-society initiatives, for innovative programs?
- Program team—program delivered mainly by dedicated staff or staff with main responsibilities elsewhere?

Satisfactorily assessing individual programs against these criteria is itself a challenge. In some cases, institutional support can be more apparent than real. Professorial appointments have been made in some British universities to chairs in the public understanding of science or science and society that appear to address the same or related agenda but have not contributed to the development of science communication as a subject of education and research (Miller 2008). The appointments have, it seems, been motivated mainly by concern for the university's prestige; the holders of such posts tend to be chosen for their high profile as science popularizers rather than for their interest in pedagogic or theoretical issues in science communication. These appointments, it has been argued, 'embodied a split within universities over the perceived role of academics with respect to public science' (Mellor et al. 2008).

One criterion that is difficult to articulate but that appears important is that of enthusiasm: the individuals and (mostly very small) teams coordinating these science communication programs have often taken on the role of champions of this new subject, in many cases being the first champions or immediate associates of the first champions of the subject in their institutions. They have had thrust upon them the task of advocating, justifying or defending—as the circumstances demand—the case for science communication.

Hong and Wehrmann (2010) have emphasized the criterion of professional education, observing on the basis of their survey of program content that 'it is still doubtful to what extent the science communication programs equip students to become professionals.' They argue that the objectives and content of these programs should centrally 'reflect on the real world of science communicators'. With colleagues at Technische Universiteit Delft, the Netherlands, they have taken initiatives to raise science communication students' awareness of career opportunities and science communication teachers' awareness of professionalization issues. While some program coordinators might choose to emphasize equally the dimension of intellectual curiosity, it is generally true that science communication programs have spread more or less in line with the spread of relevant professional employment opportunities.

As mentioned above, the first postgraduate program in Australia was explicitly linked to a practical activity and employment outlet. An internal survey in 2011 of graduates of the Dublin City University Masters in Science Communication, with responses from approximately one-third of all the 200-plus graduates, showed that

60% were working full-time or part-time in science communication.[4] Their main areas of work were journalism, informal education, public information, research in science communication and teaching or training in roughly equal proportions. A demand-oriented survey of science communication education and training would likely show a diffusion of new opportunities in science museums, science centers, science outreach, science information, science writing, science websites and so on, in a similar pattern to the diffusion of science communication programs. This linkage can be a support for the sustainability of science communication programs as long as the employment trend is upwards; it could be a handicap if the trend is reversed.

Some programs have strengthened their position through more formal links with the world of work. At the University of Western England, the Science Communication Unit provides consultancy services that include exhibit design, campaign design and evaluation. The unit also provides short communication courses for clients working in scientific institutions and science-based companies. Several science communication programs are linked with science centers, among them the Master's in Science, Media and Communication at the University of Cardiff, Wales, which is presented in collaboration with a science center, Techniquest. Among other programs that have received financial support from companies, foundations or government to support professionalization in the sector, the master's program at Universitat Pompeu Fabra in Barcelona, Spain, has had long-term financial support from the pharmaceutical company Novartis and from the philanthropic arm of a savings bank, Fundacion La Caixa.

Further supports for science communication as a university subject have come through international networking of science communication teachers in conferences, research projects and other collaborations. Of the 220 registrants for the 2010 international PCST conference in New Delhi, India, who responded to an evaluation survey (44% response rate), 21% said their main involvement in science communication was in teaching or training and 23% said it was in research.[5] Several collaborations between university-based science communication specialists, including the present volume, have grown from the biennial PCST conferences, and university lecturers in science communication are strongly represented among the PCST scientific committee that organizes the conferences.

There are also national conferences and workshops in the field, often also the initiatives of university-based science communication teams. A meeting of science communication teachers from British and Irish universities took place in London in 1997. The science communication team at Imperial College London initiated an annual conference, Science and the Public, the second of which led to a collected

[4] This survey was undertaken by the author to mark 15 years of the master's program. There were no reliable means to reach all of the approximately 225 graduates of the program; 77 responses were received. A report on the survey is posted at http://www4.dcu.ie/communications/resources/pdf/Results_of_survey_of%20graduates_of_MSc_in_Science_Communication.pdf.

[5] This survey was undertaken by the author for the guidance of the organizers of future PCST conferences; it is not published.

volume, *Science and its publics* (Bell et al. 2008). The editors' introduction to that volume reflected the contested status of science communication as a subject for formal academic study; the exchanges between science communication practitioners and academics that formed part of the conference's background 'were a sharp reminder of the tension that lies at the heart of the field of science communication: how should the relationship between theory and practice be managed?' (Mellor et al. 2008).

Another initiative of science communication specialists in universities was the formation in the early 2000s of the European Network of Science Communication Teachers (ENSCOT) as a project funded through the European Commission's Raising Public Awareness of Science and Technology program. This group of university-based lecturers in science communication and science journalism from Britain, France, Germany, Ireland and Spain developed teaching materials in various aspects of science communication (ENSCOT Team 2003). They were joined by others from several more countries and from outside universities in subsequent projects, European Science Communication Workshops and ESConet, developing and delivering short communication courses for research scientists.

The growth of research, publishing and conference activities promoted by science communication specialists in universities is a significant mark of the embedding of science communication as a university subject. Several regular publications have been initiated directly by individuals and teams associated with taught science communication programs. They include:

- *Quark*, published quarterly for over a decade from the mid-1990s by the Science Communication Observatory at Universitat Pompeu Fabra, Barcelona
- *Com Ciencia*, an online science journalism magazine published by the Labjor unit at the State University of Campinas, Brazil, which published its 100th edition in 2010
- The *Japanese Journal of Science Communication*, an open-access journal published twice a year since 2007 by Hokkaido University
- *JCom* (Journal of Science Communication), an open-access journal published quarterly since 2002 by SISSA, Trieste, Italy, where Italy's longest established master's degree in science communication is based.

University academics associated with the development of science communication programs have also been key figures on the editorial boards and in contributions to the internationally distributed peer-reviewed journals in the field: *Public Understanding of Science*, launched in 1992 with John Durant, a founder of Imperial College London's science communication master's, as first editor, and *Science Communication*, which acquired its present name (it was formerly *Knowledge*) in the 1990s, reflecting the emergence of science communication as an academic subject.

The growth of Ph.D. research in science communication may be the single most substantial mark of science communication's full emergence as a university subject. Several of the program coordinators whose correspondence is summarized above referred to the small groups of Ph.D. researchers in their teams as a significant gain; the taught master's programs surveyed here have provided many of the recruits to

Ph.D. projects. Since 2000, perhaps 100 Ph.D. theses in science communication have been completed internationally, and the indications are that there may be at least as many currently underway. Outline details of 57 Ph.D. theses collected for analysis came from 14 countries on 5 continents (van der Sanden and Trench 2010). The subject areas included media and journalism (14 of the 57 theses), means of communication (10), engagement and dialog (7), scientists' role and image (7), and roles of stakeholders (6). The diversity of the field is also represented in the broad spread of research aims, among which it was not possible to identify common categories. In a tentative general commentary, it was noted that 'science communication [as reflected in the Ph.D. theses] is much more "science" than "communication"' (van der Sanden and Trench 2010).

Here again we see that the diversity and uncertainty that characterize science communication as a field of education and research are conditions both of the subject's vitality and of its vulnerability. In Ph.D. studies, international developments mean that the individual relationship between apprentice Ph.D. student and master Ph.D. supervisor is giving way to programs of research that include required taught elements and collective supervision. In science communication, there have been new starts in organizing such programs between departments and even between universities but there have also been program closures, reflecting the difficulties of managing those relationships. For example, a joint Ph.D. program in science communication between SISSA and the University of Milan closed after several years in operation.

However, the Ph.D. scholars who have emerged from the system over the past decade in particular represent an undeniable achievement for science communication as a university subject. They also represent a second generation of science communication specialists with, in general, higher levels of formal qualification in science communication than many of the founding figures in the field, who took up teaching and research in science communication on the basis of personal interest and backgrounds in scientific research, science education, media practice or other sectors.

In terms of personnel and capacities, science communication has become ever more deeply rooted in the university system over two decades. It is no longer solely the province of scattered individual champions. But it remains marginal, often trapped uncomfortably between the major shifting blocs of natural sciences and humanities or social sciences. Despite the many inspiring initiatives aimed at dissolving the boundaries between the 'two cultures', and even to develop a 'third culture', the binary division is still strongly entrenched in higher education. The widespread trend to reorganize universities around a few major subcenters rather than myriad departments has, if anything, reinforced the two-cultures divide. So too has the research funding environment, which gives increasing weight to the large-scale and highly visible activities that are characteristic of the natural sciences and engineering and very rarely found appropriate in the humanities or social sciences. In a circular motion, these large-scale activities become defined as 'core' to the institutions, while other activities are regarded as non-core and thus as priorities for pruning when rationalization is deemed necessary. Thus, on top of inherited tribal structures and different research paradigms, financial factors and conventional wisdom on the rationalization of university structures are militating against interdisciplinary collaboration across the great divide.

In the view of E.O. Wilson (1999), 'the greatest enterprise of the mind has always been and always will be the attempted linkage of the sciences and humanities.' He has argued for the possible 'unification of knowledge', but even more modest projects can illuminate the 'big picture' of the human condition through collaborations of sciences and humanities. There have been many of those, and there will continue to be many despite the uncongenial trends. It may be that science communication can defend its position most easily in functional terms, as providing the means for transmission of scientific information to non-specialist audiences, but that limits the scope and potential of the subject. Precisely because of its position in the gaps between the sciences and the humanities, science communication has much more to offer: it can be a place of intellectual inquiry into the convergences and divergences of different disciplines; it can be an engine of institutional reflexivity, helping modern universities to examine how they manage their internal diversity and how they articulate with the wider world.

After two decades in which science communication has faced many challenges to survive and thrive in universities, it may be time to ask whether universities can face the challenge of having science communication in their midst.

Acknowledgements I thank the coordinators of science communication programs who responded to my questions. Several also generously contributed additional comments.

References

Bell, A. R., Davies, S. R., & Mellor, F. (Eds.). (2008). *Science and its publics*. Newcastle: Cambridge Scholars.

Bennett, D. J., & Jennings, R. C. (Eds.). (2011). *Successful science communication: Telling it like it is*. Cambridge: Cambridge University Press.

Brake, M. L., & Weitkamp, E. (Eds.). (2010). *Introducing science communication: A practical guide*. Houndsmills/Hants: Palgrave Macmillan.

Bucchi, M., & Trench, B. (Eds.). (2008). *Handbook of public communication of science and technology*. London: Routledge.

Cheng, D., Claessens, M., Gascoigne, T., Metcalfe, J., Schiele, B., & Shi, S. (Eds.). (2008). *Communicating science in social contexts*. Dordrecht: Springer.

Davis, L. S. (2010). Science communication: A 'down under' perspective. *Japanese Journal of Science Communication, 7*, 65–71, Retrieved from http://hdl.handle.net/2115/42663.

de Burgh, H. (2003). Skills are not enough: The case for journalism as an academic discipline. *Journalism, 4*(1), 95–112.

de Semir, V. (2009). Master in scientific, medical and environmental communication. *Journal of Science Communication*, *8*(1).

EC (European Commission). (2010). *European guide to science journalism training*. Brussels: EC. Retrieved from http://ec.europa.eu/research/conferences/2007/bcn2007/guide_to_ science_ journalism_en.pdf

ENSCOT Team. (2003). ENSCOT: The European network of science communication teachers. *Public Understanding of Science, 12*, 167–181.

Gascoigne, T., et al. (2010). Is science communication its own field? *Journal of Science Communication, 9*(3).

Greco, P. (2009). The master's degree of Trieste. *Journal of Science Communication, 8*(1).

Haynes, E. R. (2009). A graduate course for science communicators: A Mexican approach. *Journal of Science Communication, 8*(1).

Holliman, R., et al. (Eds.). (2009a). *Investigating science communication in the information age: Implications for public engagement and popular media*. Oxford: Oxford University Press.

Holliman, R., et al. (Eds.). (2009b). *Practising science communication in the information age: Theorising professional practices*. Oxford: Oxford University Press.

Hong, C.-P., & Wehrmann, C. (2010). *Do science communication university programs equip students to become professionals? A comparison of 20 university programs worldwide*. Paper presented to 11th international conference on public communication of science and technology, New Delhi, 6–9 Dec 2010.

Kahlor, L. E., & Stout, P. (Eds.). (2009). *Communicating science: New agendas in communication*. London: Routledge.

Lynch, M. (2011). Still emerging after all these years. *Social Studies of Science, 41*(1), 3–4, Editorial.

Mellor, F., Davies, S. R., & Bell, A. R. (2008). Introduction: 'Solverating the Problematising'. In A. R. Bell, S. R. Davies, & F. Mellor (Eds.), *Science and its publics* (pp. 1–14). Newcastle: Cambridge Scholars.

Miller, S. (2008). So where's the theory? On the relationship between science communication practice and research. In D. Cheng, M. Claessens, T. Gascoigne, J. Metcalfe, B. Schiele, & S. Shi (Eds.), *Communicating science in social contexts* (pp. 275–287). Dordrecht: Springer.

Mulder, H., Longnecker, N., & Davis, L. (2008). The state of science communication programs at universities around the world. *Science Communication, 80*(2), 277–287.

Priest, S. H. (2010). Coming of age in the academy? The status of our emerging field. *Journal of Science Communication, 9*(3).

Science and the Media Expert Group. (2010). *Science and the media: Securing the future*. London: Department for Business, Innovation and Skills. Retrieved from http://interactive.bis.gov.uk/scienceandsociety/site/media/files/2010/01/Science-and-the-Media-Securing-the-Future.pdf

Trench, B. (2009). MSc in science communication, Dublin City University. *Journal of Science Communication, 8*(1).

Trench, B., & Bucchi, M. (2010). Science communication, an emerging discipline. *Journal of Science Communication, 9*(3).

Turney, J. (1994). Teaching science communication: Courses, curricula, theory and practice. *Public Understanding of Science, 3*(4), 435–443.

van der Sanden, M., & Trench, B. (2010). *Analysis of doctoral research in science communication*. Paper presented to 11th international conference on public communication of science and technology, New Delhi, 6–9 Dec 2010.

Vogt, C., et al. (2009). Master's degree program in scientific and cultural communication: Preliminary reports on an innovative experience in Brazil. *Journal of Science Communication, 8*(1).

Wilson, E. O. (1999). *Consilience: The unity of knowledge*. London: Abacus.

Chapter 17
Visible Scientists, Media Coverage and National Identity: Nobel Laureates in the Italian Daily Press

Massimiano Bucchi

Abstract This chapter analyzes media coverage of the Nobel Prizes, focusing especially on the coverage in the Italian daily press of the Nobel Prize for physics received by Guglielmo Marconi in 1909. It thus offers an opportunity to explore general features and trends in the coverage of science by the press, as well as its treatment of Nobel Prizes in the sciences. Media treatment of Marconi's Nobel Prize highlights two key elements of science coverage in the Italian daily press: the media's dependence on highly prominent individuals, and the connection between science and broader social, political and cultural frames. A 'national identity and pride' frame, in particular, often emerges in media stories about Nobel laureates in the early twentieth century.

Keywords Nobel Prizes • Science in the media • Science in the daily press • Visible scientists

The phenomenon of 'visible scientists' and scientists turning into media celebrities is not entirely new, although nowadays it may be fuelled by broader mediatization processes shaping science.[1]

It is therefore interesting to look at this phenomenon in a historical perspective, using a case study that is both relevant in international terms—an early recipient of the world's most renown scientific award—and profoundly informative about the specific Italian media and cultural context.

---

[1]For a discussion of mediatization of science (that is, the increasing orientation of science institutions and dynamics to the operational logic and criteria of mass media), see Weingart (2001) and Peters et al. (2008). For a brief introduction to the 'visible scientists' theme, see Bucchi (2010).

M. Bucchi (✉)
Science and Technology in Society Programme, Università di Trento, Trento, Italy
e-mail: massimiano.bucchi@unitn.it

B. Schiele et al. (eds.), *Science Communication in the World: Practices, Theories and Trends*, DOI 10.1007/978-94-007-4279-6_17,

Analyzing the daily press coverage of the Nobel Prizes in the sciences and in particular the case of the physics prize to Guglielmo Marconi in 1909, this chapter offers an opportunity to explore general features and trends in the coverage of science by the Italian daily press, as well as its treatment of Nobel Prize assignments in the sciences.

The chapter draws on ongoing research about the public image of the Nobel Prizes in the sciences. The first section is a quick overview of the features and long-term trends of Italian daily press coverage of science and technology issues. The second section outlines some key dimensions of the coverage of Nobel Prizes in the sciences. The third section looks more specifically at how Marconi and his Nobel award were presented in the daily press. Research on the coverage of science and Nobel awards mentioned here has focused pre-eminently on *Il Corriere della Sera*, one of the leading Italian newspapers and one of the few covering the entire duration of the Nobel Prizes (from 1901 to the present).

17.1 Science and the Italian Daily Press: An Overview

Particularly since the second half of the nineteenth century, the daily press in many countries has regularly reported on events connected to science, such as conferences, meetings of scientific academies, and announcements of new results by renowned scientists (Raichvarg and Jacques 1991).

In 1881, the public experiment on anthrax vaccine conducted by Pasteur at Pouilly-le-Fort received wide and prolonged coverage in the local (*Journal de Seine et Marne)*, national (*Le Temps*) and international daily press (the London *Times* sent a correspondent at Pasteur's invitation). For several weeks, the developments of the experiment and the heated debates at the Académie des Sciences occupied the internal pages and even the front pages of the newspapers, and even spread into the 'Feuilleton des Temps' section, just like one of the many popular novels published at the time in episodes in the same newspapers (Bucchi 1997, 1998).[2]

Detailed studies have documented a true general 'flowering' in the production and consumption of popular science in Italy after the country's political unification in the nineteenth century. The efflorescence was marked in particular by the success and visibility of a few scientists as authors, as well as of book series and magazines (Govoni 2002, 2007). It was, however, a rather brief season, which had already started to fade by the end of the century.

More specifically, coverage of science and technology by the Italian daily press went through different phases and cycles of varying intensity and with different features.[3]

[2] See also Geison (1995) and Cadeddu (1987).

[3] For a description and analysis of long-term trends in postwar Italy, with a strong emphasis on international comparison, see Bucchi and Mazzolini (2003).

During the period considered here, partly because the size and structure of newspapers was quite different from those with which we are familiar today, science did not have a fixed and institutionalized space, but surfaced in different and somewhat disparate sections. For example, the very first issue of *Il Corriere della Sera* (5–6 March 1876) carried on its second page, under the heading 'Ciarle del curioso' ('The curious man's chat'), an article on carnivorous plants that drew heavily on Darwin's book *Insectivorous plants.* Less than 2 months later, science was on the front page in the form of a long and detailed review of a book by Schützenberger on fermentation (di Giorgio 1876).[4]

The following year, responding to a letter from a reader, an editorial solemnly promised the reader that, from then onward, at least on Sundays, the newspaper would offer them 'a break from the tyranny of politics' through 'popular science articles, travel reports, short stories'.[5]

As a first concrete example of what had been promised, the same issue featured a long review of the book *Lezioni di Astronomia* (*Lessons in Astronomy*) by Quirino Filopanti. The piece raised in particular the issue of potential life forms on other planets, widely citing a 'visible scientist' of the time, the astronomer Giovanni Virginio Schiaparelli, who would often be interviewed by *Il Corriere* during the following years.

Throughout the twentieth century, coverage of science in the Italian daily press went through phases and cycles with different intensity and varied features. Some of the key elements of this long-term relationship between science and the daily news included:

- An increasing proportion of space devoted to science news over time (which is also evident when one takes into account changes in newspaper format and size)
- The emergence of dedicated science/technology/medicine sections in the second half of the century
- A shifting focus from the physical and astronomical sciences to the biomedical domain, particularly since the 1980s.[6]

Evidence from comparable international studies, however limited, suggests that those trends were global rather than merely typical of Italy. Some distinctive national features nevertheless emerge: for example, the active contribution of scientists as sources and authors of articles is much more substantial in Italian compared to British daily press coverage. Many leading Italian—and also international—scientists contributed in this capacity to the presentation of science issues by *Il Corriere della Sera* (see Bauer 1998).

[4] See Caprara (2009).

[5] Editorial, *Il Corriere della Sera*, 8–9 April 1877, p. 1, cited in Caprara (2009).

[6] For more details, see Bucchi and Mazzolini (2003).

17.2 The Nobel Science Prizes in *Il Corriere della Sera*, 1901–1999

The coverage of Nobel Prizes in the sciences (physics, chemistry and medicine/physiology) offers several valuable insights into the public dimension of science and its socio-historical transformations. The announcement of the prizes often receives significant attention from the general media worldwide; it triggers particular excitement and makes headline news in the countries where the laureates reside or were born. Furthermore, Nobel laureates represent the epitome of the so-called 'visible scientists'—the scientific elite who attract key attention, resources and prestige both within the scientific community and in society at large (Merton 1973; Goodel 1977; Zuckerman 1977; Bucchi 2010). On this basis, it becomes particularly relevant to study media coverage of Nobels in the sciences as a source of information on the visibility and status of scientists and, more generally, on the public image of science in society throughout the past century.

In this section, I outline some results from wider, ongoing research on the coverage of Nobels in the sciences in the Italian daily press from 1901 to 1999. I focus in particular on one source, *Il Corriere della Sera*. Besides being the most prestigious and one of the most widely circulated Italian newspapers, *Il Corriere* also provides a solid long-term background, having been published without interruption since 1876.

Between 1901 and 1999, the paper published a corpus of 263 articles on Nobel Prizes in the sciences. Although the Nobel Prizes made news in Italy from their inception, an analysis by decades throughout the century points to peaks of attention between the 1950s and the 1980s, which can be plausibly connected, at least to some extent, to cases of Italian scientists being awarded the prize. Quite interestingly, this period also sees an increasing involvement of members of the scientific community as authors of newspaper articles about the Nobels, peaking in the 1970s, when scientific experts authored 40% of the total published articles on Nobel Prizes in the sciences.[7]

The main emphasis of the articles soon shifted from the ceremony to the announcement. After World War II, most articles were published in October rather than during the Nobel Week festivities, in conjunction with a new announcement schedule. Articles also become longer: while the Nobels were initially reported in very short news items, interviews and comments became more common across the decades.[8]

[7]This active role of scientists as commentators appears to characterize Italian daily press coverage; see Bucchi and Mazzolini (2003).

[8]This trend should obviously be read in the context of more structural transformations of the newspaper, which grew in size and developed new sections during the century; see Bucchi and Mazzolini (2003).

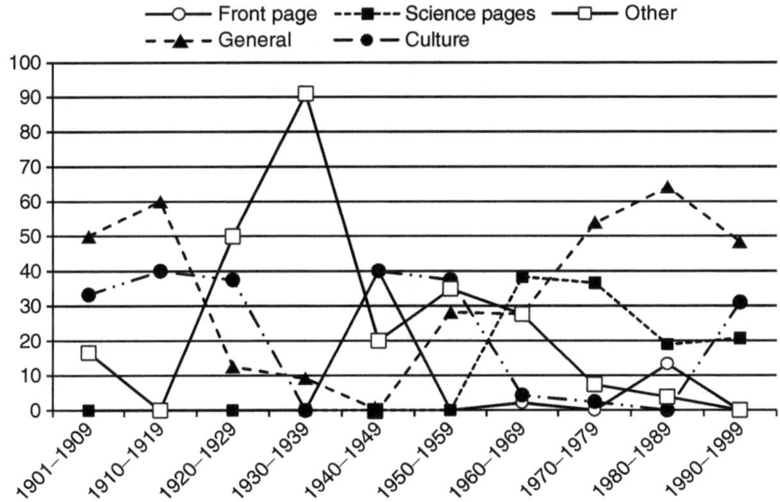

Fig. 17.1 Nobel articles, *Il Corriere della Sera*, 1901–1999, by newspaper section (Note: Lines represent the percentage of articles, per section, over each decade. Total number of articles = 263)

More detailed focus on the news context in which the Nobels are reported provides further interesting elements for reflection. Nobel articles were for many years published in 'culture' or miscellaneous newspaper sections, but from the 1950s dedicated science sections hosted a number of them. More recently, however, articles on Nobel Prizes in the sciences have increasingly 'migrated' out of those sections into general news sections (Fig. 17.1). This reflects both a general phenomenon of science-related news 'hybridizing' with political and social issues and increasing media interest in both the personal angle of Nobel scientists and the broader social/cultural implications of their activities.[9]

Another visible trend relates to an increasing prevalence, in the articles, of an image of research as a collective endeavor rather than as an individual achievement. While an emphasis on the individual strongly dominated coverage in the first part of the century, research as a collective endeavor gradually acquired salience and became a feature of most articles in the last decades.

What is the dominant scientific field in terms of media representation? This aspect can be assessed more thoroughly also in relation to other Nobel Prizes outside the natural sciences (peace, literature). In mentions of individual laureates by their fields, physics and medicine largely dominate, accounting for more than one-third of citations each. Chemistry laureates account for another 21% of mentions; all other prizes share the remainder.

[9]See, for example, Latour (1991), Bucchi (2006, 2009), Bucchi and Mazzolini (2003) and Bader (1990).

Table 17.1 Nobel laureates most often cited in articles, *Il Corriere della Sera*, 1901–1999

Nobel laureate	Field	Year	No. of Nobel articles citing laureate	% of citations
Carlo Rubbia	Physics	1984	15	5.7
Enrico Fermi	Physics	1938	11	4.2
Rita Levi Montalcini	Medicine	1986	11	4.2
Giulio Natta	Chemistry	1963	10	3.8
Emilio Segrè	Physics	1959	10	3.8
Albert Einstein	Physics	1921	10	3.4
Salvador Luria	Medicine	1969	9	2.7
Abdus Salam	Physics	1979	7	2.7
Simon van der Meer	Physics	1984	7	2.7
Jose Bovet	Medicine	1957	6	2.3
Renato Dulbecco	Medicine	1975	6	2.3
Max Planck	Physics	1918	6	2.3
Guglielmo Marconi	Physics	1909	5	1.9

Note: Total articles = 263, total Nobel laureates cited = 468, total citations = 894. Includes only laureates with five or more citations

Citations can be broken down by the country of origin of the laureate and by their country of residence at the time of the award. By both measures, the United States largely dominate Nobel coverage, accounting for 29.4% of citations by country of origin and more than 40% by country of residence. Germany and the United Kingdom follow, and less than 10% of citations were framed in an Italian context. Although this distribution could be generally described as reflecting the Nobel population during the last century, further comparative analysis of coverage in other countries could help us to understand, for example, whether this picture of limited nationalistic bias is typical of Italy or more of a global trend in Nobel coverage.

A national focus, however, more clearly emerges when one looks at the level of individual scientists. All of the 'top 10' visible scientists in terms of citations are, in some way, connected to Italy, with the exception of Albert Einstein (the sixth most widely cited Nobel laureate in *Il Corriere della Sera*). Most of them are Italian or of Italian origin; the only two non-Italians (Abdus Salam and Simon van der Meer) had institutional relationships with Italy—Salam was affiliated with the International Centre for Theoretical Physics in Trieste and van der Meer shared the Nobel Prize for physics with Carlo Rubbia in 1984 (Table 17.1). Again, the prevailing focus on physics and medicine emerges also from individual citations.

Marconi is cited in five articles connected with his Nobel Prize. However, one should put this result in the context of both general transformations in the structure of the newspaper (which was much shorter at the time of Marconi's prize) and in the coverage of the Nobels, with articles becoming longer and richer in details across time. Also, as we shall see in the next section, Marconi was already a well-known figure to Italian newspaper readers well before his Nobel award.

17.3 Marconi and His Nobel in *Il Corriere della Sera*

At the time of Marconi's Nobel, *Il Corriere* had begun to regularly follow themes and protagonists in science and technology, celebrating new discoveries and inventions, commemorating famous scientists at their deaths, and covering national and international conferences.

Already in July 1897, Il Corriere dedicated significant space and attention to Marconi, publishing a long article under the heading 'L'invenzione del giorno. Il telegrafo senza fili' ('The invention of the day: the wireless telegraph'). The article, unsigned, offers to newspaper readers a detailed description of Marconi's invention (albeit labeled in the article as a 'discovery'), making use also of illustrations of electrical schemes. The article also interestingly mixes a technical style with a more imaginative, almost fictional style:

> We don't know what ether actually is; but it is really necessary to define it? Does the human soul necessarily require a definition to understand the most minute degrees of sentiment and the strongest storms of passion?

Towards the end, the article also stages a dialog between Marconi and an English journalist, apparently drawn from the international press:

- So from this room you could dispatch a message across London? (the correspondent from *Strand* magazine asked Marconi ...)
- No doubt! Having tools with the necessary one power one certainly could.
- To all houses in London?
- To all of them ....

Marconi became more and more passionate about the discussion and eventually expressed the hope that his invention could be used as a powerful fighting device:

> ... [B]uilding an apparatus that can produce waves of different lengths, would it not be possible to make a resounding device to vibrate and ignite gunpowder? Marconi smiled and paused ... looking almost ashamed ... but that look reflected all the hopes and dreams of glory of the young inventor.[10]

Almost 2 years later, on 23 April 1899, the Sunday illustrated supplement of the newspaper, *La Domenica del Corriere*, featured on its back page a colorful illustration of Marconi describing his 'telegraph experiments' to a group of gentlemen.

At the dawn of the century, Marconi was already a familiar figure to newspaper readers: *Il Corriere della Sera* regularly reported on the successes and spread of his invention and dwelt with some nationalistic pride on his 'triumph' over Thomas Edison's initial skepticism. On 16 December 1901, a long article was devoted to the first wireless transmissions from England to North America ('I gave the world a wonderful Christmas gift,' claims Marconi proudly). On 31 December, *Il Corriere* duly informed its readers about Marconi's intention to marry in New York 'the beautiful Miss Giuseppina Holman, from Indianapolis'.

[10] Il telegrafo senza fili, *Il Corriere della Sera*, 4–5 July 1897.

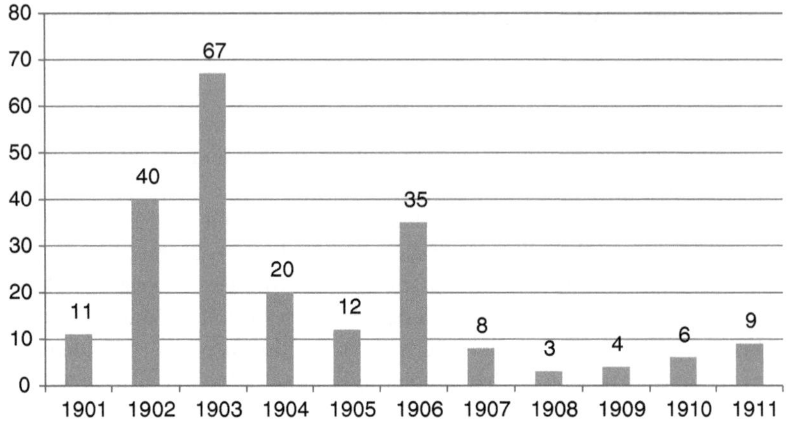

Fig. 17.2 Articles on Guglielmo Marconi, *Il Corriere della Sera*, 1901–1911 (n=215)

Media attention and emphasis on Marconi's activities continued to grow in the following years: his visits to Europe—and to Italy in particular—were greeted with enthusiasm and a series of tributes and honors were offered to him, including the honorary citizenship of Rome and the membership of several scientific and learned societies. 'Applause and gratitude to Guglielmo Marconi' were recorded in the reports of debates in the Italian Chamber of Deputies 'for making Italy's name shine in new glory'(30 January 1903), as well as in many other accounts of official occasions. Long interviews with Marconi also appeared, and reporters were sent to his home town to find out picturesque details of his adolescence and precocious talents. A few critiques from abroad—mostly about his system being not fully adequate to protect the secrecy of messages—and reports of technical failures also appear.

Figure 17.2 provides an overview of *Il Corriere*'s coverage of Marconi's activities from 1901 to 1911. An impressive total of 215 articles was published during the decade, with a peak in 1903, when 67 articles on Marconi's activities appeared.

On 16 November 1909, a short notice appeared on page 5 of *Il Corriere della Sera*:

> Svenska Dagblatt [sic] says that the Nobel prize for physics will be shared between Guglielmo Marconi, inventor of radiotelegraphy, and professor Carlo Ferdinando [sic] Braun from Strasbourg, who carried out important studies on the same invention.

A few days later, another short article confirmed the news, this time describing the other physics laureate—erroneously mentioned as 'Bauer'—as 'the one who perfected wireless telegraphy' (p. 3).

Finally, a long article published on 11 December 1909 carried official reports on the Nobel Prizes, providing short biographies of the laureates and descriptions of their achievements. Probably due to the fact that he was already very famous, however, Marconi was very briefly mentioned on this occasion; Braun's studies were contextualized 'in that wonderful field of electric waves which has been explored by Marconi and has had a successful influence on the different wireless telegraph systems' (p. 2).

In short, Marconi was a highly prominent media figure in Italy well before receiving the Nobel Prize. The relevance and practical benefits of his invention had soon become clear to the general public, and his worldwide reputation powerfully resonated with the emerging national pride of recently unified Italy. His routine familiarity with royal families and heads of state around the world was a source of excitement and curiosity about him as a member of an international proto-jetset. While his Nobel award was widely reported, it appears to have made modest difference to his already great public profile. When the Academy of Sciences conferred the prize on him, his invention had already been celebrated and made familiar in the Italian public sphere, making the award almost, to some extent, 'old news'.

In the months and years after receiving the Nobel, Marconi's new projects and personal life continued to be reported in Italy. On 22 May 1910, *Il Corriere della Sera* announced that:

> Mrs Beatrice Marconi, wife of our illustrious citizen Guglielmo Marconi, who has been living for more than a month now in their villa in Pontecchio, has given birth to a nice male child. News has immediately been sent to Marconi, who was traveling from America towards England. (p. 4)

The coverage of Marconi's Nobel Prize highlights some of the key features of science coverage in the Italian daily press in the early twentieth century: its dependence on highly prominent individuals and its connection of science to broader social, political and cultural frames—such as Italian national identity and pride.

Acknowledgements Research on media coverage of the Nobel Prizes in the sciences started within the context of a grant by the Italian Ministry of Research, under the direction of Renato G. Mazzolini. I would like to acknowledge the kind support of the Italian Culture Institute in Stockholm through the C.M. Lerici Programme. I am also grateful to Karl Grandin and the Center for History of Science at the Royal Academy of Sciences, as well as the staff at the Nobel Museum, for their help and assistance during my research in Stockholm; the Nobel Foundation for allowing access to its media archive; Kajsa Eriksson and the Swedish Research Council for inviting me to present aspects of this research; and Piero Mazzinghi and Giuseppe Pelosi. Lorenzo Beltrame, Alessia Bertagnolli, Carlo Borsatto, Silvia Giovanetti and Federico Neresini participated in data collection and analysis.

References

Bader, R. (1990). How science news sections influence newspaper science coverage: A case study. *Journalism Quarterly, 67*(1), 88–96.

Bauer, M. (1998). The medicalization of science news: From the 'rocket-scalpel' to the 'gene-meteorite' complex. *Social Science Information, 37*(4), 731–751.

Bucchi, M. (1997). The public science of Louis Pasteur: The experiment on anthrax vaccine in the popular press of the time. *History and Philosophy of the Life Sciences, 19*, 181–209.

Bucchi, M. (1998). *Science and the media: Alternative routes in scientific communication*. London: Routledge.

Bucchi, M. (2006). *When media stardom and scientific excellence combine: Nobel laureates in the Italian daily press*. Paper delivered at the 8th public communication of science and technology conference, Seoul, May 2006.

Bucchi, M. (2009). *Beyond technocracy: Citizens, politics, technoscience*. New York: Springer.
Bucchi, M. (2010). Visible scientists. In S. Hornig-Priest (Ed.), *Encyclopedia of science and technology communication* (pp. 932–933). London: Sage.
Bucchi, M., & Mazzolini, R. G. (2003). Big news, little science: Science coverage in the Italian daily press, 1946–1997. *Public Understanding of Science, 12*, 7–24.
Cadeddu, A. (1987). Pasteur et la vaccination contre le charbon: Une analyse historique et critique. *History and Philosophy of the Life Sciences, 9*, 255–276.
Caprara, G. (2009). *L'avventura della scienza: Sfide, invenzioni e scoperte nelle pagine del Corriere della Sera*. Milan: Fondazione Corriere della Sera-Rizzoli.
di Giorgio. (1876). Le fermentazioni. *Il Corriere della Sera*, 30 April–1 May 1876, p. 1.
Geison, G. (1995). *The private science of Louis Pasteur*. Princeton: Princeton University Press.
Goodel, R. (1977). *The visible scientists*. Boston: Little Brown.
Govoni, P. (2002). *Un pubblico per la scienza. La divulgazione scientifica nell'Italia in formazione*. Rome: Carocci.
Govoni, P. (2007). The rise and fall of science communication in late nineteenth century Italy. In M. Bauer & M. Bucchi (Eds.), *Journalism, science and society: Science communication between news and public relations* (pp. 21–32). London/New York: Routledge.
Latour, B. (1991). *Nous n'avons jamais été modernes*. Paris: La Découverte.
Merton, R. K. (1973). The Matthew effect in science. In R. K. Merton (Ed.), *The sociology of science: Theoretical and empirical investigations*. Chicago: University of Chicago Press.
Peters, H. P., Heinrichs, H., Jung, A., Kallfass, M., & Petersen, I. (2008). Medialization of science as a prerequisite for legitimation and political relevance. In D. Cheng, M. Claessens, T. Gascoigne, J. Metcalfe, B. Schiele, & S. Shi (Eds.), *Communicating science in social contexts: New models, new practices* (pp. 71–92). New York: Springer.
Raichvarg, D., & Jacques, J. (1991). *Savants et ignorants: Une histoire de la vulgarisation de sciences*. Paris: Seuil.
Weingart, P. (2001). *Die Stunde der Wahrheit? Zum Verhältnis der Wissenschaft zu Politik, Wirtschaft und Medien in der Wissensgesellschaft*. Weilerswist: Velbrück Wissenschaft.
Zuckerman, H. (1977). *The scientific elite: Nobel laureates in the United States*. New York: Free Press. New edition, 1996, New Brunswick/London: Transactions.

Chapter 18
Engagement: The Key to the Communicative Effectiveness of Science and Ideas

Hak-Soo Kim

Abstract Engagement is seen to be the key to understanding the process of behavior. It is a theoretical concept deriving from the universe's *partial order* condition. I argue that a problematic situation is the precondition for engagement, and that communicating is effective for enabling that engagement. Engagement is conceptualized as the act sequence of exposing, focusing attention, and cognizing (with moving to follow—or not). I then illustrate three possible types of the sequence: orienting-centered, constructing-centered or reorienting-centered. These types help explain why learning, creativity and reform are difficult to accomplish. I find that the more we are engaged with a problem, the more we are further engaged with science's potential specific contribution to solving it. Therefore, engagement seems to be the key to the communicative effectiveness of science and ideas.

Keywords Engagement • Problematic situation • Communicating • PEP/IS • Learning • Creativity • Reform

18.1 Introduction

The concept of *public engagement with science* has begun to substitute for its predecessor, *public understanding of science*. About a decade ago, the British House of Lords published a report stating that public confidence in science had been

---

An earlier draft was presented at the 2nd Venice PCST Colloquia on Quality in Science Communication and Public Engagement, hosted by the Instituto Veneto di Scienze (Venice Academy of Science), Italy, 15–16 January 2009.

H.-S Kim (✉)
College of Communication, Sogang University, CPO Box 1142, Seoul, South Korea
e-mail: hskim@sogang.ac.kr

B. Schiele et al. (eds.), *Science Communication in the World: Practices, Theories and Trends*, DOI 10.1007/978-94-007-4279-6_18,

severely eroded and that science should engage the public with direct and full-scale activities (SCST 2000). As Alan I. Leshner (2003), chief executive officer of the American Association for the Advancement of Science, editorialized in *Science*, public engagement with science seemed more urgent.

Public understanding has been a continuing concern of science, in part because public awareness of science has been modest at best. Still, public support of science is critical to the scientific enterprise. Leshner pointed out that public concern arises from scientific misconduct and potential misadventure. Examples of the former are Schön's (Service 2002) and Hwang's (Kim 2007b) data forgery cases; human cloning is an example of the latter. Despite successful scientific initiatives to prevent or cure human maladies, public engagement comes with public anxiety, but without a clear and productive destination.

When the term 'scientist' was first used in the mid-nineteenth century, scientists had already begun to show up as an intellectual force in society, and speedy scientific development brought about a cultural divide between science and the public, even between scientists and humanistic intellectuals (Snow 1993). Scientific development had social and human effects, and made necessary a communication or education program emanating from science to the public. We might come to share at least 'a feel for science' through such communication, according to Cohen (1952).

Initial efforts in communication with the public were largely communication *to* the public. Popularization of science and science education became policy, with commensurate concern for scientific literacy and the public understanding of science. Asimov (1983) argued that scientific illiterates might bring about social instability or even destruction by failing to make decisions intelligently. So many understand football, but so few understand science, he said.

Scientific literacy was also soon linked to expanding national capability. Along with the technology it fostered, it was assumed to contribute to economic prosperity. Prewitt (1983) saw this as leading to more informed democratic participation and positive support for science policy. Durant, Evans, and Thomas (1989) seconded that view, seeing science as the past century's greatest achievement, deserving of public support.

We see here the conceptual linking of literacy and (positive) attitudes, the latter of which are all too often considered a major component of understanding by those conceptualizing communication effects from the point of view of the message sender. What an 'engagement' perspective will do is to remove at least some of the misjudgments occasioned by the earlier unidirectional view.

However, as circumstances grow bleaker, the temptation for public engagement is to simply communicate science *more* and *harder* to the public. To continue a singular communication policy of information dissemination in this fashion raises the possibility of resistance instead of assistance. Nevertheless, in this 'post-academic science' epoch, according to Ziman (1996), when there is a need for more funds and for interdisciplinary teams of scientists and engineers, public understanding of science is seen to be enormously important.

The record on the achievement of scientific literacy and public understanding of science is not impressive. For example, in Britain, Miller (2001) found that the public was not much 'scientized', despite the British scientific community's intensive effort following the establishment of the Committee on Public Understanding of Science (COPUS) in 1986. Public engagement with science as a 'pushier' movement for enhancing scientific literacy and public understanding of science has not made a difference in Europe, whatever tools (festivals, interactive applications, hands-on experiences) are utilized (see Miller 2012). The same one-way hierarchical science communication from the scientists to the public was also the South Korean Government's main concern (Cho and Kim 2012). The outcome remains as ever: however high public interest in science might be, the public's knowledge of science is sparse and its attitude is amorphous.

Should we keep to the traditional assumption that effective communication requires a sequential linear relationship of interest, knowledge, attitude—and support? Should we continue to focus on measures of literacy or understanding as the key measures of the communicative effectiveness of science? Are there problems here that are theoretical and methodological, not just practical?

The purpose of this article is to address these problems, to see whether an engagement perspective might help us to do better in addressing the relationship between the public and the realm of science. I illustrate a genuine, typical engagement in order to take a fresh look at the concept of engagement, after which I explicate the concept of engagement theoretically. After illustrating types of behavioral failure of engagement, I show how communication can be used to achieve engagement with science.

18.2 Exemplifying a Genuine Engagement

What most people engage with most frequently seems to be the weather. Before we go to work, we check the weather on TV or in the newspaper. If the weather forecaster notes a high probability of rain, we decide to take an umbrella, a raincoat or a different pair of shoes. If the day is going to be sunny, we might take lighter clothes or a pair of sunglasses.

As a matter of fact, weather is an omnipresent problem for us. Because it is *consequential* to all aspects of our life, it comes to us as an ever-present problematic situation. That is why we engage with the weather so often. That is also why mass media are extremely sensitive to the weather and report it most frequently (Kim et al. 2008). In a word, weather is the best selling news, insofar as the main business of the media is selling problem-related news.

Weather is an inherently collective, shared problem. It affects the whole community, all of whom focus on it. It is also a very tangible problem, as community members can directly experience rain, snow, wind and so on. As we confront a problem such as weather, we begin to engage with it in order to solve it. Our need for survival comes from such a problematic condition. Thus, need and its affinities (for example, want, expectation, desire and motivation) arise in that problematic situation in the first place. This indicates that a hierarchy of needs (Maslow 1970) might be gratuitous, unless some consequences of problems are anticipated. The consequentiality forces our engagement.

Communicating is basically transmitting or exchanging information (Carter 1965; Kim 1986), though not necessarily accompanied by persuasion. When the information is about a problem, we envision a problematic situation (Dewey 1938) and may begin to engage with that situation. Thus mass media can contribute to our engagement with many (social) problematic situations—so much so that they are considered to be very powerful as agenda setters. Even our engagement with entertainment may be enhanced when the media invite us, vicariously, into a problematic situation (for example, of loss, conflict, illness, disaster or war).

Thus, engagement can be a genuinely theoretical concept rather than just an empirical or normative one. This means that engagement may be a key concept in explaining behavior *in general*, beyond generalizing behaviors *in particular*. An explanation of how engagement comes about and how communicating helps it to come about shows how challenging it will be to further develop public engagement with science. Some observers have presumed that communicating is all-powerful in bringing about a gain in scientific knowledge and a change in attitude towards science. They are mistaken in that presumption, but they are not mistaken in assuming that communicating seems to be the precondition for such outcomes.

18.3 Explicating the Concept of Engagement

The universe is assumed to be in partial order after the Big Bang. This condition is evidenced by omnipresent collisions. This also means that every entity is given not only incomplete instructions for survival but also the behavioral necessity to survive in potential collisions. Behavioral necessity is the source for the *ever-present* behavioral problem that every entity faces before solving a *situational* problem. Every entity ought to, somehow, solve the behavioral problem in the first place in order to avoid or arrange collisions (Carter 2010a, b; Kim in press).

Engagement must be the essential manifestation of solving the behavioral problem. Human beings, organic entities having life, have developed engagement as the key to behavior for survival. Thus, engagement should be considered as a theoretically genuine concept, not a normative one for particular movements or campaigns.

Our engagement begins with the act of exposing. We are continually exposed to many problems in our environs. This exposing is expanded by communicating (for example, interpersonally or via media). Our engagement, however slight, proceeds as we face those problems, giving them our attention. But many problems bypass us, without engagement and without this further act of focusing attention (they may be seen to be inconsequential or uncontrollable, based on past experiences). Because this focusing of attention is selective (Broadbent 1958), we cannot deal with all problematic situations at the same time. We inevitably select one problematic situation among many to focus on. This helps explain why it can be difficult to call people's attention to multiple problems, however important they may be. In a sense, genuine engagement can be said to begin with this act of focusing attention, which could also be helped by communicating.

Moreover, we sometimes attribute some consequentiality to a problem. Cognition comes into play. Communicated information could help such projection, as by *giving an impression*. Cognizing may work further to define and/or to solve the problem. We may use any or all of three cognitive modes (Carter 1978; Kim 2003) in regard to the problematic situation: by *orienting* to available information about it (perhaps to seek out more information); by *constructing* new information about it (perhaps developing one's own idea of the problem); and/or by *reorienting* via feedback information about it (perhaps seeing the problem as evidence of misadventure).

Of course, orienting is the easiest engagement mode of cognizing because it works with available information. (This is probably why observers think of 'messages' as the chief function of communicating with the public.) The others require strategists to give more consideration to how people think.

The outcome of cognizing (that is, an idea) guides us into an action. We could move our body in a direction, based on that idea. However, many (swift) acts of moving look to be automatic or habitual, because our past (safe) experiences have allowed exposing and focal attention (via recognition) to lead immediately to moving. Sometimes we have accidents because we ignore a new problem or misrecognize it (as if it were an old problem). In this way we are all handicapped, not only by limited sensory capacity for exposing, and/or by inadequate bodily limbs for moving, but also by incapacity or inability to focus attention and/or cognize.

(Communicating itself is often an act of moving, as when we give expression by crying or shouting in order to get attention, or when we assign a name to a possibility in order to think about bringing it about. It also serves to substitute for more drastic moves, like fistfighting or larger conflicts.)

This process view of engagement is based on Carter's behavioral theory (1988, 2010a, b): that body and behavior are conceived as essentially independent but functionally interdependent, and that each has structural features distinctive to it, and each has distinctive consequences. Behavior's capabilities could be more critical in producing outcomes than the body's capacities. Behavior is a molecular sequence of acts, here: exposing, focusing attention, cognizing, and moving. So, via the three distinct cognitive modes (Orienting, Constructing, Reorienting), we can conceptualize the process of engagement with a problematic situation as three kinds of molecular sequence (Kim 2003, 2007a):

1. O-Engagement [Exposing → Focusing Attention → Orienting] → Moving
2. C-Engagement [Exposing → Focusing Attention → Constructing] → Moving
3. R-Engagement [Exposing → Focusing Attention → Reorienting] → Moving

In this conceptualization, we should not forget that a problematic situation (which by definition has consequentiality) should necessarily bring forth some engagement, barring escape behavior. This general condition should also be able to guide us (see below) as to how we could engage with science by communicating.

18.4 Behavioral Failure of Engagement

Everybody talks about the consequences for science of a lack of learning, creativity or reform. Then we conduct diverse public campaigns to overcome that lack, headlined by slogans like 'Let's learn!', 'Let's be creative!', 'Let's reform!' Nobody says how that could be accomplished behaviorally.

As seen above, the first sequence of orienting-centered engagement (O-Engagement) is closely related to learning, insofar as orienting and learning are simply taking in available information. In a sense, orienting is the easiest mode of cognizing. However, we experience failure of learning all the time. Then, compulsory education, stuck in the traditional learning paradigm (McGuire 1985), pushes us to orient to established ideas (that is, existing knowledge) from the beginning. We are forced to be exposed to, focus attention on, and orient to available knowledge in a closed class or via a coercive test. We gain little knowledge beyond tests, because most of knowledge reaches us as solutions (or answers) rather than as problematic situations. Without conceiving a problem in the first place, we find it difficult to engage in its solution. That is why teaching as instruction fails to bring forth students' engagement. Little learning is achieved and a vast educational investment is wasted. Science learning or literacy is no exception, despite highly publicized campaigns (see Bauer et al. 1994, 2000; Miller 1998, 2004; Pardo and Calvo 2002; Sturgis and Allum 2004). We overlook the fact that communicating a problematic situation is the most effective first step for engagement, not only in school but also in ordinary life.

The second sequence of constructing-centered engagement (C-Engagement) is closely related to creativity. Creating is constructing new information or ideas. We tend to suppose that creativity exists as an inherited capacity (for example, we may refer to a 'genius'). However, creativity does not just occur; there is a process. When we are in a crisis that demands more serious engagement and some innovative effort, we may construct new information. We have to come up with a new idea or we cannot survive the crisis. Successful C-Engagement introduces new cognitive elements and relations (Carter and Stamm 1993; Kim 1986). This means that creative cognizing rarely arises without the thinker being exposed to and focusing attention on a problematic situation. (If only competing solutions are considered, problem becomes transformed to issue, inviting partisan behavior. Decision making displaces problem solving—placing different demands on communication. See Kim in press.) This second sequence, that is, C-Engagement, arises less frequently than the first one.

The third sequence of reorienting-centered engagement (R-Engagement) is closely related to reform. R-Engagement looks to change things, informed by the reported and/or perceived failure of problem solving. (The failure might be to not notice an emerging or enlarged problem or problematic aspect of the situation.) Reform could be pursued, as by protest. Such reform is change only to an extent. We need to keep a certain degree of stability, so as not to break down the whole system. R-Engagement also rarely occurs without exposure to and focal attention on the new problematic situation. Communicating could facilitate our sensitivity to the new situation. However, R-Engagement may emerge out of sheer desperation.

Here we have seen the three sequences of engagement behavior. None of them is easy to evoke or complete. Even O-Engagement may end up mostly being exposure to available information, not passing on to acts of focusing attention and cognizing. A Nobel laureate's distinction of perception, intuition and reasoning (Kahneman 2003), generalized from experiments, seems to touch on these sequences without explicating those three concepts or showing their theoretical base. For example, perception seems to indicate only the act of exposing, while intuition and reasoning seem to indicate, respectively, the cognitive act of orienting and that of constructing or reorienting. However, without our prior engagement in acts of being exposed to and focusing attention on a problematic situation, we do not necessarily proceed to the cognizing act of orienting, constructing or reorienting.

Communicating can help each act in and of the engagement process, but not very effectively without a basis in problem solving. Without engagement with problem solving, communication can aim only at a vaguely conceived literacy.

18.5 Engagement with Science

We want to discern how we might engage with science. Our engagement begins with a problematic situation, so, to the extent that science is dealing with problematic situations, we could find ourselves concerned and interested. This differs from more familiar formal and informal avenues of (at least partial) engagement, such as a science class in school or a public science exhibit. It requires a different kind of communicating—with special attention to what mass media can contribute.

Science itself is more or less a matter of problem solving, in addition to contributing to problematic situation definition or solution construction (see Lubchenco 1998; Pielke and Byerly 1998). A scientist can see a scientific query as a problematic situation, which could lead to the full sequence of the scientist's engagement behavior. Sometimes, a scientist's unethical conduct becomes a problematic situation for the general public and the public will then complete a full sequence of R-Engagement behavior, as in the Hwang scandal (Kim 2007b). Where there is scandal, the mass media jump to call our attention to it. Clearly, it is with the problematic situation that communicating, like engagement, must begin.

But we start from where we are. How much engagement with a problem do we have so far? And how is it related to science's contribution to problem solving? Herein comes Kim's PEP/IS model (2007a): Effective communication of science lies in the processes of public engagement with a problem or an issue relative to science.

A PEP/IS index (as evidence of communicative effectiveness) was developed by measuring the relationship between the public's engagement with a problem (P-Engagement) and the public's engagement with science as a problem solver (SPS-Engagement). We measured P-Engagement on a four-point scale as the level of engagement sequence with a problem: (1) non-exposing ('haven't heard'); (2) exposing ('heard but paid no attention'); (3) focusing attention ('have interest

Table 18.1 Correlations between P-Engagement and SPS-Engagement (PEP/IS-I Index)

	Public		Scientists		Total	
Problem	Pearson *r*	*n*	Pearson *r*	*n*	Pearson *r*	*n*
Global warming	0.135*	797	0.079	202	0.136*	999
Energy shortage	0.102*	799	0.138	202	0.120*	1,001
Pollution	0.157*	800	0.170**	202	0.164*	1,002
Aging population	0.140*	800	0.112	202	0.132*	1,002
Rich–poor gap	0.046	798	0.073	202	0.053	1,000
New virus epidemic	0.123*	798	0.199*	202	0.149*	1,000
Economic uncertainty	0.058	798	0.096	202	0.055	1,000
Food shortage	0.141*	799	0.092	201	0.136*	1,000
Cancer	0.109*	800	0.248*	202	0.134*	1,002

*$p < 0.01$ (two-tailed)
**$p < 0.05$ (two-tailed)

Table 18.2 Correlations between P-Engagement and SPS-Engagement (PEP/IS-II Index)

	Public		Scientists		Total	
Problem	Pearson *r*	*n*	Pearson *r*	*n*	Pearson *r*	*n*
Global warming	0.148*	800	0.171*	203	0.163*	1,003
Energy shortage	0.102*	798	0.203*	203	0.133*	1,001
Pollution	0.126*	800	0.198*	203	0.144*	1,003
Aging population	0.146*	799	0.052	203	0.129*	1,002
Rich–poor gap	0.084**	796	0.004	203	0.086*	999
New virus epidemic	0.104*	800	0.191*	203	0.131*	1,003
Economic uncertainty	0.102*	799	0.152*	203	0.116*	1,002
Food shortage	0.140*	799	0.140*	202	0.142*	1,001
Cancer	0.152*	800	0.173**	203	0.157*	1,003

*$p < 0.01$ (two-tailed)
**$p < 0.05$ (two-tailed)

and don't pass by relevant information given'); (4) cognizing ('have much interest and seek information to answer or solve'). We also measured SPS-Engagement on a four-point scale as the level of science's potential contribution to solving that problem: (1) non-contribution ('can't contribute at all'); (2) little contribution ('can seldom contribute'); (3) much contribution ('can contribute much'); (4) very much contribution ('can contribute very much').

We found in most cases significantly positive correlations between P-Engagement and SPS-Engagement: the more engaged the general public and the scientists are with a problem, the more engaged they are likely to be with science relevant to solving it (Kim 2012). We retested the above relationship with the general public (adults) and the scientists in another survey conducted in October 2007 (Kim et al. 2007). Table 18.1 shows the new results. Moreover, the second time, we added a new measurement of SPS-Engagement: the level (on a five-point scale) of science's potentially 'more specific contribution' to solving a problem: (1) science's impossible contribution;

(2) science's possible but unclear contribution; (3) science's contribution to problem definition; (4) science's contribution to solution construction; and (5) science's contribution to both problem definition and solution construction. Table 18.2 shows the correlations (PEP/IS-II index) between P-Engagement and the newly measured SPS-Engagement.

Comparing Tables 18.1 and 18.2, we find that the latter's PEP/IS-II indices have significantly positive correlations in more cases than the former's PEP/IS-I indices. The difference applies for both the general public and the scientists, but more apparently for the total respondents. This seems to indicate that the more engaged the general public and the scientists are with a problem, the more engaged they are likely to be with science's 'specific contribution' to solving it.

18.6 Conclusion and Discussion

As exemplified by our concern with the weather, engagement seems to be a paramount theoretical concept in behavior in general, not just to behavior about science. Life itself is full of the processes of engagement, insofar as many problems permeate our life. So, the concept of engagement can be explicated in a process perspective. A problematic situation is a precondition for engagement. Then, the processes of engagement develop from the acts of exposing and focusing attention to the act of cognizing. The act of cognizing has three modes: orienting, constructing and reorienting. All of them could be helped by communicating that basically functions as transmitting information or ideas.

Based on the explication of engagement, we found that learning, creativity or reform could be accomplished only through the full sequence of engagement behavior. That explains why not only creativity or reform but also learning is so difficult to achieve. Unless we are awakened to a problematic situation projecting its consequentiality, we are unlikely to bring along acts of cognizing for problem definition and solution construction, whatever knowledge, ideas and science are ready to serve. Communicating will be more effective when it most appropriately helps to enable the sequence of engagement behavior. So, communicate the problematic situation first. Establish relevance!

Again, we demonstrated the close relationship between public engagement with a problem and public engagement with science as a problem solver, via two PEP/IS indices. We found that the public could be 'further' engaged via science's specific contribution to solving a problem.

We seem to have lost an opportunity to enhance effective communication about science, insofar as the mass media have only brought about occasional exposure and focal attention to problems, without advancing us into the cognitive mode in regard to science. On the other hand, science journalism and classroom instruction seem to hold strongly to the traditional learning-theory paradigm that mere exposure to scientific knowledge would lead to scientific literacy and public understanding.

Our public understanding of science or public engagement with science research history has demonstrated that this is not the case. We've taken the wrong track for too long, haven't we?

References

Asimov, I. (1983). Popularizing science. *Nature, 36*(10), 119.

Bauer, M., Durant, J., & Evans, G. (1994). European public perceptions of science. *International Journal of Public Opinion Research, 6*(2), 163–186.

Bauer, M., Petkova, K., & Boyadjieva, P. (2000). Public knowledge of and attitudes to science: Alternative measures that may end the 'science war'. *Science, Technology, and Human Values, 25*(1), 30–51.

Broadbent, D. E. (1958). *Perception and communication*. New York: Pergamon.

Carter, R. F. (1965). Communication and affective relations. *Journalism Quarterly, 42*(2), 203–212.

Carter, R. F. (1978). A peculiar horse race. In G. F. Bishop, R. G. Classer, & M. Jackson-Beeck (Eds.), *The presidential debates: Media, electoral, and policy perspectives* (pp. 3–17). New York: Praeger.

Carter, R. F. (1988). Life: The double crystal. Unpublished manuscript, University of Washington School of Communications, Seattle, Washington.

Carter, R. F. (2010a). Art, art, and communication. In L. Foreman-Wernet & B. Dervin (Eds.), *Audiences and the arts: Communication perspectives* (pp. 265–277). Cresskill: Hampton.

Carter, R. F. (2010b). *Behavioral foundations of effective problem solving*. Retrieved November 1, 2010, from http://bfeps.org

Carter, R. F., & Stamm, K. R. (1993). How we thought about the Gulf War. In B. L. Greenberg & W. Gantz (Eds.), *Desert storm and the mass media* (pp. 152–165). Cresskill: Hampton.

Cho, S. -K., & Kim, O. (2012). From science popularization to public engagement: The history of science communication in Korea. In B. Schiele, M. Claessens, & S. Shi (Eds), *Science communication in the world: Practices, theories and trends*. Springer.

Cohen, I. B. (1952). The education of the public in science. *Impact of Science on Society, 3*(2), 67–100.

Dewey, J. (1938). *Logic: The theory of inquiry*. New York: Henry Holt.

Durant, J. R., Evans, G. A., & Thomas, G. P. (1989). The public understanding of science. *Nature, 340*, 11–14.

Kahneman, D. (2003). A perspective on judgment and choice: Mapping bounded rationality. *American Psychologist, 58*(9), 697–720.

Kim, H.-S. (1986). Coorientation and communication. In B. Dervin & M. J. Voigt (Eds.), *Progress in communication sciences* (Vol. 7, pp. 31–54). Norwood: Ablex.

Kim, H.-S. (2003). A theoretical explication of collective life: Coorienting and communicating. In B. Dervin & S. H. Chaffee (Eds.), *Communication, a different kind of horserace: Essays honoring Richard F. Carter* (pp. 117–134). Cresskill: Hampton.

Kim, H.-S. (2007a). PEP/IS: A new model for communicative effectiveness of science. *Science Communication, 28*(3), 287–313.

Kim, H.-S. (2007b). South Korea: The scandal of Professor Hwang Woo-Sok. In M. Bauer & M. Bucchi (Eds.), *Journalism, science and society* (pp. 255–258). New York: Routledge.

Kim, H.-S. (2012). Measuring PEP/IS, a new model for communicative effectiveness of science. In M. W. Bauer, R. Shukla, & N. Allum (Eds.), *The culture of science: How the public relates to science across the globe* (pp. 375–384). New York: Routledge.

Kim, H.-S. (in press). Climate change, science and community. *Public Understanding of Science*.

Kim, H.-S., Choi, J. -M., Park, Y. -M., & Song, J. -R. (2007). *Strengthening the scientist leadership: Improving the Science Korea Movement*. Final project report to the Ministry of Science and Technology, Korean Government, Seoul, South Korea.

Kim, H.-S., Ha, H. -S., Park, Y. -M., & Song, J. -R. (2008). *A study on upgrading the value of meteorological information toward the construction and leisure industries*. Final project report to the Korea Meteorological Administration, Seoul, South Korea.
Leshner, A. I. (2003). Public engagement with science. *Science, 299*(5609), 97.
Lubchenco, J. (1998). Entering the century of the environment: A new social contract for science. *Science, 279*, 491–497.
Maslow, A. H. (1970). *Motivation and personality*. New York: Harper & Row.
McGuire, W. J. (1985). Attitudes and attitude change. In G. Lindzey & E. Aronson (Eds.), *Handbook of social psychology* (pp. 233–346). New York: Random House.
Miller, J. D. (1998). The measurement of civic scientific literacy. *Public Understanding of Science, 7*, 203–223.
Miller, S. (2001). Public understanding of science at the crossroad. *Public Understanding of Science, 10*, 115–120.
Miller, J. D. (2004). Public understanding of, and attitudes toward, scientific research: What we know and what we need to know. *Public Understanding of Science, 13*(3), 273–294.
Miller, S. (2012). Benchmarking climate indicators for science communication and public engagement across Europe. In M. Bauer, R. Shukla, & N. Allum (Eds.), *The culture of science: How the public relates to science across the globe* (pp. 436–448). New York: Routledge.
Pardo, R., & Calvo, F. (2002). Attitudes toward science among the European public: A methodological analysis. *Public Understanding of Science, 11*, 155–195.
Pielke, R. A., Jr., & Byerly, R., Jr. (1998). Beyond basic and applied. *Physics Today, 51*(2), 42–46.
Prewitt, K. (1983). Scientific illiteracy and democratic theory. *Daedalus, 112*(2), 49–64.
SCST (Select Committee on Science and Technology) (2000). *Science and society*, House of Lords Select Committee on Science and Technology, third report. Retrieved from http://www.parliament.the-stationery-office.co.uk/pa/ld199900/ldselect/ldsctech/38/3801.htm
Service, R. F. (2002). Bell labs fires star physicist found guilty of forging data. *Science, 298*, 30–31.
Snow, C. P. (1993). *The two cultures*. London: Cambridge University Press.
Sturgis, P. S., & Allum, N. (2004). Science in society: Re-evaluating the deficit model of public attitudes. *Public Understanding of Science, 13*, 55–74.
Ziman, J. (1996). Is science losing its objectivity? *Nature, 382*, 751–754.

Chapter 19
From Public to Policy

Jan Riise

Abstract Science is not the unquestioned truth and platform for policy decisions it might have been. On the contrary, scientific results and issues are being debated, and dialog and two-way communication are key characteristics. A range of dialog events and formats has been developed, including science cafés, science parliaments and citizens' conferences. However, mechanisms for bringing public opinion and expectations to policymakers are not that well developed and evaluated. The 'mandate' to do so is not only about empowerment; it really is important, especially for adult participants, that there is an interest in the outcome of the debate or activity, although the value of mutual learning and direct interaction with scientists should not be underestimated. Science communication events such as festivals, science centers and museums provide excellent opportunities to organize dialogs. Often, formal agreements connect organized events to policymaking institutions as stakeholders or funders, and as such they ought to be able to benefit from their own networks in terms of legitimacy and the mandate for debates. Furthermore, they may provide an informal setting, a 'third place', that is a neutral ground for both scientists and members of the general public. Online activities, including individual initiatives and groups (for example, on Facebook) are briefly discussed. To some extent, science events' and science centers' internet presences ought to be an advantage and a credible starting point for online dialog development.

Keywords Public participation • Dialog • Debating science • Science events • Science museums • Science policymakers

J. Riise (✉)
Agadem AB, Dunkilihöjdsv 24, 43994 Onsala, Sweden
e-mail: jan@agadem.se

B. Schiele et al. (eds.), *Science Communication in the World: Practices, Theories and Trends*, DOI 10.1007/978-94-007-4279-6_19,

19.1 Introduction

Science has traditionally been regarded as a sound basis for policy and decision-making: reliable, objective and valid. In many cases, however, scientific fields are characterized by complex and non-linear dynamic systems. In others, research may be challenged by local knowledge or just a lack of overview. Uncertainty about what actually can be known may make science subject to different interpretations and make it problematic to use science as a basis for policies, guidelines, rules and laws. What used to be scientific consensus may well be questioned by other researchers, local knowledge or skeptics.

From this perspective, views held by the public about scientific issues have become increasingly interesting. Dialog activities and events have become regular parts of science communication community activities.

However, although some of these activities have formal links to decision-making authorities or government agencies, most do not.

In cases such as the European Commission's consultations for a green paper or a policy, or, very recently, the Royal Society of Chemistry's invitation to debate the future British chemistry landscape,[1] it is quite straightforward. Participants should have reasonable confidence that their contributions are actually read or listened to. In the European Commission case, more than 2,000 contributions were received as blog post comments, online questionnaires or written responses (EC 2011). On the other hand, it can be argued that such activities are not truly dialogical, but rather more 'market research' (see, for example, Sutcliffe 2011).

In other cases, ranging from science cafés to science parliaments and consensus conferences, the actual use of the outcome of the activity may be questioned. What happens to the resolutions of the youth parliament, when the member of the European Parliament gets back to the office after having received them at the final ceremony?

This article is written from the point of view of a practitioner—moreover, a practitioner without any scientific training in research methods, the use of references etc. The logical consequence is, I guess, that its validity and reliability may be rightfully questioned, and the whole story dismissed. Nevertheless, it is my hope that the following pages—sometimes supported by peer-reviewed articles or reports—may point to some fields and interfaces in and between science, the public and policy that could benefit from more research.

The article is about the changing role of science in society, about the increasing need and demand for public participation. It starts with a brief discussion on the relationships between citizens, scientists and policymakers. I introduce a range of formats that have been developed and used for dialog, and then discuss what happens after the talking—who cares about the resolutions and the voices heard?

[1] Royal Society of Chemistry website (n.d.). Retrieved 26 August 2011 from http://my.rsc.org/chemistrylandscape

Finally, I argue that science events and science centers could be very useful as physical places for dialog events to take place—as a kind of neutral 'third place'—and with the credibility and mandate to make them more meaningful.

All shortcomings and lack of support for arguments in this article are the responsibility of the author. Nevertheless, it is my hope that I have been able to point to some areas that could be subject of further studies, as well as initiatives.

19.2 The Double Deficit Model and Public Engagement

The 'deficit model' holds that there is a lack of knowledge among the lay public, and that, if this deficit could be eliminated with the help of science communication, the public would be more enthusiastic about science and more likely to accept or even argue for increased public funding for research.

Now, if things are not that simple, and members of the public have ideas about what research to fund and what knowledge might be reliable and valid, then there might also be a 'deficit' in places, methods and opportunities for them to say that.

This 'double deficit' situation can probably be discussed from a wide range of perspectives. In this article, the focus is on some ideas for public participation that have been tried, often with the objective of eventually influencing policymaking. The deficit model is often connected to the idea of science as 'facts' that can be communicated only if properly translated, and delivered without 'spin' (that is, without being framed for particular purposes). The article does not dive into that discussion, but presents the idea of 'joint fact finding', thus indicating that I might not entirely share that particular opinion.

It is also probably true that a substantial and constructive dialog only can take place if both, or all, sides have reasonably similar sets of factual knowledge.

The relationship between science and scientists on one side and policymakers and politicians on the other has been the subject of discussion and debate for quite some time. To an increasing extent, scientists may be dependent on public funding decisions, but on the other hand science has always had to satisfy the powers-that—be: Galileo also had to have his work accepted, in his case by the Church.

However, there is also a trend towards more emphasis on the usefulness of science, on how science can be used to solve societal problems. This means that scientists need to be accepted not only for the facts—the truths—they provide, but also for the extent to which their results are useful (see, for example, Jasanoff 2005). The ongoing transformation of the European Commission's Directorate General for Research into the Directorate General for Research and Innovation can probably be seen as one piece of evidence for that. Innovation, meaning making use of science, is the new challenge and the road to growth and welfare.

'Public engagement in science' is one of today's key expressions. It is mentioned in every conference presentation, invitation, keynote talk, article or introduction to the field of science communication.

Improved governance and the empowerment of citizens are often quoted as main reasons for public engagement and participation activities, but also for a range of other reasons:

- To provide a platform and a meeting place for discussion and debate between the public and researchers
- To facilitate mutual learning between public and researchers
- To identify public needs and concerns
- To merge citizens' values and opinions with the expertise of scientists, to create an increased acceptance and research agendas that are both scientifically interesting and socially robust.

The European Union acknowledges the need for closer relationships between science and society from a slightly more practical point of view. The Science in Society program under Framework Programme 7:

> is based on the rationale that the ability of European societies to develop themselves in a positive and sustainable way depends, to a large extent, on their capacity to create and exploit knowledge and to innovate. (EC 2011)

A Swedish study asking several thousand Swedes about their trust in different professional categories revealed slowly decreasing figures for researchers and scientists. The trend has been the same for a decade: slow but steady (Vetenskap and Allmänhet 2011). The reasons are varied, but some suspect that research is too closely connected to political and industrial agendas that may not take all factors into consideration, including public concerns about health or the environment. The idea is that the more informed and confident about research a respondent is, the more positive they are about the allocation of resources for world-class research activities.

However, public engagement development so far seems to be primarily an issue for practitioners. Academic evaluations of the outcomes are few, but there have been some.

19.3 Formats for Discussing Science

A number of formats for discussing science have been developed over the past couple of decades. The perhaps most informal and well-known are the science cafés, or *cafés scientifiques*, in many countries all over the world.

The basic idea of a science café is to use an informal venue, such as a café or a bar, for presentations and discussions concerning scientific issues. A scientist is introduced by a moderator and gives a brief presentation of their subject, followed by questions and answers and perhaps also a discussion. Depending on the subject and the moderator's ability to include and stimulate members of the audience, science cafés can be both interesting and rewarding for participating scientists as well as for the visitors.

There are no formal requirements for organizing science cafés. Where café discussions include topics that could have an impact, such as on research agendas, they produce no resolutions or other forms of documentation that are supposed to be handed over to policymakers or others. Apart from those who are actually there, no-one is listening: the systematic dissemination of the outcomes of the meeting is entirely up to the participants themselves.

Other activities include parliaments, forums, citizens' conferences, consensus conferences, citizens' exhibitions and events where the format is based on the licensed use of software and hardware, such as the twenty-first Century Town Hall.

Most have in common that there are no formal links to policymaking. The organizers, participating citizens and scientists are all there out of their own interest, and the basic aim of the activity is to produce a dialog.

Of course, events that actually connect to policymaking also exist, for example in the form of public consultations. This kind of event has been studied somewhat more, and sometimes criticized for being more like a 'public relations campaign for science' based on a 'lingering deficit model' (Davies et al. 2009).

Studies concerning dialog events that do not by definition have such direct links to policymaking are less frequent, although there are some reports that are quite positive and optimistic about the value of such events, both for scientists and for participants. Such events include student parliaments, student forums, junior science cafés, citizens' conferences, consensus conferences, citizens' exhibitions, online activities and joint fact finding.

The typology is not entirely clear, which causes some confusion among researchers trying to define and analyze the different formats, for example with regard to the formats' policymaking validity (see, for example, Rowe and Frewer 2005 for a more detailed introduction to and categorization of dialog events).

19.3.1 Student or Science Parliaments

The European Youth Parliament has developed and used the format of a parliament for a number of years, and has gained significant experience and knowledge. The basic idea is to bring together a group of 60–100 students, 16–18 years old, for 3 days to discuss scientific or societal issues. They work in committees, organize hearings and produce committee resolutions that then are debated in a plenary session. The final document is then handed over to politicians.

As part of the EU-funded 2WAYS project, coordinated by the European Science Events Association, science parliaments were arranged in 29 European cities in 2009 and 2010. The parliaments gave 18–19-year-olds an opportunity to discuss scientific issues in a political context. The format was adapted from the Young Europeans Parliaments that have been organized for several years in most European countries. In the science parliaments, the young participants discussed the use of embryonic stem cells, access to genetic information, personalized medicine, and the consequences of the possible existence of a gene for criminality and violence.

In each city, some 60 students gathered for 2 or 3 days of hearings, committee work and plenary discussions, and agreed on resolutions on all four issues. The resolutions were then handed over to local or regional political representatives.

Two students from each parliament were selected and invited to come to Brussels for the first-ever Young Europeans Science Parliament in November–December 2010.

The resolutions of the Young Europeans Science Parliament were presented to the vice president of the European Parliament, Ms Silvana Koch-Mehrin, at a ceremony in Brussels.

19.3.2 Student or Pupil Forums

A class of 25–30 students or pupils visits a research institute and analyzes a future-oriented issue with experts and scientists. They are asked to look for solutions or products that can help solve the problem and to prepare a presentation (poster, play, exhibit etc.) to visualize their ideas and to convey their views and ideas to scientists and other concerned people.

19.3.3 Junior Science Cafés

Science cafés have been organized in many places around the world for at least two decades. Increasing attention has been given to cafés organized by school pupils and students that identify the topic and invite scientists and other attendees. Support is offered (for example, by the Wissenschaft-im-Dialog) in moderation techniques and other necessary project skills.

19.3.4 Citizens' Conferences

A citizens' conference is a quite ambitious event, gathering 50–200 randomly selected participants for a 2-day meeting over a weekend. Participants discuss one controversial topic, have hearings with experts and eventually produce a 'citizen declaration', which is handed over to policymakers and/or scientists. Different opinions can be expressed. One obvious drawback is the 'randomness' of the participants: even though the initial selection may be random, attendance will be biased in favor of those who can afford to spend a weekend for this purpose.

19.3.5 Consensus Conferences

A consensus conference is similar to a citizen conference, but goes on for three weekends with 20 participants who are expected to reach consensus on a controversial topic.

The format has been used for a number of years, for example by the Danish Technology Board to collect and include public experience in technology assessments. On a random selection basis, people are asked whether they want to participate, but it goes without saying that the final group is made up of people who are committed to the issue or to the participation objectives.

However, a large majority of participants in German citizen and consensus conferences arranged during 2010 state that they learned much during the exercise (ZIRN and W-i-D 2011). The results in terms of dialog, discussion and participation are very good; the downside is the relatively small groups of people involved and the costs of arranging the events.

19.3.6 Citizens' Exhibitions

A citizens' exhibition presents the opinions of citizens on a particular issue through posters produced by the organizer and based on extensive interviews with perhaps 15–20 people selected to represent a spectrum of views. The posters are displayed and complemented with hands-on exhibits explaining the scientific background and a podium discussion with scientists and members of the public.

A citizens' exhibition has the potential to reach a large number of people, although it takes additional activities to strengthen the dialog aspect. Examples include a citizens' exhibition on 'personalized medicine' that took place in Bremen, Germany, from 12 to 26 November 2009. An evaluation of the event concluded that four out of five visitors found that the posters were understandable, that much of the science was new to them, and that the exhibition offered a wide range of perspectives, including the interviewees' fears and hopes, and also a good overview of the research field (ZIRN and W-i-D 2011).

19.3.7 Twenty-First Century Town Meetings

A somewhat more developed format is the twenty-first Century Town Meeting, a registered trademark and marketed by the non-profit organization, Global Voices. It is a format that uses technology and software to facilitate dialog in meetings with 50–5,000 participants. Polling and immediate feedback are important features used to reach agreement on priorities or build support for new proposals. The format engages citizens and deepens their commitment.

On 16–17 November 2010, more than 230 participants from 43 countries met at the Royal Belgian Institute for Natural Sciences in Brussels to create a 'Positive Vision for Biodiversity' using the twenty-first Century Town Meeting format and technology. A wide range of stakeholders was brought together for 2 days of discussions, including policymakers, civil society organizations and representatives of the business sector. The twenty-first Century format was used to develop the common vision, for voting and for sharing ideas.

19.3.8 *Joint Fact Finding*

Joint fact finding dialogs are characterized by the participation of several different stakeholders, including the public, with the objective of identifying and establishing a credible knowledge basis, and determining whether and what sort of further analyses may be needed. The joint fact finding process has proved to be valuable for societal debates involving scientific issues. It creates opportunities for insights into other stakeholders' thinking and perspectives, and it defines the knowledge that is accepted by all parties.

The method has been used to work towards common views where the validity of scientific data has been discussed for many years, such as for the governance of Baltic Sea fisheries. Environmental groups, consumer groups, fishermen, fishing industry representatives and scientists have all interpreted and used the data in different ways. A research project with the aim of creating a 'round table', particularly for the Polish fisheries sector, studied the mechanisms and issues and finally established such a forum (Stöhr and Chabay 2010).

19.4 Empowerment and Actual Influence

There are several problems connected to the participation formats described above. One of the more obvious is the fact that the participants are not necessarily representative of a larger population. From a democratic point of view, this naturally reduces the value of the opinions expressed, as they are simply not representative enough.

One of the challenges for the organizers of participation events is to mobilize a reasonable number of people and to demonstrate that the outcome of the particular event has some validity as an opinion shared by a significant proportion of the community. This has nothing to do with the potential individual or collective benefits of the training and participation in policymaking processes; that is another story.

An extensive study set up by Wissenschaft-im-Dialog and Ortwin Renn at the University of Stuttgart (ZIRN and W-i-D 2011) presented its final report in July 2011, after having studied a range of participatory activities from 2009 to 2011.

The study confirms that the mobilization of participants is a major problem. It might take 200 calls to get one participant to a consensus conference. Maybe 20 schools need to be contacted in order to find one class to participate in a student parliament.

19.4.1 *What Happens After Participation?*

One important question is of course about what happens after the parliament, consensus conference or other participatory event. The question has two aspects: what happens with the outcome of the event, and whether the participants act or think in any way differently from before the event.

Some might respond, 'It's not about validity; it's about empowerment,' pointing to the valuable training in democracy offered by, say, a student parliament.

The resolutions might be handed over to local, regional, national or European politicians to be immediately forgotten, but the real value may be the knowledge about how to produce the resolutions in the first place.

John C Besley and colleagues at the University of South Carolina set out to explore whether and how participants in a citizen engagement program afterwards became involved in or initiated discussions about the subject, in this case nanotechnology. In an article titled 'What, if anything, do they say?' (Besley et al. 2008), the authors also provide a literature review. Several authors are dealing with the exclusiveness of participatory activities, as they might turn out to be rather expensive exercises, as discussed above. Others discuss the challenges of scale and how to involve more citizens, not least to achieve better representativeness. Some have looked at how media can be used and can function as multipliers, reaching people other than those who were at the event.

Besley et al. knew of no studies on interpersonal communication after participation. They report that interpersonal communication takes place after participation in engagement activities, and is generally positive about scientific progress and economic opportunities.

Organizers of science communication and participation events may take into consideration that the event might also influence people other than those who participated, through communications of the participants. Media coverage may strengthen the effects even more, but it might not be realistic to expect a media presence other than for the biggest events. The social networks of each individual may thus also be a bridge, reaching further than the actual participants.

19.4.2 The Importance of a Mandate

The 2-year project in Germany mentioned above was recently concluded and reported. Having studied a large number of participatory activities, the research team found that the most important factor for success is the 'mandate'. That is, when the event or activity was openly connected to the interests of policymakers in finding out about people's opinions, the event was more successful in terms of participation and the development of views and opinions. Hidden agendas, with the aim of influencing opinions and resolutions in one way or another, were met with suspicion and frustration. This was more visible for adult groups than students, but the patterns were the same (ZIRN and W-i-D 2011).

19.4.3 The Informal Learning Component

During 2009 and 2010, 29 local science parliaments were organized in European cities as part of the EU-funded 2WAYS project organized and coordinated by the European Science Events Association. All the local parliaments discussed the same four issues: the use of embryonic stem cells, access to genetic data, the effects of genes that trigger violent behavior, and personalized medicine. The participants were mostly 16–19 years old. Two students were selected from each parliament to participate in the first-ever Young Europeans Science Parliament in Brussels at the European Parliament in December 2010.

The 29 local parliaments' participating students were asked a number of questions as part of the project's impact study. First of all, the students experienced the parliament event as a motivational and meaningful activity from a learning point of view. They perceived the parliament environment as encouraging them to ask questions and take part in the discussions. The students did not see the parliament event as any less formal than their everyday classroom education; it could be that the organizers' emphasis on the formal profile of a parliament (including dress codes) might have influenced that opinion (Salmi 2010).

Science cafés, parliaments and other dialog formats do not necessarily have a formal link to policymaking. Science café organizers, universities, science festivals, museums and science centers often arrange the events. There is no mandate or assignment from a policy-writing institution, such as a ministry or a non-government organization. Consequently, research on the influence on policies of dialog events not connected with policymaking has been rather limited.

However, there has been some discussion about the value of events such as science cafés for individual development. Dialog activities may be seen as opportunities for empowerment and personal benefit, which may be a step towards further involvement. The activities themselves can also be seen as parts of an incremental process. And 'symmetrical learning' during dialog should not be underestimated (Davies et al. 2009).

This is also an argument often used by event organizers when they ask scientists to participate as guests and presenters at science cafés and similar activities. This has probably not been studied in depth, but anecdotes and personal observations confirm a sometimes surprisingly positive attitude and experience—after the event.

Less is known about how to incorporate the views and knowledge that citizens may have. One study suggests that the 'experts' understanding of the public' is to a large extent dependent on who is in control of the knowledge (Young and Matthews 2007). And how do even very ambitious dialog programs actually feed into government and research policies and agendas?

Electronic communications and wide access to scientific information are significant resources needed to tackle the big challenges that the world is facing. Furthermore, cultural and language differences must be discussed—not only as they differ between countries or continents, but also in different environments: political, scientific, or civil society.

19.5 Online Activities

All of the activities and formats discussed above are real events, taking place somewhere and at some time, 'IRL' (in real life). It seems reasonable to assume that formats for public participation are being developed and will be further developed in the near future. Some attempts have been made, such as the European Commission's consultations for the new Framework Programme during 2011, in which a blog was created for discussions and debates about particular aspects of the drafts.

This will surely be the subject of other articles to come, as formats are developed and used. The German 'Wissenschaft debattieren!' project also includes a study of online activities. One of several conclusions is that the integration of scientists in online forums is necessary (ZIRN and W-i-D 2011).

However, it seems important to add the dimension of spontaneous public opinion making, using the internet as the tool and medium. The following story is an example of the use of Facebook and the unpredictability of what comes through and what does not.

In 2009, Dr Paolo Zamboni, an Italian vascular surgeon, published an article about a completely new treatment for MS (multiple sclerosis)—a condition that has mainly been regarded as an autoimmune condition and incurable. Dr Zamboni suggested a surgical widening of veins in the neck, as he had found that many MS patients suffered from 'chronic cerebrospinal venous insufficiency' (CCSVI). It was claimed that the treatment was very successful, and that three out of four patients reported an improvement after surgery.

The report to a large extent went by unnoticed, seemingly because there were also several unanswered questions about control groups and other matters. More research was needed before the CCSVI theory and treatment could be accepted.

In Canada, however, some major media made headlines of the report. Between November 2009 and January 2011, about a 100 articles were published worldwide on the treatment, including 80 in Canada and 16 in Italy. Literally hundreds of Facebook and web-based groups and pages demanded more resources for further research.

The Canadian and American multiple sclerosis societies have now funded seven studies to further investigate the connections between MS and CCSVI.

It is tempting to regard these funding decisions at least to some extent as a result of public engagement expressed through Facebook and other social media.

19.6 The Importance of the Location

It has been argued that cafés, pubs and similar places where people go to enjoy the company or the atmosphere are actually cornerstones of community vitality and democracy. Ray Oldenburg (1999) calls them the 'Great Good Places' or 'Third Places' (not at home, not at work).

Such 'great good places' are also important from a political point of view. Modern society's centers of public administration and political power are not exactly built for public participation. Politicians and policymakers need the 'third place' to meet their voters or community-based organizations.

Such places are at the same time excellent venues for science communication and public engagement in science, as they offer exactly the neutral and informal setting that is important for credibility and trust. Science cafés use the advantages of the 'third place':

> Science cafés are spreading as a grassroots movement. The casual, open format readily engages the public in conversations about science.[2]

There may be many reasons for creating and organizing science festivals, science days, science weeks or similar events. The European Science Events Association, Eusea, listed a number of its members' objectives in its 'White Book' of science events (Eusea 2005):

- Regional development
- Improved relations between schools and universities
- Improved relations between research and industry
- The generation of tourism.

Most event organizers include the mission of raising awareness of science and technology, especially among youth (Eusea 2005). Another, even broader, aim is to 'improve relations and communications between science and society' (Neresini et al. 2009).

The use of 'unusual' places, such as streets, shopping malls or railway stations, is closely connected to the idea of science festivals and constitutes in many cases a significant part of the brand and profile of the event. Such places are mainly chosen to facilitate direct interaction between the public and scientists by lowering thresholds for the public—using well-known places, a kind of 'neutral ground' for the dialog or presentation. Furthermore, this makes it possible for visitors to approach the activity at their own pace; it is possible to watch from a distance without being directly involved at once.

An important aspect of science events is the participation of 'real' scientists; presentations and activities are not only done by 'presenters', but by the people doing the research.

19.7 Conclusions

It seems that there is great interest in dialog activities and public engagement in science on the part of event organizers, policymakers, research funders and others, including the scientific community.

However, this is still a field where methods and activities are being tried out, even though some of them, such as the science cafés, have been around for quite some time. What definitely needs to be developed are the links between different actors in this field. In order to do that, some issues need to be discussed and some questions answered. One of the central issues concerns the communication between

[2] http://www.sciencecafes.org/what/.

scientists and the public on the one hand and policymakers and research funding organizations on the other. Will blogs, consultations and other forms of access to opportunities to take part and comment be enough?

First, science festivals, events, museums and centers could have an important role to play in this context. In many cases, they are established and funded by local, regional and national authorities, agencies and governments. As such, they have direct links to policy and political decision makers. This could put them in an excellent position to solve the 'mandate' problem described above. With some additional attention to the links between the public and policy, it should be possible to involve stakeholders, board members and owners in the activities.

Second, museums, science centers and events are probably just great as 'third places', at least temporarily. They provide an informal setting and usually include a café and other public spaces, and the idea of communicating science is the very basis for the venue.

Finally, their networks include schools at all levels, teachers, universities, local and regional organizations, sponsors, funders and other stakeholders. Add website visitors, and the conditions are set for the organization of dialog events for citizens, students or the public in general.

References

Besley, J. C., Kramer, V. L., & Toumey, C. (2008). Interpersonal discussion following citizen engagement about nanotechnology: What, if anything, do they say? *Science Communication, 30*(2), 209–235. Retrieved from http://scx.sagepub.com/cgi/doi/10.1177/1075547008324670

Davies, S., McCallie, E., Simonsson, E., Lehr, J. L., & Duensing, S. (2009). Discussing dialogue: Perspectives on the value of science dialogue events that do not inform policy. *Public Understanding of Science, 18*, 338–353. Originally published online, 1 October 2008.

EC (European Commission) (2011). *Green Paper on a common strategic framework for research and innovation funding: Analysis of public consultation*. Retrieved August 27, 2011, from http://ec.europa.eu/research/horizon2020/pdf/consultation-conference/summary_analysis.pdf#view=fit&pagemode=none

Eusea (European Science Events Association). (2005). *Science communication events in Europe, 2005*. Göteborg: Eusea.

Jasanoff, S. (2005). Judgment under siege: The three-body problem of expert legitimacy. In S. Maasen & P. Weingart (Eds.), *Democratization of expertise? Exploring novel forms of scientific advice in political decision-making* (Sociology of the Sciences Yearbook, Vol. 24, pp. 209–224). Dordrecht: Springer Science+Business Media B.V.

Neresini, F., Dimopoulos, K., Kallfass, M., & Peters, H. P. (2009). Exploring a black box: Cross-national study of visit effects on visitors to large physics research centres in Europe. *Science Communication, 30*, 506–533.

Oldenburg, R. (1999). *The great good place*. Cambridge, MA: DaCapo Press.

Rowe, G., & Frewer, L. J. (2005). A typology of public engagement mechanisms. *Science, Technology and Human Values, 30*(2), 251–290.

Salmi, H. (2010). *2WAYS Impact Study*. Retrieved August 25, 2011, from http://www.twoways.eu/Shared/Uploads/Files/2WAYS_ImpactSurvey_lowestres.pdf

Stöhr, C., & Chabay, I. (2010). Science and participation in governance of the Baltic Sea fisheries. *Environmental Policy and Governance, 20*(5s), 350–363.

Sutcliffe, H. (2011). *Distinguishing public engagement: When is it market research, PR and sales?* Retrieved August 30, 2011, from http://www.matterforall.org/blog/index.php/category/public-involvement/page/6/

Vetenskap & Allmänhet. (2011). Vetenskapen i Samhället—resultat från SOM-undersökningen 2010. *VA-rapport* 3. Retrieved from http://va.se/downloads/varapport2011_3.pdf

Young, N., & Matthews, R. (2007). Experts' understanding of the public: Knowledge control in a risk controversy. *Public Understanding of Science, 16*, 123–144.

ZIRN & W-i-D (ZIRN/Universität Stuttgart & Wissenschaft-im-Dialog) (2011). *Abschlussbericht Forschungsprojekt 'Wissenschaft debattieren!'* ZIRN & W-i-D.

Chapter 20
Science Culture and Its Indicators

Martin W. Bauer

Abstract The notion of 'science culture' is a conceptual as well as an empirical challenge for social researchers. This chapter suggests a language convention to distinguish 'scientific culture' from 'science culture' and reviews several efforts to conceptualize and measure 'science culture' and the difficulties of mobilizing adequate data streams for that purpose. It is time to bring these different attempts into a coherent discussion and move towards a more globally coordinated effort. To that end, the chapter outlines some of the common problems of conceptualizing and measuring science culture and suggests a way forward.

Keywords Indicator • Mass media • Science culture • Science literacy

This chapter addresses the questions: What is scientific culture? How can we measure it? Let us define 'culture' with Sorokin (1957:2) as:

> everything which is created or modified by the conscious or unconscious activity of two or more individuals interacting with one another or conditioning one another's behaviour. … science, philosophy, religion, art, technics and all the physical paraphernalia of advanced civilization are cultural phenomena.

The symbolic world of beliefs, norms and invested, often fetishized, artifacts is a condition as well as a catalyst of a society's productive activities. The symbolic world conditions how people deal with challenges as they arise. The 'subjective' constraints on human action—the dispositions, attitudes, imagination, moods and sentiments—are no epiphenomena of objective patterns of activity, but factors of action in their own right. This view is more or less taken for granted by historians

M.W. Bauer (✉)
Institute of Social Psychology and Methodology Institute, London School of Economics and Political Science, Houghton Street, London WC2A 2AE, United Kingdom
e-mail: m.bauer@lse.ac.uk

B. Schiele et al. (eds.), *Science Communication in the World: Practices, Theories and Trends*, DOI 10.1007/978-94-007-4279-6_20,

who take the long view. In their research, cultural comparisons of cultural mentalities feature prominenty in explaining the divergent paths of technological developments within the western world, and when comparing the west with other regions of the globe (see, for example, Berg and Bruland 1998; Mokyr 2000).

As human actions are constrained by both external and internal structures, the 'objective world' is no proxy for the 'subjective world', nor vice versa. When it comes to human activity, subjective resources cannot fully compensate for objective deficiencies, nor objective assets for subjective deficits. Mathematically formulated, culture is a product of the material (objective) and the symbolic (subjective): $C = O \times S$. For cultural development, individual or collective, both factors matter; culture does not arise from only one of them, and the lack of one will strongly diminish the whole culture.

Inonu (2003) put together an instructive table which shows that scientific production, expressed as the number of scholarly publications per year, is poorly explained by the economic achievement of the country (GDP or GDP per capita in purchasing power parity terms). There are poorer countries that are rich in science, and there are richer countries poor in science. This strongly suggests that non-material conditions need to be considered to understand a country's science base.

For democracy and development, similar concerns apply. The process of democratization does not simply derive from material development: we cannot simply say that the richer the country, the more democratic it will be. Research shows the importance of mediating subjective factors such as the 'emancipatory impetus', which is a value that favors autonomy and public self-expression. Such cultural features buttress a functioning public sphere that mediates the relation between economic achievement and democratization. In other words, economic development brings democracy only if certain values scaffold a functioning public sphere (see, for example, Welzel 2006). By analogy, in order to explain 'scientific creativity and productivity' as a function of economic power, we might also have to consider the public sentiment that supports these efforts in society at large.

20.1 What Are Cultural Indicators?

The term 'cultural indicator' has been used variably in the literature. Let us explore some of its different meanings. Generally speaking, the concern for cultural indicators is an extension of the social indicator movement, which since the 1960s has attempted to establish a system of societal accounting that goes beyond economic performance indicators (see Bauer 1964; Melischek et al. 1984).

20.1.1 Performance of the Culture Industry

First, the term 'cultural indicator' refers to the performance of the culture industry, the sector of the economy that includes design, architecture, advertising, cinema, the arts, music, museums, and the production and consumption of their products and performances. Culture is seen as a productive sector, the 'creative industry'. Indicators of the culture industry are added value to gross national product, visitor numbers, the industry's share of the labor market, its relative growth and export value (for the United Kingdom, see Work Foundation 2007). It is well established that the advertising sector is closely tied to the economic cycle and has a long-term constant ratio of GDP (Chang and Chan-Olmstead 2005). To consider science as part of the culture industry is likely to be controversial: science policymakers would be reluctant to be seen as part of a ministry of culture.

20.1.2 Cultural Diversity and Its Conservation

Second, UNESCO uses the term to compile statistics on cultural diversity, including of languages spoken, religions, festivals, natural and built heritage sites, museums, communication and translation efforts, and the consumption of cultural goods such as cinema, museum and concerts.[1] Diversity brings problems of equal access, but is a source of creativity and thus an intangible asset of the economy. A system of indicators is in development and aspires to global reach and consistency.

20.1.3 Local and Traditional Knowledge

Third, the Food and Agriculture Organization sponsored an initiative on 'Cultural indicators for SARD' (sustainable agricultural development) (FAO 2003). The term served as the title for a questionnaire presented to indigenous peoples and their representatives to assess the significance of local and traditional knowledge in their agricultural practices. Here, the term conferred significance on traditional knowledge as an asset of local agriculture and operationalized local knowledge for comparative purposes.

20.1.4 Mass Media Trends

Fourth, the term has a history in mass media effects research. In that field, 'cultural indicators' refers to the cultivation research program, which studies the mid-range power of the mass media to cultivate beliefs about the world, such as 'the world is generally a mean place' (see Gerbner 1969). This program combines systematic mass media scoring (the cultural indicator) with large-scale survey research

[1] See http://www.unesco.org/culture/worldreport.

(the public belief) to assess the extent to which belief is 'cultivated' as a function of exposure to television: the more hours a day a person watches TV, the more likely they are to assimilate their worldview to that of the average TV program. These studies were preoccupied with violence on TV and the resulting belief in a 'mean world'. Similar cultivation effects were observed for gender images or the framing of biotechnology (see Bauer 2005). In cultivation studies, 'culture' mostly denotes the 'unrealistic world of television', which is a de facto driver of everyday beliefs, demonstrated mainly in the United States.

20.1.5 Sociology of Culture

The term is also found in cultural sociology, which maps social change on the basis of cultural trends. Here the data stream is often mass media material coded for time-series analysis, with few variables over long time periods. The classical example of such a study is Sorokin's (1985) coding of mentalities over the past 2,500 years and his inferences about the ups and downs of ideological frameworks such as empiricism, determinism or rationalism. Those data were reanalyzed by Klingemann et al. (1982), who confirmed the longitudinal patterns of trends.

20.1.6 Social Values

Finally, the term 'cultural indicator' appears in large-scale survey research to refer to a class of questionnaire items that tap into cultural dispositions (values) with long cycles of change. By contrast, 'superficial' opinions, attitudes and beliefs have a shorter life cycle. Here the problem is to operationalize a class of value concepts with survey items, and to monitor long-term changes in and across populations. Examples are the research into 'post-materialism' (for example, Inglehart 1990) and subsequent efforts of the 'world value survey' focusing on values orientations such as survival, lifestyle, wellbeing and happiness (Schwartz 2011).[2]

20.2 Science Culture Concepts and Indicators

Let us start with a plea for a naming convention. We have two labels often used interchangeably for the phenomenon explored here: *science culture* and *scientific culture*. Researchers have been analyzing 'cultural features' in and around science inside and outside the laboratory for the core and the periphery of the thinking

[2]http://www.worldvaluessurvey.org.

collective. Without putting too much emphasis on labels, it might be useful to agree on the following: the culture of research groups, the thinking community, the lab talk and practice shall be denoted by the term 'scientific culture'. The composite word 'scienti-fic' derives from the Latin *scientia facere* (science making). Thus, the culture of science-in-the-making shall be called 'scientific culture'. On the other hand, for the purposes of examining the wider context of science-in-society, we would like to use the term 'science culture'.[3] This suggestion is consistent with the distinction between culture-as-practice and culture-as-context as it is used in organizational analysis (see Smircich 1983), or the esoteric and the exoteric spheres of science (Fleck 1935). Next, let us briefly examine several concepts of science culture that have been proposed in the literature.

20.2.1 Scientific Temper

An old idea of science culture is encapsulated in the notion of the 'scientific temper'. Psychologist E.B. Titchener (1929:29) referred to the particular frame of mind that fosters the pursuit of scientific research: disinterested curiosity, impersonal love for truth, cautious drawing of conclusions and an 'attitude of dissent against the orthodoxy of practical occupation'. The notion gained traction in the 1930s and 1940s in generalizations about public attitudes. In 1946, India's first Prime Minister, Jawaharlal Nehru, declared a commitment to diffuse the scientific temper as 'science mindedness throughout the population … measured by the extent to which ordinary people were using methods of science to solve life's problems', which was inscribed into the Indian Constitution in 1976 (Kumar 2011:266ff). The mobilization for science communication and large-scale surveys of attitudes to science in the Indian context are inspired by this notion (Raza and Singh 2012; Raza et al. 2002; Shukla 2005)

20.2.2 Civic Scientific Literacy

Jon D. Miller's well-worked idea of 'civic scientific literacy' (CSL) was developed over many years in collaboration with the National Science Foundation's science indicators module (e.g. NSF 2006) and has been copied across the globe. Its core module comprises a cognitive measure of science literacy—a quiz on timeless general science knowledge items that are not tied in with any particular contemporary controversy. An individual can be assigned a position on a standardized scale of 0–100; the validity claim is that a threshold of 70 marks the basic competence that

[3]This suggestion for language use was recently also made by Suzanne de Cheveigne (personal communication).

enables one to read and understand a *New York Times* science feature, and thus to follow and participate in public debates, not least over climate change (Miller 1983, 2012). Miller understands this notion of CSL as a contribution to the civic culture of a country. In the absence of science being a party political issue, which most of time indeed it is not, the significance of CSL lies in defining an 'issue public' that pays attention to science, the 'attentive public of science'. The pros and cons of this notion are very much part of research into the public understanding of science (PUS) over the past 30 years, which I do not rehearse here (see Bauer et al. 2007; Allum 2010)

20.2.3 The Cultural Distance Model

A model of science culture based on the idea of 'cultural distance' is offered by Gauhar Raza and colleagues in India. It takes its original motive from dissatisfaction with the CSL idea of a universal general knowledge scale and its true/false format. Attending large Hindu religious festivals and involving pilgrims in discussions about natural phenomena such as moon phases, heliocentrism, the roundness of the Earth, earthquakes and other matters, the researchers were taken by the variety of knowledge that they encountered. People were far from being ignorant about these natural phenomena. The researchers were seeking to position everyday notions of natural phenomena as more or less remote from scientific orthodoxy, and to relate that remoteness or proximity to the democratization of knowledge (that is, to basic education). To score the cultural distance, they determined how many years a population has to be schooled so that 50% can give a scientifically acceptable answer. Empirical results show that the number of school years to reach that threshold depends on how remote from everyday life the particular scientific notions are (Raza et al. 1991, 2002; Raza and Singh 2012).

20.2.4 A Multidimensional Input–Output Model

Godin and Gingras (2000) and Godin (2012) hark back to earlier notions promoted but later not sustained by UNESCO on science culture (see UN 2003): the practices of appropriation of science in society. In explicit contrast to the economic input–output model of R&D and manpower leading to patented innovations and economic growth, which is used by the Organisation for Economic Co-operation and Development (OECD), Godin and Gingras explore a larger notion of societal practice of appropriation and support for science. They distinguish three modes of appropriation: the learning mode, through which citizens acquire knowledge, know-how and attitudes; the implication mode, through which society draws benefits from science; and the socio-organizational mode, through which society develops institutions that secure scientific activities. The three modes are interrelated in their functioning. For each of the modes, the authors draw up lists of potential input, activity and output indicators.

The system offers an order of culture indicators in nine conceptual cells: learning mode input, activity and output; implication mode input, activity and output; and socio-organizational mode input, activity and output ($3 \times 3 = 9$). The system is exemplified using current Quebec data, with occasional comparisons with the whole of Canada and all OECD countries.

20.2.5 Science Culture Indices: SCI-I and SCI-S

Korean colleagues have recently developed a system of indicators for science culture from an educational perspective (Song 2010). The notion distinguishes the individual level (SCI-I) from the societal level of analysis (SCI-S) and, for each level, the modalities of potential and practice. This 2×2 logic is made concrete by three suggested indicators for each field, yielding a system of 12 indicator classes. The measures derived within the PUS tradition—attitude, interest and knowledge of science—are subsumed by one of the four quadrants, the individual-level-potential mode. The individual-level-practice mode comprises formal science education, the use of high-tech devices, information seeking and engagement with science. The societal-level-potential mode refers to infrastructure such as R&D investment, manpower, and science museum and exhibition facilities, while the societal-level-practice mode includes the mass media coverage of science, civic mobilization on scientific issues and the staging of events such as science festivals. The system has been exemplified across several Korean cities, and by attempts to compare several countries in the region (Hong Kong, mainland China, Japan, Korea, Taiwan). However, the key difficulty of the system lies in the access to data. Most of the suggested data are either not accessible or do not exist (Song 2011). However, this is a common problem in the construction of science indicator systems (see Butler 2006).

20.2.6 The Science Culture Index (SCI): Production and Appropriation

Shukla and Bauer (2012) constructed a globally valid indicator of scientific culture by considering both objective measures of performance and subjective measures of perceptions of science. They considered a micro-integrated database of perception data on knowledge, interest, informedness, attitudes and engagement with science in 32 European countries and 22 Indian states. This basis of 54 units of analysis allowed them to examine the plausibility of perception indicators (knowledge, attitude etc.) in conjunction with performance indicators (R&D spend, manpower etc.). The Science Culture Index (SCI) assumes that scientific performance and the mentality of science are mutually supportive; and this mutual support is captured by the old Chinese Yin–Yang symbol (see Fig. 20.1). Scientific research fosters a scientific

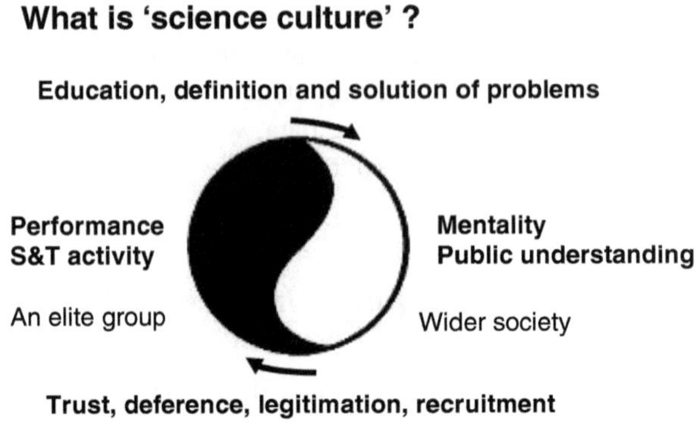

Fig. 20.1 The Yin and Yang of science: scientific performance and mentality of science in society as mutually reinforcing processes

mentality (the 'scientific temper', as Indians like to say), and that mentality in turn supports scientific research by recruiting youth into careers and by creating respect for the voice of science as a cultural authority. Five subjective and four objective indicators form weighted linear combinations; the indices for Scientific Culture (STI) and PUS Culture (PUS) combine into a final Science Culture Index: $SCI = a(STI) + b(PUS)$. More research is needed to validate this idea on a more global database, and by conducting sensitivity analyses of the SCI in relation to the different indicators.

20.2.7 The Spiral Model of Science Communication

A similarly recursive model of science culture is emerging from Brazil and from the perspective of science communication. Carlos Vogt, a sociolinguist, poet and former head of São Paulo's resourceful funding agency, FAPESP, offers a concept for the integration of several science indicators: the spiral model of science (Vogt 2012). The model is based on two fundamental dimensions of communicating science in society. On the one hand, we have the dimension of esoteric versus exoteric communities; on the other, that of monological versus dialogical exchanges. The four quadrants of this 2D model are populated by quantitative and qualitative indicators. Science education is mainly monological, socializing students and the wider public into a canon of established facts and ideas; this education has an esoteric and an exoteric angle. The esoteric side educates aspiring university students into career scientists. Here, the annual numbers of (science) Ph.D.s are a suitable indicator. The exoteric side includes formal and informal education of the wider public through school curricula and science exhibitions. Dialogical communication is esoteric as practiced in research and at the laboratory level, indicated by well-established publication and citation records. And dialogical communication also takes place exoterically in mass media debates and in forums of public engagement, such as consensus conferences or other forms of public hearings and exercises to scope public sentiment. The historical path is envisaged as 'spiraling' resonances between these four quadrants, moving through exoteric and esoteric, monological and dialogical genres of communication.

20.2.8 *Science Culture as Profiles of Educational Attainment*

Large-scale international attempts to compare educational attainment have been stimulated through initiatives such as PISA (the Programme for International Student Assessment, supported by the OECD), and international consortia such as TIMSS (the Trends in International Mathematics and Science Study, with a focus on mathematics education) or the ROSE (Relevance of Science Education) project, which is concerned with motivational factors of science education (Sjoeberg and Schreiner 2012). These efforts are not necessarily designed to assess cultural diversity, but to benchmark the performance of the education system on measures of educational attainment. PISA orders countries and regions within countries according to their average educational attainment based on a representative sample of schools. Science education has been the focus of PISA 2003 and 2009 (PISA 2009).

If it is used at all in these initiatives, the term 'culture' appears to be a secondary concern. For example, TIMSS uses a variety of scales to assess the cognitive demands of mathematical reasoning. By profiling the strengths and weaknesses of different countries on those scales, one can characterize national cultures of mathematics; for example, the United States school system focuses on declarative and procedural knowledge, France emphasizes advanced concepts, Sweden orients towards practical problem solving, and Germany excels in graphical representations of mathematical problems. These profiles reflect traditions that privilege some mathematical competencies at the expense of others (see Klieme and Baumert 2001).

20.2.9 *Science Culture as the 'Common Place' for Communication*

Another idea of science culture with a focus on communication arises from a network of researchers that Steve Miller (2012) organized around Europe. The idea was to take stock of the multitude of activities promoting public engagement and the differences in public expectations of science across Europe (the K/I, or Information Need Index). The purpose was not to evaluate public engagement or the mentality of the wider public, but to obtain a better idea of where the communication effort must start from. Science communication here follows a model of classical rhetoric that starts from an assessment of the existing 'common places' on which to build persuasive arguments that carry the day. Every rhetorical activity needs to accept the location of its audience and work with redundancies on common places. This box-ticking exercise provided an index of different contexts of science communication across Europe.

An excursion: Subjective and Objective Indicators of Science Culture

Our focus on 'culture' raises a discussion, or debate, about science indicator systems. One has to hark back to the beginnings of such debates in the 1950s to find a similar agenda of combining data on perception and performance of science (see Godin 2005).

In the United States, the National Science Foundation *Indicator reports* have continued to publish chapters on both types of data since the later 1970s, but no attempt has ever been made to bring those data into conversation with each other. Similarly, in Brazil, FAPESP (2004) includes both types of data in its reports on the science system of the state of São Paulo. In the European Union, these monitoring activities are entirely spit between the European Office of Statistics, which publishes R&D figures (see EIS 2005), and Eurobarometer, which offers the occasional perception survey. There is no coordination between these agencies. The *India science report* of 2004 started up with holistic ambitions, but did not carry them through; the final report does not map objective and subjective indicators together (Shukla 2005). In China, science literacy is a part of human resources development, the quest to improve the quality of the population, for which comprehensive input–output indicator systems are in the making. In this context, we need to remind ourselves of the old UNESCO agenda to measure science-related activities in addition to R&D investment and scientific manpower (Godin 2005, 2012).

Efforts to assess the public perception of science are also a part of the subjective social indicators movement, which since the 1970s has established monitors of the 'subjective state of the nation'. For example, the measurement of 'confidence in institutions' raises issues of measurement error: measures vary with the company that does the survey, and inferences of trends and trend changes must be based on large differences in order not to be misleading (Turner and Krauss 1978). A recent boost for these undertakings has come from social statisticians and economists who call for complementing measures of national performance (GDP) with indicators of subjective wellbeing (see ONS 2011). The United Kingdom will henceforth ask 200,000 of its citizens each year how happy they are with their life situation.

20.3 Challenges for Future Work

An intensive discussion of cultural indicators of science is overdue and will have to address issues that all the efforts outlined above have in common. Clarifying the common ground might indeed be the basis for making progress in the coming years.

20.3.1 A Campaign Guidance System or a Typology of Science Cultures?

There are pragmatic tensions between 'social' and 'cultural' indicators. Social indicators evaluate interventions as indicators-of-action. They indicate the success or failure of the handling of social affairs through outcomes such as poverty levels, mortality, crime or literacy rates. By contrast, cultural indicators are indicators-for-action. The context is not itself the target of the action, but calls for strategic adaptation. Contexts for action are out of the actor's control, at least in the first instance (see Melischek et al. 1984).

In a campaign guidance system, indicators-of-action help in designing messages (inputs), steering the choice of vectors (channels) and assessing the public understanding as outcome (output). The system is modeled on the activities of artillery gunners loading cannon with ordnance, who need a guidance system in order to aim accurately. The performance of the system is successful if it hits the target as planned, maybe after discounting collateral damage.

The alternative is the comparison of cultural symbol systems as indicators-for-action. Here the focus will be on comparing genres of science communication. Symbol systems are co-evolving processes of written and spoken references to science in many contexts of life. Different genres of communication resonate with each other; they mutually reinforce or dampen their salience and reception (for example, a reference in a newspaper or a blog entices you to read the original paper). The climate of science communication is first and foremost understood and appreciated as a guide for adaptation, and only secondarily as a target of intervention.

This duality of performance and context indicators remains fundamental to our problem of science culture, though both purposes may co-exist because they motivate the mobilization of similar data streams but with different pragmatic perspectives.

In this context, typologies are science cultures are a fertile avenue of indicator research (see Lebart 1984; OST 2000; Liu et al. 2012; Kawamoto et al. 2011; Mejlgaard and Stares 2012). On the basis of measures of literacy, interest, attitudes and engagement with science, it is possible to profile social milieus of science into socio-economic variables such as age, level of education, urban and rural habitation and others. Longitudinal and cross-sectional comparisons can be undertaken on these typologies of scientific milieus. Are these milieus expanding or contracting, merging, mutating or splitting further? This is a very promising field of enquiry. A literature review of existing typologies and their underlying methodology would make a big difference in the first instance.

20.3.2 General or Specific Indicators?

The problem of monitoring science-in-general or the study of specific developments such as nuclear power, biotechnology, nanotechnology, birth control mechanisms or synthetic biology continues to create dilemmas for researchers. The funding streams are more often than not focused on specific issues, while the interest of the researcher might well be the general science culture.

One argument holds that general attitudes do not exist, because responses to general questions are based on the respondent's perception of specific issues. Hence, it is better to know what the specific issues are, even at the cost of forfeiting time-series comparability. Also, specific indicators are more useful because they can be used as campaign guidance information (for example, one might want to increase public awareness of synthetic biology from 8% to 25% of the population within 1 year).

On the other hand, the argument for general indicators holds that we want to construct items with time-series comparability. The construction of time-series requires perennial items that do not become out of date or out of fashion. The pressing issues of the 1950s (such as water fluoridation or nuclear fallout) are non-issues in the 2010s. Assessments of cognitive competencies have to rise above such temporal concerns, and the same must apply for attitudes and interests in science.

Furthermore, a general indicator of science might be related to particular issues, but that is an empirical matter. Knowledge, interest and attitudes to science relate to specific issues in the same manner as knowledge, interest and attitudes to democracy or parliamentary procedures relate to specific policies, parties and personalities. But it would be mistake to conflate the one with the other. The rejection of a particular government or policy is not the same as the rejection of democracy as a whole, and support for a particular policy is not identical with support for the democratic system. The general and the specific attitudes need to be kept separate, not least to understand their mutual dynamic (see Easton 1976). There will be interesting variations across time and place in how specific issues aggregate to general attitudes, and that very relationship between specific and general could be a cultural indicator.

20.3.3 Developing the Indicator System Globally

International survey research encourages comparable sampling procedures, questionnaire formats and interview protocols. The imperative of comparability demands semantic equivalences of question wording and response alternatives. However, those requirements are difficult to implement. Difficulties in translating between different languages and varied speech practices (for example, the difficulties of giving 'no' or 'disagree' as an answer) are only one of many problems to solve. The biggest obstacle of all is the lack of determination and support for the global coordination of these efforts. Opportunities need to be sought to discuss these and other urgent matters in the construction of cultural indicators:

- Construct local time-series; analyze, in the local context, which items are diagnostic and drop those that are not (item response analysis).
- If items are altered from one wave to the next, consider split-half designs to calibrate the changes in the time-series.

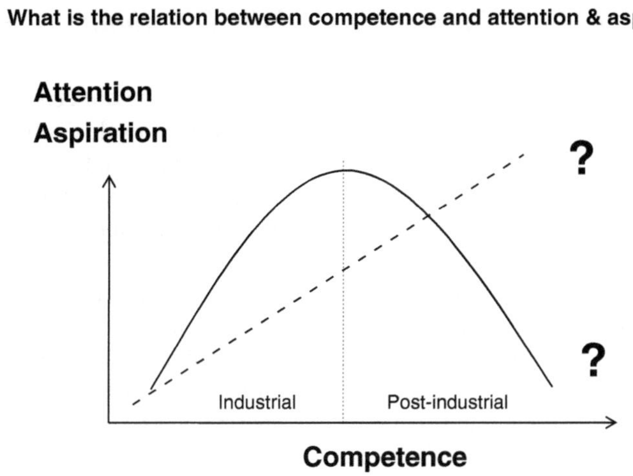

Fig. 20.2 Two hypothetical relationships between competence and attention and aspirations towards science. In an industrial context, this might be positive; in a post-industrial context, negative

- When removing survey items, stick to a set of core items for purposes of international comparison: retain core concepts of literacy, expectations, interests and engagement activities. Defining such a core set of items will be useful. We need something analogous to the *Frascati manual* used by the OECD to assess the R&D contributions of each country.
- Develop new items fit for the purposes and concerns of your local context.
- Exchange new items and new ideas through reports and meetings.
- Micro-integrate existing data. This includes micro-integration of data with a time-series and across countries. Such databases will allow for a step-change in the analysis of these indicators.

20.3.4 Non-linear Relationships

When constructing combined indicators from a set of items, it is important to examine carefully the exact relation between single items. The relationship between items might itself be an indicator. What is the relationship between literacy and attitudes? This question has preoccupied PUS researchers for quite some time (see Allum et al. 2008). Shukla and Bauer (2012) observe that the relationship between literacy and attitudes is positive in India, while across Europe it has tended to be negative, and thus overall non-linear. Whether the relationship between these indicators is positive or negative might indeed be context (time or place) dependent. In some contexts, a positive attitude to science is functional, while in other contexts, negative or more skeptical attitudes to science are functionally required. Our index constructions need to be able to take into account this non-linearity across a wide spectrum of contexts, which is by itself a cultural indicator (see Fig. 20.2).

20.3.5 Different Data Streams Beyond the Questionnaire Survey

A final aspect of cultural indicators is the kind of data that are collected. It seems obvious that we cannot assume that science culture resides exclusively in public perceptions assessed through nationally representative surveys and questionnaires. Indeed, many of the concepts of 'science culture' discussed above extend the ambition in this respect. The problem is access to, but also the existence of, such data. However, some data streams are already developed to a practical level and will need to be considered in the construction of future indicators of science culture. Discourse-based indicators of mass media coverage of science are very promising (see Bauer 2000, 2012). Tabulations of science event making in consensus conferences and other participatory forums are also useful sources of international comparisons (see Einsiedel 2008), and so are recent attempts to assess the mobilization of scientists for the purpose of public engagement at the individual level (see Bauer and Jensen 2011; Bentley and Kyvik 2011) or at the level of laboratories (Neresini and Bucchi 2011).

20.4 A New Beginning, Avoiding Old Traps

The quest for cultural indicators of science is hardly a very new endeavor—as for so many questions of the social sciences, it already has an established tradition. Past attempts did not flourish for many reasons, so it is important to learn from them to avoid old traps in the future. I suggest four principles of operation for future attempts to establish measures of science culture across the globe.

20.4.1 No One Best Way for Science Culture

If we accept that science culture has both an objective side of performance and a subjective side of mentality, we might also accept that activity and mentality do not stay in a one-to-one relationship. There is no 'one best way' of matching mentality and performance, but different mentalities allow for optimal performance.

20.4.2 A Quest for Functional Equivalences

One implication of the first principle is that we should abandon the quest for the single best mentality, and this becomes an invitation to conduct comparative research into functional equivalences. The same scientific performance is reached by different mentalities, and the same mentality gives rise to different levels of performance.

20.4.3 Comparative Analysis of Old and New Data

The quest for cultural indicators of science must be a comparative exercise; it cannot be achieved in any one context alone. However, rather than starting with the search for the perfect model, it is practical to examine existing databases and subject them to different analyses and interpretations.

20.4.4 Benchmarking 'Science' Against Other Cultural Pursuits

Finally, we need to avoid the trap of reading our own concerns with science culture into everybody's life. Science is only one among many cultural pursuits, along with, say, the arts and crafts, music or religion. It is desirable to gain a realistic picture of the position of science within people's overall portfolios of interests in everyday life.[4] That picture is not likely to emerge from research that focuses only on science. Interests and attitudes to science need to be benchmarked against other life interests and attitudes.

References

Allum, N. (2010). Science literacy. In S. Hornig Priest (Ed.), *Encyclopedia of science and technology communication* (pp. 724–727). Los Angeles: Sage.

Allum, N., Sturgis, P., Tabourazi, D., & Brunton-Smith, I. (2008). Science knowledge and attitudes across cultures: A meta-analysis. *Public Understanding of Science, 17*(1), 35–54.

Bauer, R. A. (1964). *Social indicators*. Cambridge, MA: MIT Press.

Bauer, M. (2000). Science in the media as a cultural indicator: Contextualising surveys with media analysis. In M. Dierkes & C. von Grote (Eds.), *Between understanding and trust: The public, science and technology* (pp. 157–178). Amsterdam: Harwood Academic Publishers.

Bauer, M. W. (2005). Distinguishing GREEN from RED biotechnology: Cultivation effects of the elite press. *International Journal of Public Opinion Research, 17*(1), 63–89.

Bauer, M. W. (2012). Public attention to science 1820–2010: A 'longue duree' picture. In S. Rödder, M. Franzen, & P. Weingart (Eds.), *The sciences' media connection: Communication to the public and its repercussions* (Sociology of the sciences yearbook). Dordrecht: Springer.

Bauer, M. W., & Jensen, P. (2011). The mobilization of scientists for public engagement. *Public Understanding of Science, 20*(1), 3–11.

Bauer, M. W., Allum, N., & Miller, S. (2007). What can we learn from 25 years of PUS survey research? Liberating and expanding the agenda. *Public Understanding of Science, 16*(1), 79–96.

Bentley, P., & Kyvik, S. (2011). Academic staff and public communication: A survey of popular science publishing in 13 countries. *Public Understanding of Science, 20*(1), 48–63.

[4]The author spent a pleasant afternoon conjecturing on such matters with Bruce Lewenstein while exploring the culture of Valencia in Spain back in 2010. We both agreed that this must be on the agenda, without clarifying at the time how it might be achieved.

Berg, M., & Bruland, K. (Eds.). (1998). *Technological revolutions in Europe: Historical perspectives.* Cheltenham: Edward Elgar.

Butler, D. (2006). The data gap: Statistics on science investment and performance are lacking across the Muslim world. *Nature, 444*, 26–27. 2 November.

Chang, B. H., & Chan-Olmstead, S. M. (2005). Relative constancy of advertising spending. *Gazette, 67*(4), 339–357.

Easton, D. (1976). A re-assessment of the concept of political support. *British Journal of Political Science, 5*, 435–57.

Einsiedel, E. (2008). Public participation and dialogue. In M. Bucchi & B. Trench (Eds.), *Handbook of public communication of science and technology* (pp. 173–184). London: Routledge.

EIS (European Office of Statistics). (2005). *European innovation scoreboard.* Brussels: EIS.

FAO (Food and Agriculture Organization of the United Nations). (2003). *Final report on an indigenous people's initiative to establish cultural indicators of SARD.* Rome: FAO.

Fleck, L. (1979; originally published 1935, Basel). *Genesis and development of a scientific fact.* Chicago: CUP.

Gerbner, G. (1969). Toward 'cultural indicators': The analysis of mass media mediated message systems. *AV Communication Review, 17*, 137–148.

Godin, B. (2005). *Measurement and statistics on science and technology: 1920 to the present.* London: Routledge. vol. 22 of series.

Godin, B. (2012). The culture of science and the politics of numbers. In M. W. Bauer, R. Shukla, & N. Allum (Eds.), *The culture of science: How the public relates to science across the globe* (pp. 18–37). New York: Routledge.

Godin, B., & Gingras, Y. (2000). What is science and technological culture and how it is measured? A multi-dimensional model. *Public Understanding of Science, 9*, 43–58.

Inglehart, A. (1990). *Culture shift in advanced industrial society.* Princeton: Princeton University Press.

Inonu, E. (2003). The influence of cultural factors on scientific production. *Scientometrics, 56*(1), 137–146.

Kawamoto, S., Nakayama, M., & Saijo, M. (2011). A survey of scientific literacy to provide a foundation for designing science communication in Japan. *Public Understanding of Science*, published online October 18, 2011. DOI: 10.1177/0963662511418893

Klieme, E., & Baumert, J. (2001). Identifying national cultures of mathematics education: Analysis of cognitive demands and differential item functioning in TIMSS. *European Journal of Psychology of Education, 16*(4), 385–402.

Klingemann, H. D., Mohler, P. P., & Weber, R. P. (1982). Cultural indicators based on content analysis: A secondary analysis of Sorokin's data on fluctuations of systems of truth. *Quality and Quantity, 16*, 1–18.

Kumar, P. V. S. (2011). Culture nature of scientific temper. In G. Raza, R. Ren, J. D. Hasan, & W. He (Eds.), *Constructing culture of science: Communication of science in India and China* (pp. 257–278). Delhi: CSIR-NISCAIR and CRISP.

Lebart, L. (1984). Complementary use of correspondence analysis and cluster analysis. In M. Greenacre & J. Blasius (Eds.), *Correspondence analysis in the social sciences* (pp. 162–178). London: Academic.

Liu, X., Tang, S., & Bauer, M. W. (2012). Comparing the public understanding of science across China and Europe. In M. W. Bauer, R. Shukla, & N. Allum (Eds.), *The culture of science: How the public relates to science across the globe* (pp. 139–157). New York: Routledge.

Mejlgaard, N., & Stares, S. (2012). Validating survey measures of scientific citizenship. In M. W. Bauer, R. Shukla, & N. Allum (Eds.), *The culture of science: How the public relates to science across the globe* (pp. 418–435). Routledge: New York.

Melischek, G., Rosengren, K. E., & Stoppers, J. (Eds.) (1984). *Cultural indicators: An international symposium.* Vienna: Austrian Academy of Sciences.

Miller, J. D. (1983). *Scientific literacy: A conceptual and empirical review* (pp. 29–48). Spring: Daedalus.

Miller, J. D. (2012). The sources and impact of civic scientific literacy. In M. W. Bauer, R. Shukla, & N. Allum (Eds.), *The culture of science: How the public relates to science across the globe* (pp. 217–240). New York: Routledge.

Mokyr, J. (2000). Innovation and its economics: The economic and political roots of technological inertia. In M. Olson & S. Kahonen (Eds.), *A not-so-dismal science: A broader view of economics and society* (pp. 61–91). Oxford: OUP.

Neresini, F., & Bucchi, M. (2011). Which indicators for the new public engagement activities? An exploratory study of European research institutions. *Public Understanding of Science, 20*(1), 64–79.

NSF (National Science Foundation). (2006). *Science indicators report*. Washington: NSF.

ONS (Office of National Statistics). (2011). *Measuring national well-being: Measuring what matters*. London: ONS.

OST (Office of Science and Technology). (2000). *Science and the public: A review of science communication and public attitudes to science in Britain*. London: OST & Wellcome Trust.

FAPESP (São Paulo Research Foundation) (2004). *Science indicators, São Paulo*. See Chapter 12 on public perceptions. FAPESP. Retrieved from http://www.fapesp.br/english/materia.php?data[id_materia]=463

PISA (Programme for International Student Assessment) (2009). *PISA 2009 Assessment Framework*. Paris: OECD Programme for International Student Assessment http://www.pisa.oecd.org

Raza, G., & Singh, S. (2012). The cultures of public understanding of science: Defining cultural distance. In M. W. Bauer, R. Shukla, & N. Allum (Eds.), *The culture of science: How the public relates to science across the globe* (pp. 282–299). New York: Routledge.

Raza, G., Dutt, B., Singh, S., & Wahid, A. (1991). *Prototype of the forms of scientific cognition: A survey of cultural attitudes to natural phenomena, reports I and II*. New Delhi: National Institute of Science, Technology & Development Studies.

Raza, G., Singh, S., & Dutt, B. (2002). Public, science and cultural distance. *Science Communication, 23*(3), 293–308.

Schwartz, S. H. (2011). Studying values: Personal adventure, future directions. *Journal of Cross-cultural Psychology, 42*(2), 307.

Shukla, R. (2005). *India science report: Science education, human resources and public attitudes*. Delhi: National Council for Applied Economic Research.

Shukla, R., & Bauer, M. W. (2012). The science culture index (SCI): Construction and validation. In M. W. Bauer, R. Shukla, & N. Allum (Eds.), *The culture of science: How the public relates to science across the globe* (pp. 179–199). New York: Routledge.

Sjoëberg, S., & Schreiner, C. (2012). A comparative view of adolescents' attitudes towards science. In M. W. Bauer, R. Shukla, & N. Allum (Eds.), *The culture of science: How the public relates to science across the globe* (pp. 200–216). New York: Routledge.

Smircich, L. (1983). Concepts of culture and organizational analysis. *Administrative Science Quarterly, 28*, 339–358.

Song, J. (2010). Rethinking the context and social foundation of science education. In Y. J. Lee (Ed.), *The world of science education: Science education research in Asia* (pp. 155–169). Rotterdam/Boston: Sense Publishers.

Song, J. (2011). *How to measure the social foundation of science*. Presentation to research seminar, Institute of Education, London, March 2011.

Sorokin, P. (1985, originally published 1957 by Porter Sargent). *Social and cultural dynamics: A system of change in major systems of art, truth, ethics, law and social relationships*. Revised and abridged. New Brunswick: Transaction Inc.

Titchener, E. B. (1972; originally published 1929). *Systematic psychology: Prolegomena*. London: Macmillan (1929). Cornell University Press (1972).

Turner, C. F., & Krauss, E. (1978). Fallible indicators of the subjective state of the nation. *American Psychologist, 33*, 456–470.

UN (2003). *Report on statistics of science and technology*. UN Economic and Social Council, E/CN.3/2004/15

Vogt, C. (2012). The spiral of scientific culture and cultural well-being: Brazil and Ibero-America. *Public Understanding of Science, 21*(1), 4–16.

Welzel, C. (2006). Democratisation as an emancipative process: The neglected role of mass motivations. *European Journal of Political Research, 45*, 871–896.

Work Foundation (2007). *Staying ahead: The economic performance of the UK's creative industries*. London: Work Foundation http://www.theworkfoundation.com/

Index

A
Ability, 73, 76, 104, 128, 136, 155, 183, 189, 194, 195, 198, 202, 207, 208, 232, 237, 284
Academic science, 5, 76, 161, 270
Acronym, 34, 83
Action, 6, 13, 27, 41, 52, 55, 71, 88, 89, 114, 117, 118, 133, 134, 144, 153, 187, 198, 201, 203, 229, 231, 273, 295, 305
African National Congress (ANC), 161
AlphaGalileo, 230, 231
ANC. *See* African National Congress
Anmyeondo, 184, 185
Anti-elitism, 95–97, 106
ANU. *See* Australian National University
Apartheid, 151, 152, 159–162, 166
Aspect, 8, 46, 54, 166, 250, 263, 274, 287, 292, 308
Association for Supporting Science and Technology, 183
Attitudes, 26, 28, 47, 48, 52, 74–76, 103, 165, 181, 236, 270, 295, 298–301, 305–307, 309
Australian National University (ANU), 23, 26, 244
Australian Science Communicators, 19, 20
Awareness of science popularization, 66
Axiology in science communication, 6

B
Behavioral necessity, 272
Bertelsmann foundation, 128
Big science, 5–7, 228
Branding, 95, 104, 105, 215
Bricolage, 153

C
Canadian context, 37–39
Chung-Hee, P., 183
Citizen conference, 286
Civic literacy, 89, 299–300
Cognition of science popularization, 66
Collaboration, 55, 99, 109, 115, 116, 162, 169–172, 179, 233, 249, 252, 253, 256, 299
Colonialism, 152, 155, 166
Committee on Public Understanding of Science (COPUS), 271
Commonwealth Scientific and Industrial Research Organisation (CSIRO), 22, 25
Communication
  mechanism, 75
  science, 8, 68–72, 77, 110–113, 118, 120, 144, 186, 213
Communicative turn, 3–15
Community Research and Development Information Service (CORDIS), 230, 234
Consensus conference, 13, 56, 96, 99–104, 134, 187, 240, 282, 285–288, 302, 308
Consequentiality, 271, 273, 277
Contents of communication, 74
COPUS. *See* Committee on Public Understanding of Science
CORDIS. *See* Community Research and Development Information Service
Creativity, 169–173, 188, 194, 198, 245, 269, 274, 277, 296, 297
CSIRO. *See* Commonwealth Scientific and Industrial Research Organisation

B. Schiele et al. (eds.), *Science Communication in the World: Practices, Theories and Trends*, DOI 10.1007/978-94-007-4279-6,

Cultural
distance, 139, 140, 146–149, 300
distance model, 139, 146, 300
divide, 270
indicator, 296–298, 304–309
Culture and industry, 296, 297
Culture and science, 86, 173

D
Daily Press, 259–267
Danish Board of Technology, 95, 96, 99
Danish Council of Ethics, 95
Decision making, 100, 196, 202, 236, 239, 240, 274, 282
Deficit model, 20, 28, 36, 52, 74, 77, 92, 95, 103, 137, 141, 149, 165, 184, 205, 283, 285
Degree programmes and conditions for sustainability, 129
Degree programmes and criteria for differentiation, 245, 246
Deliberation, 13, 95–106, 119, 161, 206
Democracy, 13, 46, 56, 99, 101, 119, 120, 187, 196, 208, 240, 288, 291, 296, 306
Democratic participation, 104, 133, 270
Discourse, 3, 22, 35, 36, 47, 49, 50, 52, 54, 77, 98, 109–114, 117–120, 136, 153, 157, 195, 211–214, 216, 218, 219, 221, 222, 250, 308

E
Empowerment, 281, 284, 288, 290
Engagement, 4, 8, 13, 20, 27–28, 55, 56, 85, 87, 91, 92, 95, 98, 99, 101–103, 106, 181–190, 238, 246, 247, 255, 269–278, 283, 284, 289, 291, 292, 301–303, 305, 307–309
Engagement models, 27–28
Enterprise, 11, 22, 77, 81, 82, 84, 86–88, 90, 152, 202, 234, 256, 270
Enthusiasm for science, 67, 133
Eurobarometer, 103, 198, 222, 227, 236, 237, 239, 304
EUROPA, 234
European Commission, 170, 198, 201, 203, 227–229, 231, 233–238, 247, 254, 282, 283, 290
European Science Week, 229
European Union, 47, 48, 196, 198, 207, 227–304
Evaluation, 10, 20, 26, 29, 57, 71, 99, 120, 132, 136, 151, 163, 206, 207, 215, 232, 233, 236, 248, 253, 284, 287
Evolution of science popularization studies, 66
Exhibition, 26, 42, 53, 54, 69, 90, 117, 134, 144, 181, 184–186, 200, 212, 215, 218–222, 228, 285, 287, 301, 302
Expertise, 35, 51, 55, 100–102, 105, 109–111, 116, 119, 120, 190, 284

F
Facebook, 281, 291
First courses on science communication, 20
First order and second order communication, 74
Five-year plan, 48
FP7. *See* The Seventh Framework Programme
Framework Programme (EU), 198, 204, 227–232, 234, 284, 291

G
General indicators, 306
Glassco Commission, 44
Government-driven, 169, 170, 178, 179
Grundtvig, 95–99, 101, 102, 104

H
Housewives Life Science Course, 184

I
Ideology, 8, 73, 81–82, 91, 112, 114, 161, 206, 213
Indicator, 28, 49, 54, 75, 76, 141, 146, 147, 166, 212, 295–309
Indigenous Knowledge Systems (IKS), 155–157, 161, 165, 267
Informal science education, 28, 171, 244
Information, 4, 7, 9, 13, 14, 24, 27, 29, 30, 34, 41–43, 47, 48, 50–52, 54–56, 71, 72, 91, 99, 101, 109, 110, 115, 116, 119, 120, 126, 141, 145, 152, 160, 162, 163, 166, 171, 173, 174, 176, 184–190, 193–195, 198, 201, 204–207, 227, 229–238, 241, 246, 247, 253, 256, 262, 272–274, 276, 277, 285, 290, 301, 303, 306
Infrastructure, 86, 87, 89–91, 93, 162, 183, 195–197, 212, 301

Initiatives in Germany, 127
Input–output model, 300–301
Institutionalization, 9, 11, 39, 40, 96, 98–102
Interaction, 5, 9, 10, 38, 43, 48, 54, 55, 72–74, 77, 83, 99, 100, 113, 116, 197, 206, 207, 221, 237, 249, 281, 292
Interactive museums, 200, 215
Interest, 4, 6, 8, 11, 13, 14, 21–23, 25–28, 30, 34–37, 43, 48, 51–54, 56, 67, 75, 82, 83, 96, 98, 99, 101, 103, 104, 110, 111, 115, 120, 130, 133, 134, 140, 158, 176–178, 182, 184, 185, 189, 200, 206, 212, 213, 216, 218, 221, 231, 232, 234, 236, 237, 239, 244, 245, 250–252, 255, 259, 262, 263, 265, 271, 275, 276, 281, 282, 284, 285, 289, 292, 299, 301, 305–307, 309

K
Kepu, 83
Knowledge, 3, 21, 34, 65, 82, 97, 109, 126, 141, 151–166, 171, 181, 193, 211, 233, 243, 267, 271, 282, 297
Korea Foundation for the Advancement of Science and Creativity (KFASC), 182, 187, 188
Korea Science Foundation (KSF), 170, 171, 173, 174, 177, 178
Kumbh Mela, 150

L
Latin America, 3–14, 248
Law, 13, 68, 75, 83, 85–89, 91, 100, 102, 104, 161, 185, 205, 214, 282
Legal, 82, 85, 88, 93, 161, 162, 175, 206
Living science, 76
Low-level radioactive waste disposal facility, 184

M
Mass-media, 112
Matrix, 91, 157
Measure, 28, 29, 43–45, 47, 49, 54, 75, 76, 86, 89, 119, 134, 135, 145, 147, 152, 155, 159, 177, 203, 205, 218, 233, 239, 250, 264, 271, 275–277, 295, 299, 301, 303–305, 308
Mechanics institutes, 21
Mechanism, 11, 13, 14, 56, 71, 73–75, 77, 90, 93, 114, 206, 238, 281, 288, 305
Media, 3, 20, 34, 69, 102, 110, 126, 160, 169–179, 187, 197, 211, 230, 244, 259–267, 271, 289, 297–298
Media coverage of science, 51, 301, 308
Mediation, 34, 36–39, 42, 43, 109, 110, 112, 114, 116–119, 128, 131, 221
Mentality, 153, 158, 159, 204, 301–303, 308
Milestone, 22, 68, 75
Ministry of Science and Technology, 86, 88, 90, 175, 176, 187
Mode 2 knowledge production, 5
Models in science communication, 7, 27–28
Motivation, 69, 133, 169, 179, 271
Multidisciplinary, 20, 25–27, 36, 69, 76, 77, 242
Museums
  and education, 200, 213, 218, 290
  and identity,
  of science and society, 213
  in Spain, 211–222

N
National Academy of Sciences, 135
National culture and globalization, 303
National Research Council of Canada (NRCC), 40, 44
National System of Innovation, 162, 163
Nobel prizes, 259, 260, 262–264, 266, 267
Non-profit organization (NPO), 170, 179
Nuclear armament, 160

O
Objective indicator, 302, 304
Organisation for Economic Co-operation and Development (OECD), 33, 43–47, 53, 170, 300, 301, 303, 307
Organized, 12, 13, 34, 36, 50, 67, 81, 82, 84, 90–93, 100, 114, 118, 129, 135, 143, 152, 164, 216, 219, 227, 231, 235, 236, 281, 285, 286, 289, 303
Output, 90, 128, 250, 300, 301, 304, 305

P
Participation, 4, 11–14, 49, 55, 56, 74, 77, 87, 90, 98, 99, 102, 104, 113, 133, 136, 181, 200, 204, 229, 232, 234, 238, 249, 270, 281–284, 287
PCST conferences, 252, 253
Pet phrase, 83
Physics on stage, 230, 231

Policy, 8, 11, 24, 26, 28, 34, 36, 42–46, 56, 69, 73, 91–93, 98, 101, 103, 144, 145, 161, 165, 166, 182–184, 187–189, 203, 214, 215, 228, 233–235, 270, 281–293
Policy-making, 34, 232, 235
Popularization, 7, 20, 33, 65–77, 81–93, 111, 125, 139, 152, 170, 181–190, 198, 245, 270,
Popular science, 10, 11, 14, 83, 200, 201, 204, 260, 261
Post-academic science, 5, 76, 270
Post-colonial, 152, 155, 157–159
Practical solutions, 19
Practice influences research, 39
Practices, 3–7, 13, 14, 29, 36, 39, 46, 52, 53, 66, 68, 70–73, 76, 77, 88, 104, 109, 110, 112–118, 120, 153, 155, 156, 161, 166, 219, 227, 229, 231, 235, 297, 300, 306
Press release, 131, 206, 233
Problem solving, 133, 182, 190, 274, 275, 303
Professional approach to science communication, 25, 105, 136, 187
Profession of science communication, 22
Public attitude toward S&T, 75
Public awareness of S&T, 162, 169
Public communication of science and technology (PCST), 3, 4, 20, 33, 34, 81, 83, 96, 171, 193, 199, 211, 227, 235, 242, 245
Public engagement, 13, 28, 55, 56, 98, 102, 106, 181–190, 269–272, 275, 277, 278, 283–284, 291, 292, 302, 303, 308
  in science., 181–183, 187, 291, 292
  with science, 98, 102, 106, 270–272, 277
Public participation, 4, 56, 98, 102, 282, 283, 290, 291
Public perception, 11, 206, 304, 308
Public perception of S&T, 11
Public sphere, 109, 117, 127, 296
Public's science literacy, 47–50
Public understanding of science, 20, 34, 50, 66, 68, 72–74, 77, 81, 83, 92, 95, 102, 103, 106, 125–137, 139–149, 152, 162–164, 171, 181, 182, 184–189, 193, 195, 199, 205, 206, 231, 239, 244, 252, 254, 269–271, 300
Public Understanding of Science and Humanities (PUSH), 125, 127, 128, 133
Purpose of communication, 75

Q

Qualification program science journalism, 128
Quiet revolution, 40, 49

R

Rationality, 5, 105, 128, 135, 151–166, 173
Reason, 39, 45, 46, 54, 101, 106, 114, 132, 136, 141, 143, 152–155, 158, 161, 165, 188, 190, 194, 197, 206, 213, 215, 219–221, 229, 250, 275, 282–284, 288, 290, 292, 303, 308
Relevance, 29, 84, 133, 145, 214, 267, 277, 303
Research being applied, 21, 28, 77
Research managers, 104
Research on science communication, 19–31, 33, 35–37, 39, 43–57, 103, 110, 140, 142, 248
Research on science popularization, 66
Role of science museums, 211
Ruberti, A., 228

S

SCI. *See* Science culture index
Science
  All, 186
  café, 34, 186, 281, 282, 284–286, 290, 292
  Campaign, 143–145
  centers, 24, 29, 53, 87, 90, 132, 134, 200, 202, 205, 211–213, 239, 244, 245, 253, 283, 290
  dissemination, 102
  events, 229, 230, 283, 285, 289, 292, 308
  indicator, 299, 301, 302, 304
  journalism, 7, 14, 34, 41, 43, 46, 50, 53, 57, 72, 73, 77, 127–132, 171, 176, 199, 201, 204, 231, 244, 245, 247, 254, 277
  literacy, 20, 37–57, 152–157, 163, 165, 244, 299, 304
  movement, 125–137, 142, 144–146
  museums, 20, 23, 53, 69, 86, 87, 163, 173, 184, 200, 204, 211–222, 230, 239, 245, 253, 301
  and national identity, 259–267
  parliament, 282, 285–286, 289, 290
  in society, 39, 72, 193–208, 230, 231, 235, 262, 282, 284, 299, 300, 302
Science communication
  in Canada, 57
  and contemporary capitalism, 5–6
  courses, 20, 23–26, 176, 246, 253, 254
  and institutional reflexivity, 256

in Latin America, 4, 7–14
and peer-reviewed journals, 254
and postgraduate programmes, 12, 25, 244–247
and professionalisation, 7, 9, 11, 23, 25, 31, 41, 43, 136
and relations between theory and practice, 243, 247, 254
in Spain, 211–222
and status as discipline, 243
and undergraduate programmes, 245–247
Science communication research (SCR), 19–31, 34–39, 43, 47, 49, 52–55, 57
Science culture index (SCI), 301–302
Science culture in Korea, 169–179
Science Korea Movement, 170, 173, 187
Science of science popularization, 68–71
Science, technology engineering and mathematics (STEM), 169, 171, 175
Scientific culture, 49, 65, 66, 69, 83, 193, 196–199, 201–204, 211, 213, 221–222, 228, 295, 298, 299, 301, 302
Scientific heritage, 212
Scientific knowledge, 3–5, 7, 44, 67, 69, 71–74, 82, 83, 92, 111, 117, 126, 129, 131, 134, 141, 142, 146, 152–155, 171, 173, 176, 181–184, 187–190, 194, 195, 197–199, 207, 211, 212, 216, 218, 221, 222, 239, 246, 272, 277
Scientific literacy, 66, 68, 69, 71, 73–75, 83, 85, 89–90, 92, 140, 141, 145, 165, 166, 170, 186, 189, 239, 270, 271, 277, 299
Scientific temper, 142, 143, 299, 302
Scientists and journalists, 52
SCR. *See* Science communication research
Securing social acceptance, 126, 128, 132
The Seventh Framework Programme (FP7), 228–235, 238
Sience popularization policy, 82–86
Social
actors, 35, 36, 115
movement, 4, 7, 12, 13, 152, 155
sciences, 19, 24, 26, 27, 31, 49, 109–120, 128, 130, 132, 134, 187, 244, 246, 247, 249–252, 255, 308
sciences perspective, 19, 26
Socialization of science, 149
Societal impact of popularisation, 125
Sociology, 4, 10, 27, 35, 69, 70, 72, 74, 75, 77, 99, 111, 157, 298
Space-sharing with newspaper, 178
Specific indicator, 305–306
Spiral model, 302
STEM. *See* Science, technology engineering and mathematics
Stifterverband für die Deutsche Wissenschaft, 127
Subjective indicator, 304
Summer of science, 134
Survey, 10, 20, 23, 25, 27–31, 44, 47–50, 75, 76, 83, 89, 92, 103, 140, 141, 145–147, 149, 163–166, 173, 182, 188, 227, 236–239, 242, 246, 251–254, 276, 297–299, 304, 306–308
Survey results, 25
Sustainability, 169, 179, 251–256

T
Technoscience, 3–14
Theoretical integration, 68–76
Theoretical studies, 66
Third place, 283, 291–293
Three-stage argument, 74
Three strands of science communication, 20
Training in science communication, 57, 232
Two cultures, 38, 195, 208, 255

U
Understanding, 3, 20, 34, 65, 83, 95, 111, 125, 139, 152, 171, 182, 194, 218, 228, 244, 269, 290, 300
Usage, 54–55, 112

V
Visible scientists, 259–267

W
Wide variety of disciplinary backgrounds, 19
Wissenswerte Conference, 130

Y
Years of Science, 69, 134

MIX
Papier aus verantwortungsvollen Quellen
Paper from responsible sources
FSC® C105338

If you have any concerns about our products,
you can contact us on
ProductSafety@springernature.com

In case Publisher is established outside the EU,
the EU authorized representative is:
Springer Nature Customer Service Center GmbH
Europaplatz 3, 69115 Heidelberg, Germany

Printed by Libri Plureos GmbH
in Hamburg, Germany